Public School Emergency Preparedness

Public School Emergency Preparedness

Second Edition

Don Philpott

Bernan Press
Lanham • Boulder • New York • London

Published by Bernan Press
An imprint of The Rowman & Littlefield Publishing Group, Inc.
4501 Forbes Boulevard, Suite 200, Lanham, Maryland 20706
www.rowman.com
800-865-3457; info@bernan.com

6 Tinworth Street, London SE11 5AL, United Kingdom

ISBN: 978-1-64143-320-4
E-ISBN: 978-1-64143-321-1

Contents

Acknowledgments

The authors would like to thank the Akron Public School System for permitting us to use details from their safety plan for this publication. The Akron Public School System Safety Plan team was led by Robert Boxler, Manager Health & Safety, and Dave Nunley, Grant Coordinator. In addition, the Ohio Red Cross, the Ensafe Corporation, and the University of Akron/Emergency Management Program contributed to the Akron Plan.

This book has also drawn heavily on the authoritative materials published by a wide range of sources. These materials are in the public domain, but accreditation has been given both in the text and in the reference section if you need additional information.

Introduction

In the last few months there have been several fatal shooting incidents in schools including the tragedy at Marjory Stoneman Douglas High School in Parkland, Florida.

While our schools are generally safe, there have, unfortunately, been similar events over the past few years such as the multiple shootings at Columbine High School, Virginia Tech, Oikos College in Oakland, and Santa Fe High School in Texas.

These incidents have attracted international attention and renewed calls at home for stricter gun control. New York became the first state to pass comprehensive gun control laws in January 2013. Other states have taken similar measures including Florida following the Parkland shooting and while tighter gun control legislation has been put forward in Congress it continues to face a tough uphill battle.

Gun control may be one way of trying to tackle the problem, but it is not likely to deter an aggressor determined to inflict maximum damage on others. After all, someone who is sufficiently deranged to want to kill and maim young children has many options other than guns and bullets—from homemade bombs to releasing toxic chemicals.

However, in order to put things into perspective, school shootings are still a rare occurrence and while every death and injury is one too many, more than 800 children die every year traveling to and from school.

In an ideal world, there would be no school crime and the safety of each staff member and student would be ensured. Unfortunately, in reality so many things are outside the control of the school administrator that such a task is virtually impossible. To attempt to predict the disruptive behavior of a student, a staff member, an intruder, or a terrorist is unrealistic, said Ronald D. Stephens, executive director of the National School Safety Center which was established by the Departments of Education and Justice to address the growing problem of violence in US schools.

> This nation's educational system has its foundation at the local level. Education is a federal concern, a state function and a local responsibility. Assigning such responsibility to parents, students, educators and other citizens closest to the need for schooling is both appropriate and demanding.
>
> No greater challenge exists today than creating safe schools or restoring schools to tranquil and safe places of learning. The challenge requires a major strategic commitment

> and involves placing school safety at the top of the educational agenda. Without safe schools, teachers cannot teach and students cannot learn. A safe school is foundational to the success of the academic mission. A safe school is also one that is prepared to respond to the unthinkable crisis.
>
> —Stephens

The question facing schools and communities alike is this: How far must schools go to create a safe and welcoming environment without turning our nation's schools into armed camps?

For the most part, schools were not designed to be defended. They were designed as open places of learning. Today security is an integral part of new school building plans, but contingency planning and preparedness are still the keys to secure and safe school environments. There are, however, a number of initial decisions that have to be made. These fall into two areas—school design and school security.

Regardless of the size of a school construction program, certain steps are necessary and certain procedures must be followed. These will vary greatly in scope between the design of a small elementary school and the development of a multi-school program of new and remedial construction. Review and regulation procedures by outside agencies will also vary. Internal district decisions as to the design and construction process (e.g., conventional architect design and competitive construction bid, design/build, or construction manager) will affect the scope and timing of some of the activities.

However, regardless of the size and scope of the project, the following steps should be taken; for a small project, they may entail relatively informal meetings among a few district staff, the school board, and others; for a large program, formal procedures must be established.

- Conduct an in-house assessment of the educational needs, with the assistance of a public education committee and consultants. Public committees continue throughout the programming and design process, acquiring specialist members as necessary at different stages for a large program.
- Determine the size and scope of the proposed program. (In a small district, an architect may be employed to assist the school district with this task, who may later become the design architect.)
- Conduct an assessment of the site needs to determine the size and availability of sites (and lease/purchase as necessary).
- Develop educational specifications, both in-house and/or consultants.
- Conduct an assessment of financial needs.
- Identify financial resources, including alternative sources of funding (e.g., state and federal programs, local taxes, bond issues).
- Ensure funding (e.g., pass bond issue).
- Appoint a district building program management staff (appointed officials or a committee).
- Determine the design and construction process (i.e., conventional design and bid, design/build or construction management).
- Select and hire architects and other special design consultants or design/build team members; the timing of hiring will vary, depending on number of projects, whether programming is involved, and other variables.

- Develop building programs, including building size, room size, equipment, and environmental requirements; this may be done in-house and/or architects or independent program consultants may assist.
- Appoint the district staff and public stakeholders committee for the design phase.
- Develop designs (architects), together with cost estimates. Hold public meetings with architects and encourage public input into the design, together with district progress reviews.
- Design completion, district review of contract documents.
- Submit construction documents to the district and any permitting agencies for review and approval.
- Submit documents to building department and other required agencies.
- Select the contractor (bidding) or finalize design/build or construction management contracts.
- School construction.
- School district administration of construction contract.
- Observation by architect and inspection as required.
- School completed by contractor.
- School inspected and accepted by architect.
- School inspected and accepted by school district.
- School commissioned and occupied.

The sequence of the above steps may vary, depending on the complexity of the program; some steps may be implemented simultaneously.

The aim of this handbook is to assist you in determining the risks that your facility faces and then to help you develop the appropriate strategies to overcome or mitigate them. While this book focuses primarily on man-made disasters—especially acts of violence—basic strategies for coping with natural disasters are also covered.

There can't be a school administrator in the country who still thinks "It won't happen to us" but what to do about it is a huge dilemma. While the safety of students and staff is and must always be the top priority, the reality is that security enhancements cost money. Most states have been slashing spending on education because of the harsh economic climate and this means less money for security and necessary facility safety improvements. If more money is made available for security, it means less is available for teaching and staff.

During the past decade, more than 300 violent deaths have occurred in or near school campuses. The incidents don't fit any neat patterns, they have occurred at one-classroom rural schools, inner-city high schools, and suburban elementary schools. While the risks do remain small they cannot be overlooked. The purpose of risk analysis is to discover all possible vulnerabilities and weaknesses and then to determine what measures need to be taken to overcome them or mitigate their effects. Normally when conducting a risk assessment you come up with a matrix of what needs fixing and what can reasonably be left unfixed because the risk is so small. When it comes to our schools you have to ask yourself "no matter how small the risk, what is the risk of doing nothing?"

SOME STATISTICS

- Approximately fifty-five million kindergarten to grade 12 students
- Six million adults working as teachers or staff
- About 117,000 public and private schools
- Sixteen million students in higher education
- Over 140,000 education complexes (schools, universities and administrative offices)

WHAT CAN WE DO?

First we need to be better able to identify those people who might likely commit acts of violence. Of course this is no easy thing but there are ways to tackle it.

As a society we need to be better at understanding the scope and symptoms of mental health. According to the National Institute of Mental Health, mental disorders are common throughout the United States, affecting tens of millions of people each year, and that only a fraction of those affected receive treatment.

The overwhelming majority of people with mental health disorders are non-violent, and those that have the potential to be violent stick to their medications and don't pose a problem.

It is those people who don't take their medications, or worse still, who haven't even been to a doctor at all that may pose a real threat.

We have to develop systems that encourage those with mental health problems to get treatment and receive the medications they need to lead productive lives. We need protocols in place to alert the authorities when we suspect something is wrong. Many people with mental health issues are loners, but most have families and acquaintances somewhere who might notice unusual or dangerous behavior.

Following many school shooting incidents people reported that they had suspected something was wrong because of the shooter's behavior or actions prior to the shooting. It might have been the posts on Facebook or violent statements made in front of others. In many of these cases, if people had acted on their suspicious earlier, the shootings might have been avoided.

We should do something about gun control. Everyone who buys a gun through a legitimate gun dealer has to go through a waiting period (unless they have a concealed weapons license) before they can pick it up. Millions of people who buy weapons at gun shows do not always have to follow this procedure as gun laws vary from state to state. There is also the issue of whether automatic weapons should be banned along with high-capacity magazines.

In reality, there is no one solution to this problem and the way to make our schools safe and to protect our children and those who teach them is to come up with a matrix of solutions.

Some of the solutions already put forward include having an armed deputy at every school in the country (as they now do in Florida following the Parkland shooting), and training teachers how to use firearms which they would carry in class. Another is to have a national register of people with mental health disorders. Some or all of these might be helpful but the most effective way to protect our children at school is to have practiced emergency procedure plans in place that can be implemented at a second's notice and to improve school design and layout. If a suspicious person is seen approaching the main entrance, the alarm can be raised and, hopefully, the suspect apprehended before gaining access. Point of sight surveillance over all main areas together with closed circuit television is critical in identifying and delaying the attacker. Bullet proof glass protecting the main point of entry, usually the administration building, is one solution. If the building is breached, building design can be used to delay the attacker's progress and protect those children and staff who are still inside. This involves safe routes for evacuation, safe rooms, and comprehensive building monitoring by closed circuit television which can, when necessary, be relayed to law enforcement agencies.

Surveillance and building design are also critical in protecting students and staff from other violent crimes. Better lighting and alarm buttons and just two of the many design modifications can be implemented to make people safe in vulnerable areas such as walking to car parks on dark nights. These and many other design features are discussed in detail later in this handbook.

The politicians may or may not come up with solutions. The bottom line is that you need to do something. You need to know all the risks you face before you can develop and implement an effective emergency action plan.

This book does the following:

- It outlines programs and procedures that can be applied to any US School System to address hazard mitigation and prevention, emergency preparedness and response, and recovery and restoration of school functions to an effective learning environment.
- It predetermines, to the extent possible, the operational procedures across any US School System and cooperating governmental, private, and volunteer agencies for responding to and recovering from any and all types of natural, human, or technology-based emergencies that may occur directly within school system operations, or that may occur outside the jurisdiction of the school system but nonetheless causes/could cause collateral impact to school system operations.
- It describes specific actions and assigns responsibilities and response roles to district and individual school staff emergency teams and cooperating agencies and community response partners that have agreed to share responsibilities and resources as defined in this plan.
- It outlines, in the event of an emergency involving response by fire and/or law enforcement, the district/school site personnel who should establish an Incident Command System-based response organization in accordance with procedures outlined in the National Incidental Management System.

Section I

Security Basics

FIRST STEPS

The Department of Education identifies four phases of an emergency: mitigation and prevention, preparedness, response and recovery. Emergency action plans are different from facility plans in that they take into account actions and needs of students and teachers, whereas a facility plan focuses on restoring the physical plant to its pre-disaster condition and performance. Facility planning is discussed in section four. Emergency planning allows school facilities to be prepared for all eventualities, have the appropriate contingencies in place for when they happen, and return to normal as soon as possible afterward.

The basic planning processes are as follows:

- **Mitigation/Prevention:**
  - Conduct an assessment of each school building. Identify those factors that put the building, students, faculty, and staff at greater risk, such as proximity to rail tracks that regularly transport hazardous materials, or facilities that produce highly toxic material, or propane gas tanks, and develop a plan for reducing the risk. This can include plans to evacuate students away from these areas in times of crisis and to reposition propane tanks or other hazardous materials away from school buildings.
  - Work with businesses and factories in close proximity to the school to ensure that the school's crisis plan is coordinated with their crisis plans.
  - Ensure that a process is in place for controlling access and egress to the school. Require all persons who do not have authority to be in the school to sign in.
  - Review traffic patterns and, where possible, keep cars, buses, and trucks away from school buildings.
  - Review landscaping, and ensure that buildings are not obscured by overgrowth of bushes or shrubs where contraband can be placed or persons can hide.
- **Preparedness:**
  - Have site plans for each school building readily available and ensure they are shared with first responders and agencies responsible for emergency preparedness.

- Ensure that there are multiple evacuation routes and rallying points. First or second evacuation site options may be blocked or unavailable at the time of the crisis.
- Practice responding to crisis on a regular basis.
- Ensure a process is established for communicating during a crisis.
- Inspect equipment to ensure it operates during crisis situations.
- Have a plan for discharging students. Remember that, during a crisis, many parents and guardians may not be able to get to the school to pick up their child. Make sure every student has a secondary contact person and contact information readily available.
- Have a plan for communicating information to parents and for quelling rumors. Cultivate relationships with the media ahead of time, and identify a Public Information Officer (PIO) to communicate with the media and the community during a crisis.
- Work with law enforcement officials and emergency preparedness agencies on a strategy for sharing key parts of the school crisis plans.

- **Response:**
  - Identify the type of crisis that is occurring and determine the appropriate response.
  - Develop a command structure for responding to a crisis. The roles and responsibilities for educators, law enforcement and fire officials, and other first responders in responding to different types of crisis need to be developed, coordinated, reviewed, and approved.
  - Maintain communications among all relevant staff.
- **Recovery:**
  - Return to the business of teaching and learning as soon as possible.
  - Identify and approve a team of credentialed mental health workers to provide mental health services to faculty and students after a crisis. Understand that recovery takes place over time and that the services of this team may be needed over an extended time period.
  - Ensure that the team is adequately trained.
  - The plan needs to include notification of parents on actions that the school intends to take to help students recover from the crisis.

When developing an emergency action plan, there are several things to identify:
Establish the authority under which the plan is developed.

- **Determine:**
  - What to do prior to the event?
  - What to do during the event?
  - What to do after the event?
- **Identify:**
  - Critical functions
  - Resources required to support critical functions
  - Timetable to implement critical functions
- **At a minimum, the plan should address:**
  - Coordination
  - Communications
  - Logistics
  - Who is to perform each task?
  - Who is to report and when?

  - Where task will be done?
  - How task will be done?
  - Facilities' vulnerability to disasters
  - Risk probability of each type of disaster
  - Assign risk probability to each type of disaster
  - Assign resources accordingly
  - Appropriate resource assignment
  - Chain of command setup
- **Chain of command:**
  - Define specific roles and responsibilities.
  - Assign director/administrator.
  - Assign alternates.
  - Identify individuals and alternates who are authorized to activate emergency plan.
  - Determine policy of when and how chain of command will be activated.
  - Make quick decisions related to construction, building materials, and safety.
  - Assign personnel to disaster and recovery teams.
  - Assign duties to team members.
  - Train personnel in expected duties and responsibilities to ensure implementation.

EMERGENCY MANAGEMENT PLANNING

A disaster is any natural, technological, or civil emergency that causes damage of sufficient severity and magnitude as to result in a declaration of a state of emergency by a county official, the governor, or the president of the United States.

A program of emergency preparedness must include general emergency management principles and specific plans tailored for a particular school. This includes the following:

- Preparation of emergency supplies.
- Familiarization with the types of disasters likely to impact the geographical location of the school.
- Methods of self-protection during and after a disaster or act of violence (e.g., the identification of safe zones and alternative routes of escape in the event of a terrorist or sniper attack, and identification of the safest structural areas of the school in the event of a tornado).
- Well-rehearsed evacuation protocols, including the assignment of staff to specific tasks and duties, as well as the institution of methods of early warning and prompt evacuation (e.g., intercom, "walkie-talkie," or cellular systems that permit school-wide and between-site communication during any telephone outage).
- Institution of procedures to enlist aid following a catastrophic event (including telephone procedures and the use of social media).
- A system of tracking the location and safe dissemination of children and personnel during a rescue effort.
- A plan for restoration of the school facility and removal of trauma-invoking imagery.
- A mechanism for directing media response, which can become intrusive and distressing (e.g., directing news media personnel to a central administration representative).

Facility managers should be familiar with these general principles and have a working knowledge of the school emergency preparedness and response plan. This guidebook specifically addresses these facility preparedness and safety issues.

Disaster Classifications

The three classifications of disasters are:

1. *Minor*

A disaster that is likely to be within the response capabilities of local government. Minimal need for state or federal assistance. Example: tropical storm, local flooding.

2. *Major*

A disaster that is likely to exceed local capabilities. Requires a broad range of state and federal assistance. Example: Category One, Two, or Three hurricane.

3. *Catastrophic*

A disaster that will require massive state and federal assistance, including military involvement. Example: Category Four or Five hurricane that hits a densely populated area. The core role of facility managers in preparing for a disaster is to identify areas of vulnerability and past history of different types of disasters, including the following:

- Projected storm surges by hurricane category
- Past hurricane occurrences
- Past tornado events
- Past flood events
- History of freeze
- History of droughts
- History of forest fires
- History of civil disturbance
- Potential toxic spill locations

CONDUCTING A NEEDS AND RISK-THREAT ASSESSMENT

In order to achieve the greatest risk reduction at the least cost, identifying and prioritizing a school's critical assets is a vital first step in the process to identify the best mitigation measures to improve its level of protection.

Recognizing that people are a school's most critical asset, the process described below will help identify and prioritize school infrastructure where people are most at risk and require protection.

Identifying a school's critical assets is accomplished in a two-part process:

1. Define and understand the school's *core functions* and processes.
2. Identify school *infrastructure*.
    - Critical components/assets
    - Critical information systems and data

- Life safety systems and safe haven areas
- Security systems

Identifying School Core Functions

The initial step of an asset value assessment is the determination of core functions and processes necessary for the school to continue to operate or provide services after an attack. The reason for identifying core functions/processes is to focus the design team and school administrators on what a school does, how it does it, and how various threats can affect the school. This provides more discussion and results in a better understanding of asset value. Factors that should be considered include the following:

- What are the school's primary services or outputs?
- What critical activities take place at the school?
- Who are the school's occupants or visitors?
- What inputs from external organizations are required for a school's success?

Identifying School Infrastructure

After the core functions and processes are identified, an evaluation of school infrastructure is the next step. To help identify and value rank infrastructure, the following should be considered, keeping in mind that the most vital asset for every school is its people:

- Identify how many people may be injured or killed during an incident—earthquake, terrorist attack, and so on—that directly affects the infrastructure.
- Identify what happens to school functions, services, or student satisfaction if a specific asset is lost or degraded. (Can primary services continue?)
- Determine the impact on other organizational assets if the component is lost or cannot function.
- Determine if critical or sensitive information is stored or handled at the school.
- Determine if backups exist for the school's assets.
- Determine the availability of replacements.
- Determine the potential for injuries or deaths from any catastrophic event at the school's assets.
- Identify any critical faculty, staff, or administration whose loss would degrade or seriously complicate the safety of students, faculty, and staff during an emergency. [Consider first responders or the personnel responsible for shelter operations at a school that is a designated shelter for natural hazards.]
- Determine if the school's assets can be replaced and identify replacement costs if the school building is lost.
- Identify the locations of key equipment.
- Determine the locations of personnel work areas and systems within a school.
- Identify the locations of any personnel operating "outside" a school's controlled areas.
- Determine the location, availability, and readiness condition of emergency response assets, and the state of training of school staff in their use.

- Determine, in detail, the physical locations of critical support architectures:
  - Communications and information technology (IT—the flow of critical information)
  - Utilities (e.g., facility power, water, air conditioning, etc.)
  - Lines of communication that provide access to external resources and provide movement of students and faculty (e.g., road, rail, air transportation)

ASSET IDENTIFICATION

Critical Assets

An essential element of the planning and design process is to identify your critical assets.

An asset is a resource of value requiring protection. An asset can be tangible (e.g., students, faculty, staff, school buildings, facilities, equipment, activities, operations, and information) or intangible (e.g., processes or a school's reputation). In order to achieve the greatest risk reduction at the least cost, identifying and prioritizing a school's critical assets is a vital first step in the process to identify the best mitigation measures to improve its level of protection whether from a natural disaster or terrorist attack.

For instance, a school's main function is to teach, and its critical assets are the teachers and students. The computers and other teaching aids that are used as well as the ancillary staff are also important—but these are not as important as the teachers themselves. So you have to not only identify your critical assets but you have to prioritize them as well.

Recognizing that people are a school's most critical asset, the process described below will help identify and prioritize school infrastructure where people are most at risk and require protection. In order to conduct productive interviews, a list of areas to be covered should be generated and prioritized prior to the actual interviews.

Thorough planning and research to generate relevant questions will aid the process and yield better results. Identifying a school's critical assets is accomplished in a two-step process:

1. Define and understand the school's core functions and processes
2. Identify school infrastructure
   - Critical components/assets
   - Critical information systems and data
   - Life safety systems and safe haven areas
   - Security systems

BASIC QUESTIONS

1. What is your current security situation?
   You must find out about the crime rate in your area and any current terrorist threats or alerts. You must know if your facility has been burgled or if threats have been made by current or former employees.
2. Site—Neighbors

You must review all the facilities in your neighborhood to see which, if any, pose a threat to you. Are there high-risk targets which might be attacked, and if so, what would the impact be on your facility and your ability to continue operating.

3. Site—Location
   Walk around your perimeter and determine if your immediate surrounds are secure.
4. Site—Perimeter
   Focus on your perimeter—does it offer adequate protection, what are the vulnerabilities?
5. Perimeter Security
   Check out your perimeter physical security systems—are they adequate?
6. Building Usage
   It is essential to know everything about your building—who uses it, the main access points, traffic flow, different security levels, portals, and so on.
7. Building Structure
   Consult with your experts to determine if the building structure is safe or if physical hardening is needed.
8. Utility Systems
   Utility systems are critical but often overlooked. You must know what utilities are used, how they enter the building, where they are vulnerable, if there is on-site storage, backup systems, and so on.
9. People
   In almost all cases people are your most critical asset, whether they are employees, tenants, or visitors. You must know who uses the building and when, how they access and exit it, where they go when they are inside and whether their movement is controlled. You must also know what procedures are in place to screen staff and to train them in the event of an emergency.
10. Communications and IT
    A key element of your integrated physical security plan is the protection of your IT, communications, and information systems. Even though communications is a critical part of IPS, it is often covered in a separate plan. That is why integrated physical security is so important because it must incorporate all these separate threads into one cohesive and all-encompassing plan.
11. Equipment Operations and Maintenance
    You must know all about the equipment used in the building, how it operates, and what the maintenance schedules are. If you do not know this information, make sure that the people who do have it are on your planning team.
12. External Security Systems
    You must examine all your external security systems—are they adequate, are they integrated? Do they support your deter, detect, and delay objectives?
13. Internal Security Systems
    As with external security systems above, you have to do a comprehensive evaluation of all internal security systems. Are they adequate for the job in hand, are there newer systems that could be implemented, are there areas that should be protected that are not currently covered?

14. Security System Documents
An essential part of your IPS is to ensure that all security system documents are kept up-to-date, and that they are in a secure environment yet are available to the proper authorities in the event of an emergency.
15. Security Master Plan
As with your security system documents, you have to ensure that your security master plan is regularly reviewed and revised to accommodate changing circumstances and new threats. It should be a constantly evolving and improving document.
16. Compliance
You must ensure that all modifications comply with all relevant codes.

Identifying School Core Functions

The initial step of an asset value assessment is the determination of core functions and processes necessary for the school to continue to operate or provide services after an incident. The reason for identifying core functions/processes is to focus the design team and school administrators on what a school does, how it does it, and how various threats can affect the school. This provides more discussion and results in a better understanding of asset value. Factors that should be considered include the following:

- What are the school's primary services or outputs?
- What critical activities take place at the school?
- Who are the school's occupants or visitors?
- What inputs from external organizations are required for a school's success?

After a list of a school's assets or resources of value requiring protection have been identified, they should be assigned a value. Asset value is the degree of debilitating impact that would be caused by the incapacity or destruction of the school's assets.

Asset Value Scale

Very High	10
High	8-9
Medium High	7
Medium	5-6
Medium Low	4
Low	2-3
Very Low	1

Very high—Loss or damage of the school's assets would have exceptionally grave consequences, such as extensive loss of life, widespread severe injuries, or total loss of primary services and core functions and processes.

High—Loss or damage of the school's assets would have grave consequences, such as loss of life, severe injuries, loss of primary services, or major loss of core functions and processes for an extended period of time.

Medium high—Loss or damage of the school's assets would have serious consequences, such as serious injuries, or impairment of core functions and processes for an extended period of time.

Medium—Loss or damage of the school's assets would have moderate to serious consequences, such as injuries, or impairment of core functions and processes.

Medium low—Loss or damage of the school's assets would have moderate consequences, such as minor injuries or minor impairment of core functions and processes.

Low—Loss or damage of the school's assets would have minor consequences or impact, such as a slight impact on core functions and processes for a short period of time.

Very low—Loss or damage of the school's assets would have negligible consequences or impact.

Obviously, the key asset for every school is its people (e.g., students, faculty, and staff). They will always be assigned the highest asset value. See example that follows.

Asset	Value	Numeric Value
Students	Very High	10
Faculty	Very High	10
Staff	Very High	10
Designated Shelter (safe haven)	Very High	10
Main School Building	High	9
Teaching Functions	High	9
IT/Communications Systems	High	8
Utilities association with Shelter	Medium High	7
Utility Systems	Medium High	7
Nurses Station	Medium High	7
School/Student Records	Medium High	7
Transportation (buses/parking)	Medium High	7
Security Equipment (metal detectors, CCTV, badge equipment)	Medium High	7
Administrative Functions	Medium	5
Temporary Classrooms (trailers)	Medium Low	4
Food Services (cafeteria/kitchens)	Medium Low	4
Library	Low	3
Custodial Functions	Low	3
Science Laboratories	Low	3
Vocational Equipment (shops)	Low	3
Indoor Sports Facilities	Low	2
Outdoor Sports Facilities	Very Low	1

Threat/Hazard Assessment

After identifying asset value, the next step in the assessment process is to conduct a threat/hazard assessment wherein the threats or hazards are identified, defined, and quantified. Within the DoD, intelligence community, and law enforcement, the term "threat" is typically used to describe the design criteria for terrorism or manmade disasters. Within the FEMA and other civil agencies, the term "hazard" is used in several different contexts. "Natural hazard" typically refers to a natural event, such as an earthquake, a flood, or a wind disaster. "Manmade hazards" are "technological hazards" and "terrorism." These are distinct from natural hazards primarily in that they originate from human activity. Furthermore, "technological hazards" are generally assumed to be accidental, and their consequences are considered unintended. For the sake of simplicity, this primer will use the terms "threat" and "hazard" when referring to terrorism and manmade disasters, respectively.

The following methodology is used to conduct a threat/hazard assessment against a possible terrorist attack, but it is equally applicable for all threat assessments whether they be natural or manmade.

For terrorism, the threat is from aggressors (those people with intent to do harm) that are known to exist, have the capability for hostile actions, and have expressed intentions for using hostile actions. They may seek publicity for their cause or political gain through their actions to injure or kill people, and destroy or damage facilities, property, equipment, or resources.

Aggressor tools can be forced-entry tools, vehicles, or surveillance (visual/audio). Their weapons can be incendiary devices; small arms (rifles and handguns); stand-off military-style weapons (rocket propelled grenades or mortars); explosive devices; and CBR agents. Their tactics run the gamut: moving vehicle bombs; stationary vehicle bombs; exterior attacks (thrown objects like rocks, Molotov cocktails, hand grenades, or hand-placed bombs); stand-off weapons attacks (small arms, military or improvised direct and indirect fire weapons); covert entries (gaining entry by false credentials or circumventing security with or without weapons); mail bombs (delivered to individuals or institutions); airborne contamination (CBR agents used to contaminate the air, water, or food supply to a school); and waterborne contamination (CBR agents injected into the water supply of a school facility).

A threat assessment is a continual process of compiling and examining all available information concerning potential threats and manmade hazards. It can be broken down into two processes: (1) defining threats and (2) identifying threat event profiles and tactics.

Threat Identification

The beginning point for security design is to define threats (hazards) and tactics that may be employed. From a physical attack viewpoint, schools may be susceptible to attack by a number of different threats and tactics especially in areas of high risk. Schools are typically site constrained, have well-defined traffic control and entry points, and operate on standard schedules. School administrators need to evaluate attack objectives, threat event profiles, and the effects or impact of the attack on the school and its occupants. It should also be noted that weapons and tactics change faster than the construction of schools.

A threat (hazard) is any indication, circumstance, or event with the potential to cause loss of or damage to an asset. It is important to understand identify the people with the intent to cause harm; or who, by process, materials, or proximity, can cause indirect harm to a school building. With the goal of reducing the potential risk of a school building, the design team and school administration should seek threat assessment information from local law enforcement, the local office of the FBI, State Health Departments, the DHS, and the Homeland Security Offices (HSOs) at the state level. In many areas of the country, there are threat coordinating committees that facilitate the sharing of information. Local fire departments and HazMat units will frequently understand the threat of technological hazards due to HazMat on school grounds, as well as those in surrounding industries that could cause a collateral threat to schools. In many jurisdictions, the HazMat unit is part of the fire department.

After information on potential aggressors is gathered, it should be analyzed. A common method to evaluate terrorist threats uses five factors: existence, capability, history, intention, and targeting.

Existence addresses the questions: Who is hostile to our school building or community of concern? Are they present or thought to be present? Are they able to enter the country or are they readily identifiable in a local community upon arrival?

Capability addresses the questions: What weapons have been used in carrying out past attacks? Do the aggressors need to bring them into the area or are they available locally?

History addresses the questions: What has the potential threat element done in the past and how many times? When was the most recent incident and where, and against what target? What tactics did they use? Are they supported by another group or individuals? How did they acquire their demonstrated capability?

Intention addresses the questions: What does the potential threat element or aggressor hope to achieve? How do we know this (e.g., published in books or news accounts, speeches, letters to the editor, informant)?

THREAT IDENTIFICATION

Now you know what you are protecting, you must identify what threats you face and where these threats might come from. This means identifying all internal and external threats and hazards and understanding why they might affect you. These threats include everything from internal pilfering and violence from a disgruntled student to external threats such as theft, trespass, natural disasters, and terrorism. When you have identified your various adversaries, you can determine the probability of attack and the degree of threat that each one poses.

Schools are particularly vulnerable to attack because they are site constrained, perimeters can be easily penetrated, there are clearly defined heavy traffic patterns and students and staff arrive and leave school, and often very little security during the hours of darkness.

Information you need to define threat includes:

- Type of adversary and tactics
- Potential actions of adversary
- Motivations of adversary
- Capabilities of adversary

Defining a School's Threats

Any security strategy must incorporate the constraints of the facility so that all strengths, weaknesses, and idiosyncrasies are realized and provided for. How risks are approached will largely be driven by facility constraints. If theft and vandalism are primary risks for your school, answers to questions regarding the physical plant will determine the optimal security measures. Is the school new or old? Are the windows particularly vulnerable? Does everyone who ever worked at the school still have keys? What is the nighttime lighting like? Does the interior intrusion sensor system work well, or do the local police ignore the alarms due to a high false-alarm rate? Are visitors forced or merely requested to go through the front office before accessing the rest of the school?

If outsiders on campus are a primary concern, it will be necessary to recognize the facility's ability to control unauthorized access. How many entry points are there into the buildings? Are gangs present in the area? Are the school grounds open and accessible to anyone, or do fences or buildings restrict access? Is there easy access to the school roof? Where are the hiding places within the building or on the premises? Is the student population small enough so that most of the staff would recognize most of the students and parents?

If issues of violence are a major concern, a thorough understanding of employees, student profiles, and neighborhood characteristics will be necessary. What is the crime rate in the neighborhood? Is the school administration well-liked by the students? Are teachers allowed access to the school at night? Are students allowed off campus at lunch time? How much spending money do students generally have? Are popular hangouts for young people close by and, for business establishments, does management collaborate with the school? Are expelled or suspended students sent home or to an alternative school? How many incidents of violence have occurred at the school over the past four years? What is the general reputation of the school, and how does it appear to an outsider? Are your most vocal parents pro-security or pro-privacy? Do your students like and respect your security personnel well enough to pass them pieces of information regarding security concerns? Once the school's threats, assets, and environmental constraints are understood, the security needs can be prioritized such that the school's security goals are understood by all those involved.

Identifying security needs and then securing the funding to pay for them are usually unrelated at most schools. Schools have to have a "Plan B" for program design which may be the perfect "Plan A"—but spread out over several years of implementation. If the desirable strategies (e.g., fencing, sensors, locker searches, speed bumps) are too costly or unpalatable to the community, a school may then need to modify the facility constraints (e.g., back entrances locked from the outside, no open campus for students, no teacher access after 10 p.m., all computer equipment bolted down, no lockers for students, and so on).

Most school districts or school boards will be more supportive of security measures and the requested funding if they are well educated about the most likely risks faced each year and the options available. A security staff should not have the wide-open charter to "keep everything and everybody safe." A school board should be briefed as often as once a month as to what the current security goals are and what strategies are recommended, realizing that these will and must continue to evolve. If a school board member is clearly aware of a school's most important concerns and what is required to achieve them, then he or she is less likely to be swayed by an irate parent into making a decision that will handicap reasonable security efforts.

Targeting addresses the questions: Do we know if an aggressor (we may not know which specific one) is performing surveillance on our school, nearby facilities, or facilities that have much in common with our school? Is this information current and credible, and indicative of preparations for terrorist operations (manmade hazards)?

The threat/hazard analysis for a school can range from a general threat/hazard scenario shared by all members of a community to a very detailed examination of specific groups, individuals, and tactics that must be repelled or defended against by means of school design. The Homeland Security Advisory System has five threat levels that provide a general indication of risk of terrorist attack. If the anticipated threat or projected character/use of the facility warrant, a detailed threat analysis should be developed in coordination with local law enforcement, intelligence, and civil authorities in order to more quantitatively determine the vulnerability or risk. All schools should identify actions to be taken for each threat level.

Threat Assessment Products

A threat assessment is a continual process of compiling and examining all available information concerning potential threats and manmade hazards. The product of a threat assessment is a list of threats and hazards with a threat rating assigned. The threat rating is a subjective judgment based on existence, capability, history, intention, and targeting. Often, information is sketchy and analysts must rely more on the judgment of experts, statistical probability, and occasionally assumptions to help quantify and qualify the threat (all assumptions should be documented). The same combination of linguistic scale and numerical scale used in the asset value assessment can be used for the threat assessment. Assessing terrorist threats is much more difficult than assessing the risk from natural hazards, such as earthquakes, floods, and winds. Historical data form the basis of threat, and locality indicates vulnerability to a great extent in regard to natural hazards. For terrorist threats, the likelihood of occurrence is less defined and the associated vulnerabilities have many considerations that impact making good risk management decisions.

In this case, the threat rating scale would be:

Very high—Known aggressors or hazards highly capable of causing loss of or damage to the school exist. One or more vulnerabilities are present. The aggressors are known or highly suspected of having intent to exploit the school's assets and are known or highly suspected of performing surveillance on a facility.

High—Known aggressors or hazards capable of causing loss of or damage to the school exist. One or more vulnerabilities are present and the aggressors are known or reasonably suspected of having intent to exploit the school's assets.

Medium high—Known aggressors or hazards capable of causing loss of or damage to the school exist. One or more vulnerabilities are present and the aggressor is suspected of having intent to exploit the school's assets.

Medium—Known aggressors or hazards that may be capable of causing loss of or damage to the school exist. One or more vulnerabilities may be present; however, the aggressors are not believed to have intent to exploit the school's assets.

Medium low—Known aggressors or hazards that may be capable of causing loss of or damage to the school exist. Aggressors have no intent to exploit the school's assets.

Low—Few or no aggressors or hazards exist. Their capability of causing damage to the school's assets is doubtful.

Very low—No aggressors or hazards exist.

Design Basis Threat

Traditionally, the building regulatory system has addressed natural disaster mitigation (hurricane, tornado, flood, earthquake, windstorm, and snowstorm) through prescriptive building codes supported by well-established and accepted reference standards, regulations, inspection, and assessment techniques. Some manmade risks (e.g., HazMat storage) and specific societal goals (energy conservation and life safety) have also been similarly addressed. However, the building regulation system has not yet fully addressed all manmade hazards or terrorist threats.

Identifying design basis threats (e.g., threat tactics, weapons, tools, or explosives against which a building must be protected) should be considered as part of a school's threat assessment to facilitate the work of designers during new construction or rehabilitation of an existing school building. The DoD, GSA, and DOS all have established processes to identify design basis threats for their facilities.

The typical building design and construction process is sequential, progressing from identifying building use and design goals through actual construction.

In every school design and renovation project, there are ultimately three choices of how to address the risk posed by terrorism:

1. Do nothing and accept the risk.
2. Perform a risk assessment and manage the risk by installing reasonable mitigation measures to achieve a desired level of protection.
3. Harden the building against all threats to achieve the least amount of risk.

As stated above, a risk and vulnerability assessment is essential as part of any emergency and preparedness plan. The cost of implementing changes to protect against all eventualities is both prohibitive and impractical, which is why the vulnerability assessment is necessary. There is little point in spending hundreds of thousands of dollars to protect against a risk that is highly unlikely to happen. The focus has to be on threats that could happen, that is, if you are in California, your threat mitigation must include plans to protect against earthquakes, in Florida against hurricanes, and so on.

THREATS AND POSSIBLE SOLUTIONS

Outsiders on campus

- Posted signs regarding penalties for trespassing
- Enclosed campus (fencing)
- Guard at main entry gate to campus
- Greeters in strategic locations
- Student I.D.s or badges
- Vehicle parking stickers
- Uniforms or dress codes
- Exterior doors locked from the outside
- A challenge procedure for anyone out of class
- Cameras in remote locations
- School laid out so all visitors must pass through front office
- Temporary "fading" badges issued to all visitors

Fights on campus

- Cameras
- Duress alarms
- Whistles

Vandalism

- Graffiti-resistant sealers
- Glass-break sensors
- Aesthetically pleasing wall murals (these usually are not hit by graffiti)
- Law enforcement officers living on campus
- Eight-foot fencing
- Well-lit campus at night

Theft

- Interior intrusion detection sensors
- Property marking (including microdots) to deter theft
- Bars on windows
- Reinforced doors
- Elimination of access points up to rooftops
- Cameras
- Doors with hinge pins on secure side
- Bolting down computers and TVs
- Locating high-value assets in interior rooms
- Key control
- Biometric entry into rooms with high-value assets
- Law enforcement officer living on campus

Drugs

- Drug detection swipes
- Hair analysis kits for drug use detection (intended for parental application)
- Drug dogs
- Removal of lockers
- Random searches
- Vapor detection of drugs

Alcohol

- No open campus at lunch
- Breathalyzer® test equipment
- No access to vehicles
- No lockers
- Clear or open mesh backpacks
- Saliva test kits

Weapons

- Walk-through metal detectors
- Hand-held metal detectors
- Vapor detection of gun powder
- Crimestopper hotline with rewards for information
- Gunpowder detection swipes
- Random locker, backpack, and vehicle searches
- X-ray inspection of bookbags and purses

Malicious acts

- Setback of all school buildings from vehicle areas
- Inaccessibility of air intake and water source
- All adults on campus required to have a badge
- Vehicle barriers near main entries and student gathering areas

Parking lot problems

- Cameras
- Parking decals
- Fencing
- Card I.D. systems for parking lot entry
- Parking lots sectioned off for different student schedules
- Sensors in parking areas that should have no access during schoolday
- Roving guards
- Bike patrol

False fire alarms

- Sophisticated alarm systems that allow assessment of alarms (and cancellation if false) before they become audible.
- Boxes installed over alarm pulls that alarm locally (screamer boxes).

Bomb threats

- Caller ID on phone system
- Crimestopper program with big rewards for information
- Recording all phone calls, with a message regarding this at the beginning of each incoming call
- All incoming calls routed through a district office
- Phone company support
- No pay phones on campus
- Policy to extend the school year when plagued with bomb threats and subsequent evacuations

Bus problems

- Video cameras and recorders within enclosures on buses
- IDs required to get on school buses
- Security aides on buses
- Smaller buses
- Duress alarm system or radios for bus drivers

Teacher safety

- Duress alarms
- Roving patrols
- Classroom doors left open during class
- Cameras in black boxes in classrooms
- Controlled access to classroom areas

VULNERABILITY ASSESSMENT

A vulnerability assessment evaluates vulnerability of critical assets, or any weaknesses that can be exploited by an aggressor, across a broad range of identified threats and provides a basis for determining mitigation measures for protection of people and critical assets.

The assessment of any vulnerability of a school building should be done within the context of the defined threats and the value of the school's assets. That is, each element of the school building should be analyzed for vulnerabilities to each threat and a vulnerability rating should be assigned. The same combination of linguistic scale and numerical scale used in the asset value and threat assessments can also be used for the vulnerability assessment. It should be noted that a vulnerability assessment may change the value rating of assets due to the identification of critical nodes or some other factor that makes the school's assets more valuable.

Very high—One or more major weaknesses have been identified that make the school's assets extremely susceptible to an aggressor or hazard.
High—One or more significant weaknesses have been identified that make the school's assets highly susceptible to an aggressor or hazard.
Medium high—An important weakness has been identified that makes the school's assets very susceptible to an aggressor or hazard.
Medium—A weakness has been identified that makes the school's assets fairly susceptible to an aggressor or hazard.
Medium low—A weakness has been identified that makes the school's assets somewhat susceptible to an aggressor or hazard.
Low—A minor weakness has been identified that slightly increases the susceptibility of the school's assets to an aggressor or hazard.
Very low—No weaknesses exist.

Major Steps in Conducting a Vulnerability Assessment

- Characterize facility
- Identify assets or "targets"
- Identify credible threats against assets
- Analyze undesired events
- Analyze likelihood of occurrence
- Identify consequence of loss
- Evaluate ranked list of vulnerabilities
- Identify impact on facility mission—low, medium, high

Determining the Vulnerability Rating

This task involves determining a vulnerability rating that reflects the weakness of functions, systems, and sites in regard to a particular threat. Weakness includes the lack of redundancies that will make the building system operational after an attack.

Redundancy Factor

A terrorist selects the weapon and tactic that will cause harm to people, destroy the infrastructure, or functionally defeat the target. The function and infrastructure vulnerability analysis will identify the geographic distribution within the building and interdependencies between critical assets. Ideally, the functions should have geographic dispersion as well as a recovery site or alternate work location. However, some critical functions and infrastructure do not have a backup, or will be determined to be collocated and create what are called single-point vulnerabilities. Identification and protection of these single-point vulnerabilities are key aspects of the assessment process. Concerns related to common system vulnerabilities are:

- No redundancy
- Redundant systems feed into single critical node
- Critical components of redundant systems collocated
- Inadequate capacity or endurance in post-attack environment

Identification and protection of these single-point vulnerabilities will help you to determine a more accurate vulnerability rating for your assessment.

The results of the vulnerability assessment are then analyzed in conjunction with the results of the asset value and threat assessments developed earlier. Each asset/ threat pair is then assigned a vulnerability rating and forms the basis for identifying measures to mitigate threat vulnerability and improve protection of the building and its occupants. See example that follows.

Table 1.1 Asset Threat/Hazard Assessment

	THREAT/HAZARD									
	Terrorist Act		Armed Attack		Unauthorized Entry		Low Level CBR Attack		Cyber-terrorism	
	Stationary Vehicle Bomb		Attack by Small Arms		Forced Entry at Night to Damage School Property		Hydrogen Sulfide "Stink Bomb"		Electronic Attack to Destroy or Alter School Academic Records	
ASSET	Vulnerability Rating		Vulnerability Rating		Vulnerability Rating		Vulnerability Rating		Vulnerability Rating	
Students	VH	10	H	9	VL	1	MH	7	VL	1
Faculty	VH	10	H	9	VL	1	MH	7	ML	4
Staff	VH	10	H	9	ML	4	MH	7	ML	4
Designated Shelter (safe haven)	H	9	L	3	ML	4	L	3	VL	1
Main School Building	VH	10	ML	4	M	5	L	3	VL	1
Teaching Functions	VH	10	VH	10	ML	4	H	9	L	3
IT/Communications Systems	VH	10	M	5	MH	7	L	2	MH	7
Utilities Associated with Shelter	H	8	ML	4	L	2	VL	1	VL	1
Utility Systems (gas, electrical, sewer/water)	MH	7	ML	4	M	5	L	2	VL	1
Nurses Station	H	9	H	8	M	6	L	3	VL	1
School/Student Records	H	9	L	3	M	6	L	2	MH	7
Transportation (buses and parking)	VH	10	H	9	M	6	L	3	VL	1
Security Equipment (metal detectors, badge equipment)	H	9	MH	7	M	6	L	2	VL	1
Administrative Functions	VH	10	H	8	M	6	MH	7	M	5
Temporary Classrooms (trailers)	VH	10	VH	10	MH	7	ML	4	VL	1
Food Service (cafeteria/kitchen)	VH	10	H	8	M	5	MH	7	VL	1
Library	VH	10	VH	10	MH	7	MH	7	ML	4
Custodial Functions	VH	10	ML	4	M	6	ML	4	VL	1
Science Laboratories	VH	10	M	5	H	8	L	3	VL	1
Vocational Equipment (shops)	VH	10	ML	4	H	8	L	3	VL	1
Indoor Sports Facilities	VH	10	ML	4	H	8	M	6	VL	1
Outdoor Sports Facilities	M	5	L	2	MH	7	L	2	VL	1

VH = Very High; H = High; MH = Medium High; M = Medium; ML = Medium Low; L = Low; VL = Very Low

Risk Assessment

Risk is the potential for a loss of or damage to an asset. It is measured based upon the value of the asset in relation to the threats and vulnerabilities associated with it. Risk is based on the likelihood or probability of the hazard occurring and the consequences of the occurrence. A risk assessment analyzes the threat (probability of occurrence), and asset value and vulnerabilities (consequences of the occurrence) to ascertain the level of risk for each asset against each applicable threat/hazard. Thus, a very high likelihood of occurrence with very small consequences may require simple, low-cost mitigation measures, but a very low likelihood of occurrence with very grave consequences may require more costly and complex mitigation measures. The risk assessment provides school administrators with a relative risk profile that defines which assets are at the greatest risk against specific threats.

Risk Rating System

Risk	Risk Rating
≥ 261	Very High
201 - 260	High
141 - 200	Medium High
101 - 140	Medium
61 - 100	Medium Low
31 - 60	Low
1 - 30	Very Low

Very high—The potential for loss or damage of the school's assets is so great as to expect exceptionally grave consequences, such as extensive loss of life, widespread severe injuries, or total loss of primary services, and core functions and processes.

High—The potential for loss or damage of the school's assets is so great as to expect grave consequences, such as loss of life, severe injuries, loss of primary services, or major loss of core functions and processes for an extended period of time.

Medium high—The potential for loss or damage of the school's assets is such as to expect serious consequences (e.g., as serious injuries, or impairment of core functions and processes for an extended period of time).

Medium—The potential for loss or damage of the school's assets is such as to expect serious consequences (e.g., injuries, or impairment of core functions and processes).

Medium low—The potential for loss or damage of the school's assets is such as to expect only moderate consequences (e.g., minor injuries, or minor impairment of core functions and processes).

Low—The potential for loss or damage of the school's assets is such as to expect only minor consequences or impact (e.g., a slight impact on core functions and processes for a short period of time).

Very low—The potential for loss or damage of the school's assets is so low that there would only be negligible consequences or impact.

As a minimum, mitigation measures to reduce risk and create an acceptable level of protection should be considered for those critical assets determined to be at highest risk.

RISK ASSESSMENT

The risk assessment uses the threat, asset value, and vulnerability assessments to determine the level of risk for each critical asset against each applicable threat. This takes into account the likelihood of the threat occurring (probability) and the effects (consequences) of the occurrence if the critical asset is damaged or destroyed.

The risk assessment provides security consultants, engineers, designers, and architects with a relative risk profile that defines which assets are at the greatest risk against specific threats.

Any threat subjects an organization to risk. Therefore, when a threat is exhibited a risk rating needs to be applied in order to understand how to manage the risk. A threat management team should have processes and procedures in place for measuring the probability of loss and the severity of loss, in light of a threat.

The following primary risk types should be considered in determining risk exposure:

- Mission or function risks
- Asset risks
- Security risks

Remember that the aim in developing your secure facility design is to identify all critical assets—tangible and intangible—and reduce the risks to them to an acceptable level. You must take into account the following crucial elements—deterrence, detection, delay, and response, and then recovery and reassessment. All are mitigation measures. These are the foundations on which any physical security plan must be built.

- Deterrence—provides countermeasures such as policies, procedures, and technical devices, and controls to defend against attacks on the assets being protected.
- Detection—monitors for potential breakdowns in protective mechanisms that could result in security breaches.
- Delay—provides measures that in the event of a breach delay intruders long enough to allow them to be apprehended before they achieve their objective.
- Response—procedures and actions for responding to a breach. Note: Because total protection is almost impossible to achieve, a security program that does not also incorporate detection and delay is incomplete. To be effective, all three concepts must be elements of a cycle that work together continuously.

- Recovery—your plan to continue operations as normally as possible following an incident. Mitigation planning is part of your response and recovery with the aim of minimizing the effects of any incident.
- Reassessment—crucial and an ongoing process. Before implementing any changes, you need to revisit your strategic plan to ensure that goals and objectives will be met. Whenever there are changed circumstances or when new threats are identified, revisit your strategic plan and conduct a reassessment to see what additional measures, if any, are needed.

After identifying asset value, the next step in the assessment process is to conduct a threat/hazard assessment wherein the threats or hazards are identified, defined, and quantified.

The Risk Management Process

Risk management is the process of selecting and implementing mitigation measures to achieve an acceptable level of risk at an acceptable cost. Because it is cost-prohibitive to protect against the entire range of possible threats, it is important to develop a realistic prioritization of mitigation measures. When considering mitigation measures, the following factors should be considered:

- Results of the risk assessment, including asset value and asset vulnerabilities
- Costs of the mitigation measures
- The value of risk reduction to the school
- Frequency with which the benefits of the mitigation measures will be realized
- The deterrence or preventive value of the mitigation measures
- The expected lifespan of the mitigation measures and the time value of money

The final step is to perform a benefit/cost analysis to determine which mitigation measures will produce the greatest reduction of risk at an acceptable cost.

When dealing with manmade hazards and terrorism, it is much more difficult to predict how often an event will occur and the deterrent value of mitigation measures. Although there are historical data to help predict how often natural hazards, such as floods or tornadoes, occur in various regions, the probability or frequency of manmade hazards/threats is not known. Therefore, subjective approaches for frequency must be combined with quantitative estimates of cost-effectiveness.

Additionally, the deterrent or preventive value of a mitigation measure is also difficult to quantify. Deterrence, in the case of terrorism, may also have a secondary impact in that, once a school is "hardened," a terrorist may turn to a less-protected building, changing the likelihood of an attack for both targets. For example, the Murrah Federal Building in Oklahoma City became the target of an aggressor when he was deterred from attacking his primary target, the FBI building, because it was too difficult to get the attack vehicle close to the target. He was able to park immediately adjacent to the Murrah Federal Building and successfully target the ATF.

All these factors should be considered when calculating the value of mitigation measures, and weighing their value against their cost. Ideally, sufficient resources would be available to achieve a desired level of protection against design basis threats through mitigation measures. This is not always the case, so it is also important that every school identify or designate an appropriate authority that is authorized to accept risk on behalf of the school. Sometimes when decisions are left up to committees or personnel at an inappropriate level, poor choices or decisions can be made.

The risk management process is a benefit/cost analysis to decide and prioritize which mitigation measures to implement to achieve the desired level of protection with available resources.

For more information about design and building strategies see Section IV: "Facility Design."

PLANNING AND PREPAREDNESS

Natural disasters, such as floods, earthquakes, fires, and tornadoes, can strike a community with little or no warning. An influenza pandemic, or other infectious disease, can spread from person to person causing a major health issue in a very short time. School shootings, threatened or actual, are rare but horrific and chilling when they occur. The harrowing events of September 11 ushered in a new age of terrorism. We must all be vigilant to avert acts of terror.

Knowing what to do when faced with a crisis can be the difference between calm and chaos, between courage and fear, between life and death. There are thousands of fires in schools every year, yet there is minimal damage to life and property because staff and students are prepared. This preparedness needs to be extended to all the risks that schools face.

The time to plan is now. If you do not have an emergency preparedness plan in place, develop one. If you do have a plan in place, review, update and practice it regularly.

Taking action now will save lives, prevent injury, and minimize property damage in the event of a crisis.

The importance of reviewing and revising school and district plans cannot be underscored enough, and this handbook is designed to help you navigate this process. It is intended to give schools, districts, and communities the critical concepts and components of good crisis planning, stimulate thinking about the crisis preparedness process, and provide examples of promising practices.

Crisis plans need to be customized to communities, districts, and schools to meet the unique needs of local residents and students. What is effective for a large inner-city school district where the population is concentrated may be ineffective for a rural community where schools and first responders are far apart. Crisis plans also need to address state and local school safety laws.

WHAT IS A CRISIS?

Crises range in scope and intensity from incidents that directly or indirectly affect a single student to ones that impact the entire community. Crises can happen before, during, or after school, and on or off school campuses. The definition of a crisis varies with the unique needs, resources, and assets of a school and community. Staff and students may be severely affected by an incident in another city or state. The events of Columbine, Virginia Tech, and Sandy Hook left the entire nation fccling vulnerable.

In essence, a crisis is a situation where schools could be faced with inadequate information, not enough time, and insufficient resources, but in which leaders must make one or many crucial decisions.

All districts and schools need a crisis team. One of the key functions of this team is to identify the types of crises that may occur in the district and schools and define what events would activate the plan. The team may consider many factors, such as the school's ability to handle a situation with internal resources and its experience in responding to past events.

Plans need to address a range of events and hazards caused by both nature and people, such as:

- Natural disasters (earthquake, tornado, hurricane, flood)
- Severe weather
- Fires
- Chemical or hazardous material spills
- Bus crashes
- School shootings
- Bomb threats
- Medical emergencies
- Student or staff deaths (suicide, homicide, unintentional, or natural)
- Bullying (physical and cyber)
- Sexual harassment
- Acts of terror or war
- Outbreaks of disease or infections

EMERGENCY PLANS AND TRAINING

Every school should have a school safety emergency management plan developed in partnership with public safety agencies, including law enforcement, fire, public health, mental health, and local emergency preparedness agencies. The plan should address fire, and natural and manmade disasters. A school's plan should be tailored to address the unique circumstances and needs of the individual school, and should be coordinated and integrated with community plans and the plans of local emergency preparedness agencies.

These plans should also consider CBR attack scenarios and the associated procedures for communicating instructions to building occupants, identifying suitable shelter-in-place areas (if they exist), identifying appropriate use and selection of personal protective equipment (i.e., clothing, gloves, respirators), and directing emergency evacuations. Individuals developing emergency plans and procedures should recognize that there are fundamental differences between chemical, biological, and radiological agents. In general, chemical agents will show a rapid onset of symptoms, while the response to biological and radiological agents will be delayed. Issues, such as designated areas and procedures for chemical storage, HVAC control or shutdown, and communications with school occupants and emergency responders, should all be addressed. The plans should be as comprehensive as possible, but, as described earlier, protected by limited and controlled access. When appropriately developed, these plans, policies, and procedures can have a major impact upon school occupant survivability in the event of a CBR release.

Effective crisis planning begins with leadership at the top. Every governor, mayor, legislator, superintendent, and principal should work together to make school crisis planning a priority. Top leadership helps set the policy agenda, secures funds, and brings the necessary people together across agencies. Other leadership also needs to be identified—the teacher who is well loved in her school, the county's favorite school resource officer, or the caring school nurse.

Leaders at the grassroots level will help your school community accept and inform the planning process.

Leadership is the key to crisis preparedness. An organized management structure will be needed to respond to any crisis, and this structure begins with strong leadership.

Major Elements

Leadership should start at the top. An effective crisis plan requires strong leadership from state, district, school, and community leaders. Leadership should start at the state level and continue down to the district and school levels. In selecting team members, remember natural leaders at the grassroots level.

Districts should be at the forefront in the creation of crisis plans for all of their schools. Schools should then tailor plans to fit their needs.

At the school level, the principal serves as a leader. He or she should do the following:

- *Identify stakeholders who need to be involved in crisis planning,* such as community groups, emergency responders, families, and staff. Cultivate relationships with these groups.
- *Establish a crisis planning team.*
- *Secure commitment to crisis planning* within the school and the larger community.
- *Create an incident management structure.*

The structure should provide a comprehensive organizational structure designed for all types of emergencies. It is based on the premise that every crisis has certain major elements requiring clear lines of command and control.

Know available resources. This activity includes identifying and becoming familiar with resources in the school, such as staff members certified in cardio-pulmonary resuscitation (CPR); in the community, including everyone from emergency responders to counselors; and in organizations, such as the parent-teacher association.

Set up time to train and practice with staff, students, and emergency responders. Training is multifaceted and can include drills, in-service events, tabletop exercises, and written materials. It also includes time to review and evaluate the plan.

In times of crisis, the principal serves as the manager and a leader. This does not always equate with being the person in charge of the entire crisis response; see the closer look on ICS for more details. During a crisis, a principal should perform the following tasks:

- Respond within seconds and lead with a serious, calm, and confident style.
- Implement the crisis plan.
- Yield authority, when appropriate, to others in the plan's designated command structure.
- Facilitate collaboration among school staff and emergency responders.
- Remain open to suggestions and information that may be critical in adjusting the response.

Crisis plans should not be developed in a vacuum. They are a natural extension of ongoing school and community efforts to create safe learning environments. Good planning can enhance all school functions.

Needs assessments and other data should feed into a crisis plan. Crisis plans should address incidents that could occur inside school buildings, on school grounds, and in the community. Coordination will avoid duplication and mixed messages, as well as reduce burden on planners.

Schools and districts should open the channels of communication well before a crisis. Relationships need to be built in advance so that emergency responders are familiar with your school. Cultivate a relationship with city emergency managers, public works officials, and health and mental health professionals now, and do not overlook local media. It is important that they understand how the districts and schools will respond in a crisis.

Crisis plans should be developed in partnership with other community groups, including law enforcement, fire safety officials, emergency medical services, as well as health and mental health professionals.

Do not reinvent the wheel. These groups know what to do in an emergency and can be helpful in the development of the plan. Get their help to develop a coordinated plan of response.

A common vocabulary is necessary. It is critical that school staff and emergency responders know each other's terminology.

Work with emergency responders to develop a common vocabulary. The words used to give directions for evacuation, lockdown, and other actions should be clear and not hazard specific. The FEMA recommends using plain language to announce the need for action, for example, "evacuate" rather than "code blue." Many districts note that with plain language everyone in the school building including new staff, substitute teachers, and visitors will know what type of response is required.

However, some districts have found it useful to use, but streamline, codes. Rather than a code for each type of incident they use only one code for each type of response. With either approach, it is critical that terms and/or codes are used consistently across the district.

Schools should tailor district crisis plans to meet individual school needs. In fact, a plan should not be one document. It should be a series of documents targeted to various audiences. For example, a school could use detailed response guides for planners, flipcharts for teachers, a crisis response toolbox for administrators, and wallet cards containing evacuation routes for bus drivers. Plans should be age appropriate. Elementary school children will behave much differently in a crisis than high school students.

Plan for the diverse needs of children and staff. Our review of crisis plans found that few schools addressed children or staff with physical, sensory, motor, developmental, or mental challenges. Special attention is also needed for children with limited English proficiency. Outreach documents for families may be needed in several languages.

Include all types of schools where appropriate. Be sure to include alternative, charter, and private schools in the planning process, as well as others who are involved with children before and after school.

Provide teachers and staff with ready access to the plan so they can understand its components and act on them. People who have experienced a crisis often report that they go on "autopilot" during an incident. They need to know what to do in advance not only to get them through an incident but also to help alleviate panic and anxiety.

Training and practice are essential for the successful implementation of crisis plans. Most students and staff know what to do in case of a fire because the law requires them to participate in routine fire drills, but would they know what to do in a different crisis? Many districts now require evacuation and lockdown drills in addition to state-mandated fire drills. Drills also allow your school to evaluate what works and what needs to be improved.

COMMUNICATION

Clear lines of communication are crucial to a successful response to a crisis. During the planning process, it will be important to establish effective lines of communication among and within the state, district, school, and community groups. When creating a crisis plan, there are several communication needs that should be addressed.

Communication is essential before crises occur. As already stated, plain language terminology and codes should be the same across schools in a district.

Identify several modes of communication for both internal and external communication. Keep in mind that in times of crisis, computers, intercoms, telephones, and even cell phones may not work or may be dangerous to use. Plan for several methods of communication in a crisis.

Make sure that schools have adequate supplies of communication gear and that the appropriate individuals have access to it. One school's crisis plan, for example, calls for the principal to immediately grab a backpack containing a cell phone and a walkie talkie. Communication gear is of no use if no one can access it.

Verify that school communication devices are compatible with emergency responder devices. A cell phone or two-way radio is of no use if it cannot be used with the emergency responder's phone or radio. Also, check to see that the school's communication devices do not interfere with the emergency responder's equipment.

Create communication plans to notify families that a crisis has occurred at their child's school. These pathways should include several modes of communication, including notices sent home and phone trees, so the pathways can be tailored to fit the needs of a particular crisis. For example, it may be appropriate in some crises to send a notice home, while other crises require immediate parental notification. Use these pathways throughout the planning process to encourage parental input and support.

Establish communication pathways with the community. This may be in the form of a phone or e-mail tree, a community liaison, or media briefings. It is crucial to keep the community informed before, during, and after a crisis.

Designate a PIO to deal with families, the community, and the media. The designation of one individual will help all parties stay informed with identical information.

Good communication during a crisis is also crucial. Below are some key points to keep in mind:

Keep staffs who are managing the students informed. Regardless of the amount of training staff members have received, there is going to be chaos and fear

Communication mitigates those reactions and helps regain a sense of calm and control

Notify families of action being taken. Understand that parents are going to want immediate access to their children. Safely begin reunification procedures as soon as possible. Keep families informed as much as possible, especially in the case of delayed reunification

Communication often stops after a crisis subsides. However, during the recovery phase, keeping staff and community informed remains critical.

PREPAREDNESS

Crises have the potential to affect every student and staff member in a school building. Despite everyone's best efforts at crisis prevention, it is a certainty that crises will occur in schools. Good planning will facilitate a rapid, coordinated, effective response when a crisis occurs. Being well prepared involves an investment of time and resources, but the potential to reduce injury and save lives is well worth the effort.

Every school needs a crisis plan that is tailored to its unique characteristics. Within a school district, however, it is necessary for all plans to have certain commonalities.

Also, it is impractical for all schools to work individually with emergency responders and other local agencies, although school staff should meet the people who will respond to a crisis before one happens. It is important to find the right balance and to assign district and school roles early.

Set a realistic timetable for the preparation process. While it is reasonable to feel a sense of urgency about the need to be prepared for a crisis, a complete, comprehensive crisis plan cannot be developed overnight. Take the time needed for collecting essential information, developing the plan, and involving the appropriate people.

Crisis management is a continuous process in which all phases of the plan are being reviewed and revised. Good plans are never finished. They can—and should—always be updated based on experience, research, and changing vulnerabilities.

Start by identifying who should be involved in developing the crisis plan. Include training and drills. Delegating responsibilities and breaking the process down into manageable steps will help planners develop the plan.

Action Checklist

- Determine what crisis plans exist in the district, school, and community.
- Identify all stakeholders involved in crisis planning.
- Develop procedures for communicating with staff, students, families, and the media.
- Establish procedures to account for students during a crisis.
- Gather information that exists about the school facility, such as maps and the location of utility shutoffs.
- Identify the necessary equipment that needs to be assembled to assist staff in a crisis.

Action Steps

Start by identifying who should be involved in developing the crisis plan. Include training and drills. Delegating responsibilities and breaking the process down into manageable steps will help planners develop the plan.

Section II

The Crisis Management Plan

Every school should have an emergency preparedness and crisis management plan based on the following simple premise:

> *We will develop and maintain crisis management programs to ensure we are prepared to deal with incidents and crises that could affect our students, teachers, staff, visitors, and other persons involved in school-related operations.*

THE SEQUENCE OF CRISIS MANAGEMENT

There are four phases of crisis management:

1. Mitigation/prevention addresses what schools and districts can do to reduce or eliminate risk to life and property.
2. Preparedness focuses on the process of planning for the worst-case scenario.
3. Response is devoted to the steps to take during a crisis.
4. Recovery deals with how to restore the learning and teaching environment after a crisis.

The US Department of Education recommends that each school crisis plan should address these four major areas.

Mitigation/Prevention

- Conduct an assessment of each school building. Identify those factors that put the building, students, faculty, and staff at greater risk, such as proximity to rail tracks that regularly transport hazardous materials or facilities that produce highly toxic material or propane gas tanks and develop a plan for reducing the risk. This can include plans to evacuate students away from these areas in times of crisis and to reposition propane tanks or other hazardous materials away from school buildings.

- Work with businesses and factories in close proximity to the school to ensure that the school's crisis plan is coordinated with their crisis plans.
- Ensure that a process is in place for controlling access and egress to the school. Require all persons who do not have authority to be in the school to sign in.
- Review traffic patterns, and where possible, keep cars, buses, and trucks away from school buildings.
- Review landscaping and ensure that buildings are not obscured by overgrowth of bushes or shrubs where contraband can be placed or people can hide.

Preparedness

- Have site plans for each school building readily available and ensure they are shared with first responders and agencies responsible for emergency preparedness.
- Ensure there are multiple evacuation routes and rallying points. First or second evacuation site options may be blocked or unavailable at the time of the crisis.
- Practice responding to crisis on a regular basis.
- Ensure a process is established for communicating during a crisis.
- Inspect equipment to ensure it operates during crisis situations.
- Have a plan for discharging students. Remember that, during a crisis, many parents and guardians may not be able to get to the school to pick up their children. Make sure every student has a secondary contact person and contact information readily available.
- Have a plan for communicating information to parents and for quelling rumors. Cultivate relationships with the media ahead of time, and identify a public information officer (PIO) to communicate with the media and the community during a crisis.
- Work with law enforcement officials and emergency preparedness agencies on a strategy for sharing key parts of the school crisis plans.

Response

- Identify the type of crisis that is occurring and determine the appropriate response.
- Develop a command structure for responding to a crisis. The roles and responsibilities for educators, law enforcement, fire officials, and other first responders in responding to different types of crisis need to be developed, coordinated, reviewed, and approved.
- Ensure all relevant staff maintains communications.

Recovery

- Return to the business of teaching and learning as soon as possible.
- Identify and approve a team of credentialed mental health workers to provide mental health services to faculty and students after a crisis. Understand that recovery takes place over time and that the services of this team may be needed.
- Ensure that the team is adequately trained.
- The plan should include notification of parents about actions that the school intends to take to help students recover from the crisis.

TEN COMMON MISTAKES WHEN DEVELOPING A PLAN

1. Not Consulting With Local Emergency Response Officials
2. Not Customizing Prepackaged Plans
3. Plan is Difficult to Read Under Stress
4. Plan Components aren't Consistent
5. Using Coded Announcements, or too many of them
6. Not Conducting Drills of Plan
7. Not Telling Parents about Plan
8. Not Planning for Power Failure or Phone System, Communications Failure
9. Not Training All Staff on Plan
10. Not Giving Administrators Decision Making Authority

Crisis Planning Essentials

Crisis management is a continuous process in which all phases of the plan are being reviewed and revised. Good plans are never finished. They can always be updated based on experience, research, and changing vulnerabilities. Districts and schools may be in various stages of planning. These guidelines, developed by the US Department of Education, provide the resources needed to start the planning process and can be used as a tool to review and improve existing plans.

Crisis planning may seem overwhelming. It takes time and effort, but it is manageable.

- Effective crisis planning begins with leadership at the top. Every governor, mayor, legislator, superintendent, and principal should work together to make school crisis planning a priority. Top leadership helps set the policy agenda, secures funds, and brings the necessary people together across agencies. Other leadership also needs to be identified—the teacher who is well loved in her school, the county's favorite school resource officer, or the caring school nurse. Leaders at the grassroots level will help your school community accept the planning process.
- Crisis plans should not be developed in a vacuum. They are a natural extension of ongoing school and community efforts to create safe learning environments. Good planning can enhance all school functions. Needs assessments and other data should feed into a crisis plan. Crisis plans should address incidents that could occur inside school buildings, on school grounds, and in the community. Coordination will avoid duplication and mixed messages, as well as reduce the burden on planners.
- School and districts should open the channels of communication well before a crisis. Relationships need to be built in advance so that emergency responders are familiar with your school. Cultivate a relationship with city emergency managers, public works officials, and health and mental health professionals now, and do not overlook local media. It is important that they understand how the district and schools will respond in a crisis.
- Crisis plans should be developed in partnership with other community groups, including law enforcement, fire safety officials, emergency medical services, as well

as health and mental health professionals. Do not reinvent the wheel. These groups know what to do in an emergency and can be helpful in the development of your plan. Get their help to develop a coordinated plan of response.

- A common vocabulary is necessary. It is critical that school staff and emergency responders know each other's terminology. Work with emergency responders to develop a common vocabulary. The words used to give directions for evacuation, lockdown, and other actions should be clear and not hazard specific. The FEMA recommends using plain language to announce the need for action, for example, "evacuate" rather than "code blue." Many districts note that with plain language everyone in the school building, including new staff, substitute teachers, and visitors, will know what type of response is called for. However, some districts have found it useful to use—but streamline—codes. Rather than do a code for each type of incident they use only one code for each type of response. With either approach, it is critical that terms and/or codes are used consistently across the district.
- Schools should tailor district crisis plans to meet individual school needs. In fact, a plan should not be one document. It should be a series of documents targeted to various audiences. For example, a school could use detailed response guides for planners, flipcharts for teachers, a crisis response toolbox for administrators, and wallet cards containing evacuation routes for bus drivers. Plans should be age appropriate. Elementary school children will behave very differently in a crisis from high school students.
- Plan for the diverse needs of children and staff. Our review of crisis plans found that few schools addressed children or staff with physical, sensory, motor, developmental, or mental challenges. Special attention is also needed for children with limited English proficiency. Outreach documents for families may be needed in several languages.
- Include all types of schools where appropriate. Be sure to include alternative, charter, and private schools in the planning process, as well as others who are involved with children before and after school.
- Provide teachers and staff with ready access to the plan so they can understand its components and act on them. People who have experienced a crisis often report that they go on "autopilot" during an incident. They need to know what to do in advance not only to get them through an incident but also to help alleviate panic and anxiety.
- Training and practice are essential for the successful implementation of crisis plans. Most students and staff know what to do in case of a fire because the law requires them to participate in routine fire drills, but would they know what to do in a different crisis? Many districts now require evacuation and lockdown drills in addition to state-mandated fire drills. Drills also allow your school to evaluate what works and what needs to be improved.

Crisis plans are living documents. They need to be reviewed and revised regularly. Analyzing how well a crisis plan worked in responding to an incident, whether a drill or a real event, is crucial. Documenting all actions taken while, during and after an event helps in identifying the strengths and weaknesses of a plan. Use this information to strengthen the plan.

SCHOOL SAFETY PLANNING AND PREPAREDNESS

> Families trust schools to keep their children safe during the day. Thanks to the efforts of millions of teachers, principals and staff across America, the majority of schools remain a safe haven for our nation's youth. The unfortunate reality is, however, that school districts in this country may be touched either directly or indirectly by a crisis of some kind at any time.
>
> —The Office of Safe and Drug-Free Schools, US Department of Education

Natural disasters, such as floods, earthquakes, fires, and tornadoes, can strike a community with little or no warning. While still rare, school shootings, threatened or actual, are horrific and chilling when they occur. The harrowing events of September 11 and subsequent anthrax scares have ushered in a new age of terrorism. Communities across the country are struggling to understand and avert acts of terror.

Children and youth rely on and find great comfort in the adults who protect them. Teachers and staff must know how to help their students through a crisis and return them home safely. Knowing what to do when faced with a crisis can be the difference between calm and chaos, between courage and fear, between life and death. There are thousands of fires in schools every year, yet there is minimal damage to life and property because staff and students are prepared. This preparedness needs to be extended to all risks that schools face. Schools and districts need to be ready to handle crises, large and small, to keep our children and staff out of harm's way and ready to learn and teach.

The time to plan is now. If you do not have a crisis plan in place, develop one. If you have one, review, update and practice it.

Crises range in scope and intensity from incidents that directly or indirectly affect a single student to ones that impact the entire community. Crises can happen before, during, or after school, and on or off school campuses. The definition of a crisis varies with the unique needs, resources, and assets of a school and community. Staff and students may be severely affected by an incident in another city or state. The events of Columbine and September 11 left the entire nation feeling vulnerable.

The definition of a crisis is: "An unstable or crucial time or state of affairs in which a decisive change is impending, especially one with the distinct possibility of a highly undesirable outcome" (*Webster's Ninth Collegiate Dictionary*, 1987). Additionally, *Webster* notes that "crisis" comes from the Greek word meaning "decision." In essence, a crisis is a situation where schools could be faced with inadequate information, not enough time, and insufficient resources, but in which leaders must make one or many crucial decisions.

All districts and schools need a crisis team. One of the key functions of this team is to identify the types of crises that may occur in districts and schools and define what events would activate the plan. The team may consider many factors, such as the school's ability to handle a situation with internal resources and its experience in responding to past events. Plans need to address a range of events and hazards caused by nature and people.

Planning Considerations for Safe Schools and Crisis Response

In the school's mission statement, identify the context for which the school wishes the academic learning to take place, using phrases like "to learn in a safe and secure environment free of violence, drugs, and fear." Such phrases enhance the school's legal position to create and enforce policies that promote a safe, caring, and disciplined school climate.

Identify a specific procedure for evaluating and responding to threats. Every campus should have a series of threat assessment protocols so that school officials can effectively work with mental health and law enforcement professionals in handling circumstances that could result in potential violence or harm. Also, make certain that students are on your team. For the most part, students are the best information resources for inside threats. Recent studies by the Secret Service revealed that in the vast majority of student shootings, other students on the campus were aware of the event before it occurred. Having a tip line or safe reporting mechanism in place for students is critical.

Identify the potential disasters that could occur based on the school's setting and climate. Such disasters may include:

- Civil unrest/demonstrations/rioting
- Bomb threats/explosions
- Intruders/unauthorized visitors
- Hostage takings
- Sniper attacks
- Extortion
- Assault/battery/rape
- Weapons possessions
- Drug abuse/trafficking
- Gang-related violence/drive-by shootings
- Kidnappings/abductions
- Child abuse/neglect/molestation
- Medical emergencies
- Life-threatening illness
- Accidental injury or death
- Intentional injury or death
- Utility failures
- Chemical spills
- Fire
- Automobile accidents
- Severe weather
- Natural disasters: earthquake, flood, tornado, fire, hurricane, tsunami
- Mass-transit disasters: falling aircraft/train derailment/bus accidents

Control campus access. Minimize the number of campus entrance and exit points used daily. Access points to school grounds should be limited and supervised on a regular basis by individuals who are familiar with the student body. Campus traffic, both pedestrian and vehicular, should flow through areas that can be easily and naturally supervised. Delivery entrances used by vendors should also be checked regularly. Parking lots often have multiple entrances and exits, which contribute to the vandalism and defacement of vehicles and school property. Vehicular and pedestrian access should be carefully controlled. Perimeter fencing should be considered. Bus lots should be secured and monitored. Infrequently used rooms and closets should be locked. Access to utilities, roofs, and cleaning closets should be secured.

Identify specifically assigned roles and responsibilities. Specific policies and procedures that detail staff members' responsibilities for security should be developed. These responsibilities may include monitoring hallways and restrooms, patrolling parking lots, and providing supervision at before-school and after-school activities. Specific roles and responsibilities should also be assigned for times of crisis, including the appointment of a crisis team.

Identify whom to call in a crisis. Maintain an updated list of whom to call in case of various kinds of crises. Develop a close working partnership with these emergency responders. Know the extent of services offered by these agencies. Find out what to do when an emergency responder is not immediately available. Develop a close working partnership with law enforcement officials. Determine in advance who will lead, who will follow, and how searches, interrogations, and other issues will be handled. The counselor or psychologist can also be an important partner in the aftermath of a crisis.

Provide training for all members of the school community regarding cultural awareness and sensitivity. It is important to consider the impact of cultural influences on a school community's ability to create and maintain safe, secure, and peaceful schools. Cultural influences will directly affect the information, strategies, and resources that will be used in safe school planning. The sensitivity to cultural influences also applies to creating a plan to manage and respond to a crisis.

Establish an emergency operation communication system. In addition to campus intercoms and two-way radios, school officials must be able to communicate with law enforcement and outside telephone providers. This includes the use of cell phones.

Implement a uniform school crime reporting and record-keeping system. When school administrators know what crimes are being committed on their campus, when these are committed, and where and who is involved, they are better able to develop effective strategies and supervision procedures. In addition, it is important to conduct some level of crime analysis to determine what, if any, linkages exist among various aspects of criminal activity on the campus.

Preparing for National Emergencies

Schools should stay informed about potential national security threats, be prepared for emergencies, and know how to react during an attack.

Identify potential and reliable sources of information to be accessed once a crisis situation develops. Prepare a plan that identifies the first and subsequent contacts you will make to access credible information and appropriate direction for action.

Perform an assessment of your school's risk during a national crisis. This includes:

- Evaluating health and medical on-site preparedness.
- Checking the availability and accessibility of local emergency services, including

HazMat (hazardous materials), fire, emergency medical, law enforcement, and local and federal emergency management agencies.
- Identifying potential terrorist targets in your local community.
- Identifying and taking inventory of potential informal contacts in the community who could provide food, water, shelter, medical aid, power sources, and other forms of emergency support.
- Reviewing viable communications plans, such as phone chains and parent contact information.

Be observant of the things transpiring on campus. During periods of high alert, an additional level of vigilance must be in place. Everyone who comes onto the campus must have a legitimate purpose. It is important to have a uniform screening policy for all visitors, including volunteer workers, vendors, and delivery/service personnel. All visitors should be required to sign in at the school office, state their specific business, and wear or visibly display a visitor's badge. Keep records. Watch for suspicious people, packages, and activities on or near your campus. Notify authorities of these observations. *In view of Homeland Security recommendations, it is especially important to pay added attention to the possibility of the following kinds of disasters. Review and revise your crisis plans accordingly.*

- Bomb threats/explosions
- Suicide bombings
- Intruders/unauthorized visitors
- Biological/radiological attacks
- Utility failures
- Mass-transit disasters: falling aircraft/train derailment/bus accidents

Assemble your crisis teams to reacquaint members with each other and with crisis plans and procedures. Enlist new team members as needed.

Provide training to all school staff members regarding crisis preparation. Detail each person's responsibilities during a crisis. Train them to be observant and watchful of suspicious or out-of-the-ordinary activity; how to identify and what to do with suspicious packages; how to turn off utilities and heating or air systems and who would be responsible for doing so; and basic first aid.

Conduct an inventory of campus provisions, including food, water, alternative power sources, materials for sealing doors and windows, and medical and first aid supplies that would need to be available during an emergency. Consider assembling emergency kits and food and water supplies for every classroom. Maintain an updated record of any hazardous chemicals and cleaning agents that may be stored on campus. Be sure that such materials are securely stored according to local and federal regulations.

Review and, if needed, revise existing plans for evacuation, alternative shelter, temporary lockdown, and shelter-in-place. Be prepared to put any combination of these into effect. It is possible that a situation would call for simultaneous lockdown of one section of the building while evacuating other parts of the school.

Keep parents informed of your crisis plans, procedures, and protective policies, particularly with regard to reuniting parents and children after a crisis event.

As appropriate to your community and degree of risk, conduct emergency drills for evacuation, lockdown, or shelter-in-place procedures so that staff and students are familiar with the appropriate response in an actual emergency.

> Creating safe schools is a joint responsibility involving students, parents, teachers, school officials, local law enforcement, judges, emergency personnel, social service and a variety of other youth-serving professionals. The bottom line is if we are going to require young people to attend school, then it is our responsibility to provide them with a safe, secure and welcoming environment. In this time of heightened potential for terror, including the use of chemical, biological and radiological weapons in our community, it is more important than ever for our schools to be prepared to respond in appropriate ways. Ronald Stephens, executive director of the National School Safety Center.

Planning for the Worst

Every school should have a school safety emergency management plan developed in partnership with public safety agencies, including law enforcement, fire, public health, mental health, and local emergency preparedness agencies. The plan should address fire and natural and manmade disasters. A school's plan should be tailored to address the unique circumstances and needs of the individual school and should be coordinated and integrated with community plans and the plans of local emergency preparedness agencies.

Because of the ever-present threat of terrorism, these plans should also consider CBR attack scenarios and the associated procedures for communicating instructions to building occupants, identifying suitable shelter-in-place areas (if they exist), identifying appropriate use and selection of personal protective equipment (i.e., clothing, gloves, respirators), and directing emergency evacuations. This is dealt with in more detail later on.

Identify and involve stakeholders. Identify the stakeholders to be involved in developing the crisis management plan (the people who are concerned about the safety of the school and the people who will call assist when a crisis occurs). Ask stakeholders to provide feedback on sections of the plan that pertain to them. For instance, ask families to comment on procedures for communicating with them during a crisis.

During this process, create working relationships with emergency responders. It is important to learn how these organizations function and how you will work with each other during a crisis. Take time to learn the vocabulary, command structure, and culture of these groups. Some districts have found it useful to sign MOUs with these agencies that specify expectations, including roles and responsibilities.

It is essential to work with city and county emergency planners. You need to know the kinds of support municipalities can provide during a crisis, as well as any plans the city has for schools during a crisis. For example, city and county planners may plan to use schools as an emergency shelter, a supply depot, or even a morgue.

Reviewing this information in advance will help you quickly integrate resources. Participating in local emergency planning gives school and district administrators insight into all the problems they might face in the event of a community-wide crisis and will help school efforts.

Consider existing efforts. Before jumping in to develop your crisis plan, investigate existing plans (such as those of the district and local government). How do other agencies' plans integrate with the school's plan? Are there conflicts? Does the comprehensive school safety plan include a crisis plan? What information from the district's crisis plan can be used in the school's crisis plan?

If the school recently completed a crisis plan, efforts may be limited to revising the plan in response to environmental, staff, and student changes:

- Has the building been renovated or is it currently under renovation?
- Is the list of staff current?
- Have there been changes in the student population?
- Have other hazards revealed themselves?

Determine what crises the plan will address. Before assigning roles and responsibilities or collecting the supplies that the school will need during a crisis, define what is a crisis for your school based on vulnerabilities, needs, and assets.

Describe the types of crises the plan addresses, including local hazards and problems identified from safety audits, evaluations, and assessments conducted during the mitigation/prevention phase. Consider incidents that may occur during community use of the school facility and prepare for incidents that occur while students are off-site (e.g., during a field trip).

Define roles and responsibilities. How will the school operate during a crisis? Define what should happen, when, and at whose direction, that is, create an organizational system. This should involve many of the school staff—important tasks will be neglected if one person is responsible for more than one function.

School staff should be assigned to the following roles:

- School commander
- Liaison to emergency responders
- Student caregivers
- Security officers
- Medical staff
- Spokesperson

During the planning process, both individuals and backups should be assigned to fill these roles.

If the district has not already appointed a PIO, it should to do so right away. Some large school districts have staff dedicated solely to this function. Many smaller districts use the superintendent, school security officers, or a school principal as their PIO.

Work with law enforcement officers and emergency responders to identify crises that require an outside agency to manage the scene (fire, bomb threat, hostage situations). Learn what roles these outsiders will play, what responsibilities they will take on, and how they will interact with school staff. Especially important is determining who will communicate with families and the community during an incident.

Many schools and emergency responders use the ICS, to manage incidents. ICS provides a structured way for delegating responsibilities among school officials and all emergency responders during crisis response. An ICS and/or other management plan needs to be created with all emergency responders and school officials before a crisis occurs.

Develop methods for communicating with the staff, students, families, and the media. Address how the school will communicate with all of the individuals who are directly or indirectly involved in the crisis. One of the first steps in planning for communication is to develop a mechanism to notify students and staff that an incident is occurring and to instruct them on what to do. It is critical that schools and emergency responders use the same definitions for the same terms.

Don't create more confusion because terms do not mean the same to everyone involved in responding to a crisis. It is important to determine how to convey information to staff and students by using codes for evacuation and lockdown, or simply by stating the facts. FEMA recommends simply using plain language rather than codes.

If students are evacuated from the school building, will staff use cell phones, radios, intercoms, or runners to get information to the staff supervising them? Be sure to discuss the safest means of communication with law enforcement and emergency responders. For example, some electronic devices can trigger bombs.

Plan how to communicate with families, community members, and the media. Consider writing template letters and press releases in advance so staff will not have to compose them during the confusion and chaos of the event. It's easier to tweak smaller changes than to begin from scratch.

Often the media can be very helpful in providing information to families and others in the community. Be sure to work with local media before a crisis occurs to help them understand school needs during an incident.

Obtain necessary equipment and supplies. Provide staff with the necessary equipment to respond to a crisis.

Consider whether there are enough master keys for emergency responders so that they have complete access to the school. Get the phones or radios necessary for communication.

Ask for contact information for families. Maintain a cache of first aid supplies. What about food and water for students and staff during the incident?

Prepare response kits for secretaries, nurses, and teachers so they have easy access to the supplies. For example, a nurse's kit might include student and emergency medicines ("anaphylaxis kits," which may require physician's orders, for use in breathing emergencies, such as severe, sudden allergic reactions), as well as first aid supplies. A teacher's kit might include a crisis management reference guide, as well as an updated student roster.

Prepare for immediate response. When a crisis occurs, quickly determine whether students and staff need to be evacuated from the building, returned to the building, or locked down in the building. Plan action steps for each of these scenarios.

Evacuation requires all students and staff to leave the building. While evacuating to the school's field makes sense for a fire drill that only lasts a few minutes, it may not be an appropriate location for a longer period of time.

The evacuation plan should include backup buildings to serve as emergency shelters, such as nearby community centers, religious institutions, businesses, or other schools. Agreements for using these spaces should be negotiated or reconfirmed prior to the beginning of each school year. Evacuation plans should include contingencies for weather conditions, such as rain, snow, and extreme cold and heat. While most students will be able to walk to a nearby community center, students with disabilities may have more restricted mobility. Your plan should include transportation options for these students.

If an incident occurs while students are outside, you will need to return them to the building quickly. This is a *reverse evacuation.* Once staff and students are safely in the building, you may find the situation calls for a lockdown.

Lockdowns are called for when a crisis occurs outside of the school and an evacuation would be dangerous. A lockdown may also be called for when there is a crisis inside, and movement within the school will put students in jeopardy. All exterior doors are locked and students and staff stay in their classrooms. Windows may need to be covered. The steps in determining which action is most appropriate for each situation are illustrated below.

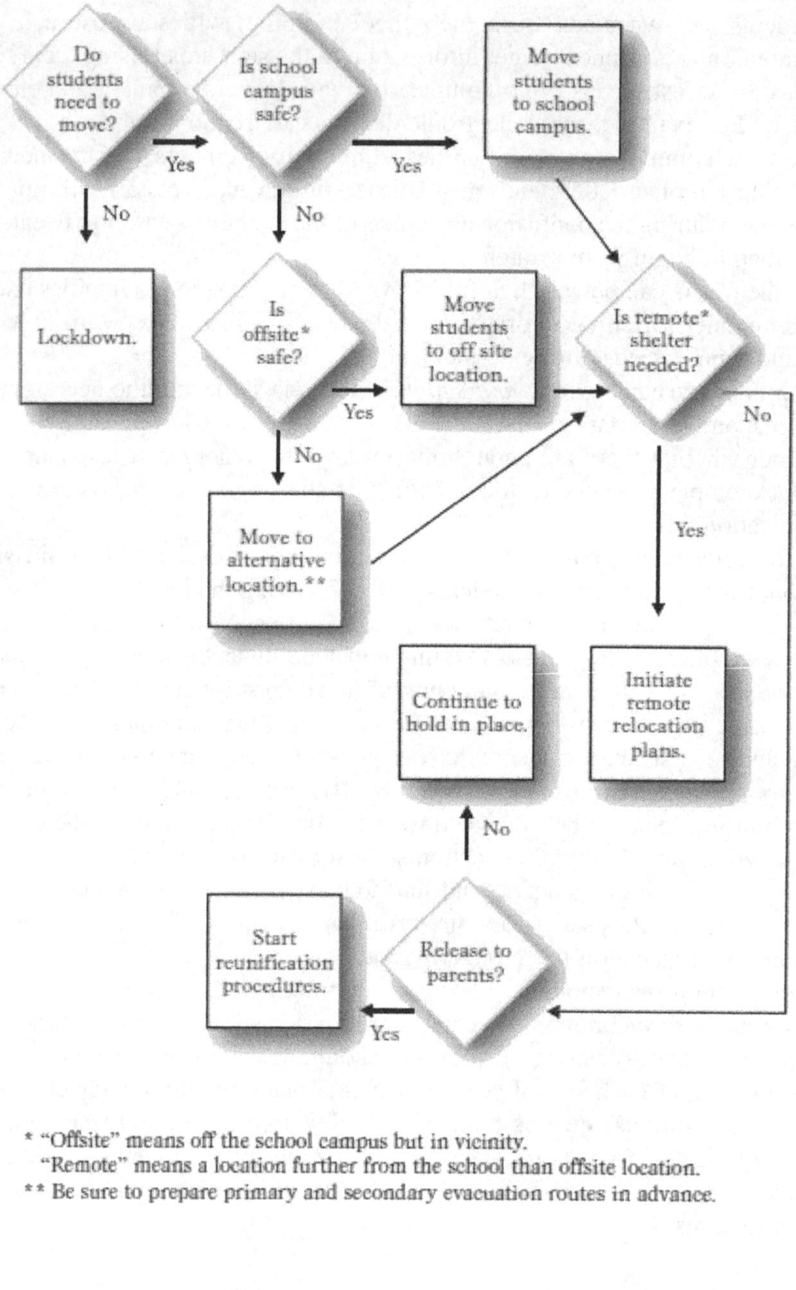

* "Offsite" means off the school campus but in vicinity.
"Remote" means a location further from the school than offsite location.
** Be sure to prepare primary and secondary evacuation routes in advance.

Shelter-in-place is used when there is no time to evacuate or when it may be harmful to leave the building. Shelter-in-place is commonly used during hazardous material spills. Students and staff are held in the building and windows and doors are sealed. There can be limited movement within the building.

Create maps and facilities information. In a crisis, emergency responders need to know the location of everything in a school. Create site maps that include information about classrooms, hallways, and stairwells, the location of utility shutoffs, and potential staging sites. Emergency responders need copies of this information in advance. During a crisis designate locations—staging sites—for emergency responders to organize, for medical personnel to treat the injured, for the public information officer to brief the media, and for families to be reunited with their children. Student reunification sites should be as far away from the media staging area as possible. Law enforcement will help determine the plans needed to facilitate access of emergency responders and to restrict access of well-wishers and the curious.

Develop accountability and student release procedures. As soon as a crisis is recognized, account for all students, staff, and visitors. Emergency responders treat a situation very differently when people are missing. For example, when a bomb threat occurs, the stakes are substantially higher if firefighters do not know when they are trying to locate and disarm a bomb whether students are in the school.

Be sure to inform families of release procedures before a crisis occurs. In many crises, families have flocked to schools wanting to collect their children immediately. A method should be in place for tracking student release and ensuring that students are only released to authorized individuals.

Practice. Preparedness includes emergency drills and crisis exercises for staff, students, and emergency responders. Many schools have found tabletop exercises very useful in practicing and testing the procedures specified in their crisis plan. Tabletop exercises involve school staff and emergency responders sitting around a table discussing the steps they would take to respond to a crisis. Often, training and drills identify issues that need to be addressed in the crisis plan and problems with plans for communication and response. Teachers also need training in how to manage students during a crisis, especially those experiencing panic reactions.

Careful consideration of these issues will improve your crisis plan and better prepare you to respond to an actual crisis.

Address liability issues. Consideration of liability issues is necessary before crisis planning can be completed and may protect you and your staff from a lawsuit.

Situations where there is a foreseeable danger can hold liability if the school does not make every reasonable effort to intervene or remediate the situation. A careful assessment of the hazards faced by the school is critical.

FACILITY SAFETY AND SECURITY

Facility disaster preparedness focuses on protecting educational facilities so that they can be used for their intended purposes. Preparedness activities vary depending on the disaster and whether there is warning. A hurricane provides warning only as its approach path can be estimated. A bomb going off on the roof of a school gives little or no warning.

The type of disaster will determine your preparedness activities. For all disasters, the following actions should be taken:

- If the school emergency plan contains a Facility Hazard Analysis, review and understand the analysis.
- Review insurance policies and keep paper copies in a safe, watertight place, perhaps off school grounds.
- Ensure communication systems have redundant backup.
- Clearly post and practice evacuation procedures.
- Gather a Go Kit (detailed in following pages).
- Create backups of important computer files.
- Gather other facility-relevant documents or equipment.
- Review and update mutual aid agreements.

Specific facility hazard preparedness activities include:

Potential Manmade Hazards

- Identify potential threats or targets near the school, along with their impact (e.g., chemical plants, gas lines, heavy truck traffic, and railroad tracks).
- Identify hidden areas adjacent to the school that might provide offenders with cover or provide students with a location for illicit activities.
- Maintain natural surveillance from the surrounding neighborhood, allowing neighbors and passing patrol cars to help serve as guardians of the school.

Seismic and Wind Hazards

- Identify alternate routes into and out of the site to avoid potential fallen trees, buildings, utility lines, or other hazards.
- Ensure backup emergency power and communication sources have been incorporated into the design.
- Ensure building setbacks are adequate to prevent damage from falling trees.

Flood Hazard

- Is the site located in a floodplain or at high risk if nearby water sources flood?
- Can emergency vehicles access the site during high water conditions?

Storm Hazard

- If there is construction on the campus, contact the facilities project manager to implement contractor tiedown procedures.
- All objects that may be blown by the wind should be taken into the building (e.g., garbage cans, lawn furniture). Anchor objects that cannot be brought inside. Also, check the roofs of buildings for loose objects.

- Clean out all gutters and roof drains. One gallon of water weighs 8.35 pounds; the weight from backed-up water adds up quickly and can easily cause a roof collapse.
- Remove and relocate all portable physical education equipment to the interior of the school.
- Lower and secure window storm awnings (if applicable).
- Close all windows.
- Take down and properly store all cloth coverings from exterior of the site.
- Pools:
  - Lower the water level of swimming pools one foot to accommodate heavy rains, but do not drain them completely.
  - Add extra chlorine to pools to prevent contamination.
  - Turn off electricity to pool equipment.
- Check generator and emergency lights (if applicable).
- Media Area/Computer Labs—If possible, relocate all equipment to tabletops and cover with plastic to reduce water damage.
- Media Area—Relocate any valuable books in the collection to the upper shelves.

Site Access Control

Ensure site perimeter is secured at a level that prevents unauthorized vehicles or pedestrians from entering.

- In high-threat areas, manholes, utility tunnels, culverts, and similar unintended access points to the school property should be secured with locks or other appropriate devices without creating additional entrapment hazards.
- Perimeter fences, walls, and vegetation may be used to provide sufficient access control, surveillance, and territoriality.

Comprehensive school crisis prevention planning includes coordinating with all individuals who will be involved if a crisis occurs, conducting practice drills, and identifying security needs.

Some of the tools that will be needed can be found in a **Go Kit**, a portable box that contains important information, such as written procedures for an emergency or disaster. The following components of a Go Kit make possible a thorough school and community response to a crisis.

Aerial Photos of Campus

An aerial perspective of the campus and the surrounding area is very helpful to all agencies involved in a critical incident, including police, fire, paramedic, and rescue personnel. Your local municipality may be able to provide you with an aerial photo of your school and surrounding campus. Because of FEMA and insurance documentation requirements, extensive photos and digital images of damage prior to cleanup and repair are critical. If a FEMA or insurance claim is challenged, a picture is worth a thousand words. After the storm, be sure to take pictures and document damage before recovery efforts begin. Additionally, photos should be taken of facilities and grounds prior to an event to document conditions prior to the occurrence of damage.

Maps

Crisis response planners need to review the traffic patterns and intersections that will be affected in a major crisis. You should identify locations where parents or guardians can retrieve their children after an incident and determine traffic safety issues your school and law enforcement officials will have to consider when directing personnel to safe areas. Each school should keep as many as twenty copies of the map available, preferably laminated, for emergency personnel. Establish an emergency traffic plan capable of protecting emergency response routes and accommodating traffic and parking needs for parents, students, and the media. Maps should illustrate these planned routes, as well as:

- Streets surrounding the school
- Intersections near the school
- Vacant lots near the school
- Location of major utilities

Campus Layout

It is important to maintain current, accurate blueprints, classroom layouts, and floor plans of the building and grounds, including information about main leads for water, gas, electricity, cable, telephone, alarm and fire sprinkler systems, hazardous materials location, elevators, and entrances. This information is extremely helpful for damage assessments and "shelter-in-place" situations when students are safely locked in classrooms.

Information should be available on the layout of the building, including room numbers and availability of phones, cable television, e-mail, computers, and cell phones in the classroom. On the campus layout diagram, it is also helpful to highlight areas that could pose a possible threat, for example, the chemistry laboratory, biology laboratory, or any welding and wood shop areas that could also become a potential source for weapons.

It is also helpful to show the location of the fire alarm turnoff, fire sprinkler system turnoff, utility shutoff valves and switches, cable television shutoff, and first aid supply boxes. These items can be color-coded on the campus layout.

Blueprint of School Buildings

Architectural blueprints of the school building(s) are important to SWAT teams, and provide additional—and more detailed—information other than the simple classroom layout diagram. This information may be critical, especially in the event of a bomb threat. The principal for the school site should be the custodian for the blueprints. Grounds and maintenance staff of the school should be familiar with these blueprints and their location.

Teacher/Employee Roster

A teacher/employee roster should be in the Go Kit, including, if available, a classroom layout diagram with teachers' names. If not, be sure to match up each teacher's name with his or her classroom and identify whether or not each teacher has a cell or land phone. This roster should identify any teacher or employee with special medical needs (e.g., diabetes) who will require medications during a prolonged period, and individuals with a disability who may require assistance in an evacuation. Someone in the front office should be assigned to take the visitor/volunteer/substitute teacher list with him/her in the event of a critical incident.

Keys

The Go Kit should also contain a master key and an extra set of keys for those rooms for which a master key cannot be used. The keys must be clearly tagged. Consider placing the keys in a locked container within the Go Kit for added security, in case the box should end up in the wrong hands. Some schools have found it advantageous to keep the master key in a lockbox (rapid entry system) outside of the school. This is a secured metal box that can easily be accessed by a code or a key without having to enter the building. This can prove especially helpful when it is not safe to enter the school.

Further information on a lockbox can be obtained from your local fire department.

Another option is the Dual Key Box, which can be useful to school districts and community colleges wanting to provide keys to various entities (e.g., sheriff's department, fire department, or board employees).

Fire Alarm Reset Procedures

One of the lessons learned from Columbine was to make it easier to turn off the alarm than it was during the incident affecting that high school. The loud alarm made it very difficult for responders to hear directions. It took considerable time before someone who knew how to turn the fire alarm off was able to do so. School officials learned that you cannot assume that the person who knows how to turn off the alarm will be available. If that person is inside the building, he or she might not be able to get to the shutoff valve or switch; if that person is outside, it is possible that he or she might not be able to safely reenter the school. Multiple people need to know how to shut off the alarm. Providing such information on where shutoff valves or switches are located in the building and the procedures for shutoff in the lockbox could prove vital. In addition, though somewhat costly, some schools have installed a secure alarm shutoff system outside the school that can control the fire alarm and fire sprinklers.

Fire Sprinkler System Reset Procedures

Fire sprinkler systems may activate during an emergency. During the incident at Columbine, no one was readily available who knew how to immediately turn off the fire sprinkler system. As a result, hallways quickly filled with water, making it difficult to escape. In some places, the water reached dangerous levels in proximity to the electrical outlets—water reaching such outlets could have caused many more injuries and possibly additional deaths. At least two people need to be trained and assigned responsibility for turning off the fire sprinkler system. As backup, the Go Kit needs to provide information on where shutoff valves or switches are located in the building and the necessary procedures for shutoff.

Utility Shutoff Valves or Switches

Shutoff and access points for all utilities—gas, electric, water, phone, and data network—need to be clearly identified and their locations listed so they can be quickly shut off in a crisis; however, consideration should be given to not shutting off certain lines, that is, gas lines to generators. If there is no fire, the water should be shut off immediately to prevent flooding from the fire sprinkler system. Unless open electric or gas lines pose an immediate threat to life, the decision whether to shut off these lines should be made by the Incident Command Officer.

Gas Line and Utility Line Layout

Include a diagram that shows where gas and other utility lines are located throughout the campus.

Cable Television Satellite Feed Shutoff

If your school has a satellite feed for a cable television system, you should also provide directions on how to shut down that feed. Several of the police officers involved in nationally televised shootings recommend that the cable television feed be shut off so that the perpetrators on the inside will not be able to view the whereabouts of the SWAT team by tuning into live coverage of the scene on the outside. On the other hand, in a natural disaster, the television system can be helpful (if working) to provide individuals who are sheltered-in-place with up-to-date information.

Checklist

To review the comprehensiveness of the crisis plan, consider the items on the checklist below:

- Determine what crisis plans exist in the district, school, and community.
- Identify all stakeholders involved in crisis planning.
- Develop procedures for communicating with staff, students, families, and the media.
- Establish procedures to account for students during a crisis.

- Gather information that exists about the school facility, such as maps and the location of utility shutoffs.
- Identify the necessary equipment that needs to be assembled to assist staff in a crisis.

MITIGATION AND PREVENTION POLICIES AND PROCEDURES

School safety and emergency management experts often use the terms "prevention" and "mitigation" differently. Crises experts encourage schools to consider the full range of what they can do to avoid crises (when possible), or lessen their impact. Assessing and addressing the safety and integrity of facilities (window seals, HVAC systems, building structure), security (functioning locks, controlled access to the school), and the culture and climate of schools through policy and curricula are all important for preventing and mitigating possible future crises.

Mitigation and prevention require taking inventory of the dangers in a school and community and identifying what to do to prevent and reduce injury and property damage. For example:

- Establishing access control procedures and providing IDs for students and staff might prevent a dangerous intruder from coming onto school grounds.
- Conducting regular evacuation drills can reduce injury to students and staff because they will know what to do and how to avoid harm.
- Planning responses to and training for incidents involving hazardous materials is important for schools near highways. There are resources in every community that can help with this process. Firefighters, police, public works staff, facilities managers, and the district's insurance representative, for example, can help conduct a hazard assessment. That information will be very useful in identifying problems that need to be addressed in the preparedness process. Rely on emergency responders, public health agencies, and school nurses to develop plans for and provide training in medical triage and first aid.

The FEMA has done considerable work to help states and communities in the area of mitigation planning. It notes that *the goal of mitigation is to decrease the need for response as opposed to simply increasing response capability.*

Mitigation is any sustained action taken to reduce or eliminate long-term risk to life and property from a hazard event. Mitigation encourages long-term reduction of hazard vulnerability (FEMA, 2002).

Mitigating emergencies is also important from a legal standpoint. If a school, district, or state does not take all necessary actions in good faith to create safe schools, it could be vulnerable to a suit for negligence. It is important to make certain that the physical plant is up to local codes, as well as federal and state laws.

Mitigating or preventing a crisis involves both the district and the community. Contact the regional or state emergency management office to help get started and connect to efforts that are underway locally.

MITIGATION STEPS CHECKLIST

Mitigation and Prevention Requires Assessment and Action

□ Take inventory of potential dangers in a school community
□ Identify and implement action items to prevent injury and property damage

Stick to the Basics (*A.V.E.*)

A—Access Control-Limit Access to Facility
V—Visitor Control-Sound Procedures
E—Employee Involvement-Empowerment
□ Employ Security Technologies
□ Physical Plant Maintenance
□ Safety and Fire Prevention (include chemical storage and disposal)
□ Environmental Healthy Programs
□ Special Needs Students (including evacuation procedures)
□ Life Safety and Security Systems
□ Inspections
□ School Resource Office Program
□ Violence Prevention Programs
□ Confidential Tip Line (anonymous reporting)

Preparedness Checklist

□ Crisis and Emergency Operation Plans
- Base your plans on incident command because no one person can do it all.
- There must be plan consistency and flexibility.
- Take the all hazards approach—is there a better place to be.

Options

□ Secure the Building
□ System Wide Secure the Building
□ Lockdown
□ Shelter-in-Place
□ Evacuate
- on-site
- off-site

Prevention

Creating a safe and orderly learning environment should not be new to any school and district. Identifying students (or in some cases staff) who may pose a danger to themselves or to others is sometimes called "threat assessment." The US Department of Education and US Secret Service have released a guide, *Threat Assessments in*

Schools: A Guide to Managing Threatening Situations and to Creating Safe School Climates, that may be useful in working through the threat assessment process. The results of a threat assessment may guide prevention efforts, which may help avoid a crisis.

Many schools have curricula and programs aimed at preventing children and youth from initiating harmful behaviors. Social problem-solving or life-skills programs, anti-bullying programs, and school-wide discipline efforts are common across the nation as a means of helping reduce violent behavior. The staff in charge of prevention in a school (counselors, teachers, health professionals, administrators) should be part of the crisis planning team. Information on effective and promising prevention programs is on the Office of Safe and Drug-Free Schools website.

Know the school building. Assess potential hazards on campus. Conduct regular safety audits of the physical plant. Be sure to include driveways, parking lots, playgrounds, outside structures, and fencing. A safety audit should be part of normal operations. This information should feed into mitigation planning.

Know the community. Mitigation requires assessment of local threats. Work with the local emergency management director to assess surrounding hazards. This includes the identification and assessment of the probability of natural disasters (tornadoes, hurricanes, earthquakes) and industrial and chemical accidents (water contamination or fuel spills). Locate major transportation routes and installations. For example, is the school on a flight path or near an airport? Is it near a railroad track that trains use to transport hazardous materials? Also address the potential hazards related to terrorism. Schools and districts should be active partners in community-wide risk assessment and mitigation planning.

To help agencies work together they may want to develop a MOU that outlines each agency's responsibility.

Bring together regional, local, and school leaders, among others. Given that mitigation/prevention are community activities, leadership and support of mitigation and prevention activities are necessary to ensure that the right people are at the planning table. Again, leadership begins at the top. Schools and districts will face an uphill battle if state and local governments are not supportive of their mitigation efforts.

Make regular school safety and security efforts part of mitigation/prevention practices. Consult the comprehensive school safety plan and its needs assessment activities to identify what types of incidents are common in the school.

Establish clear lines of communication. Because mitigation and prevention planning requires agencies and organizations to work together and share information, communication among stakeholders is critical. In addition to communications within the planning team, outside communications with families and the larger community are important to convey a visible message that schools and local governments are working together to ensure public safety. Press releases from the Governor and chief state school officer, that discuss the importance of crisis planning, can help open the channels of communication with the public.

Implementation

Partnering with local first responders. Conducting drills and exercises to test and fine-tune the plan. Implementing the plan may need structural changes, as well as attitudinal changes on the part of custodians, faculty and other members of staff. Everyone must understand why changes are needed and support them if the plan is to be successful.

Although schools have no control over some of the hazards that may impact them, such as earthquakes or plane crashes, they can take actions to minimize or mitigate the impact of such incidents. For instance, schools in earthquake-prone areas can mitigate the impact of a possible earthquake by securing bookcases, and training students and staff on what to do during tremors.

Schools cannot always control fights, bomb threats, and school shootings. However, they can take actions to reduce the likelihood of such events. Schools may institute policies, implement violence prevention programs, and take other steps to improve the culture and climate of their campuses.

According to FEMA, the benefits of effective mitigation include the following:

- Saving lives and reducing injuries
- Preventing or reducing property damage
- Reducing economic losses
- Maintaining critical facilities in functioning order
- Lessening legal liability
- Providing positive consequences

THREATS

Natural Disasters: Earthquakes, Hurricanes, Tornadoes, Thunderstorms, Floods, and Fire

The general response to most disasters should be the same—to ensure the safety of students and staff. In the event of a natural disaster, activate your emergency plan and call the police and other emergency services, if necessary.

- Assess any injures and attend to them and discover the extent of the damage to buildings.
- Evacuate if there is no external danger and the building is sound, or stay in place until told it is safe to leave.
- Designate one or more members of staff to meet and assist the first responders (fire, police, and medics).
- Activate emergency communications to keep students, staff, parents and the media informed.
- Send students and staff home if safe to do so.
- Work with the media to keep people informed about how long the school will be closed and when it plans to reopen.

The School as an Emergency Shelter

Emergency managers regularly identify schools to serve as short-term and/or long-term shelters. They are attractive sites for shelters because they have kitchen facilities that are designed to serve many people, restroom facilities that are likely to be adequate for many people, and space for cots in gymnasiums, cafeterias, and wide corridors.

New schools that are to be used for emergency sheltering are appropriately designed as essential or critical facilities that warrant a higher degree of protection than other schools.

Manmade Disasters

Power Outages

A power outage is the loss of electricity supply to an area. The reasons for a power failure include a defect in a power station, damage to a power line or other part of the distribution system, a short circuit, or the overloading of electricity mains.

A power outage may be referred to as a "blackout" if power is lost completely, or as a "brownout" if some power supply is retained but the voltage level is below the minimum level specified for the system. Some brownouts, called "voltage reductions," are initiated intentionally to prevent a full power outage.

Power failures are particularly troublesome for schools as they can impact the safety of students. For this reason, and because of their potential use as shelters, some schools have emergency power generators that are configured to start automatically as soon as a power failure occurs.

Fire

Sound the alarm, evacuate the building if necessary and call 911.
Ensure all staff and students are accounted for.
Return to classrooms when told it is safe to do so.
If the building is damaged, relocate classes or send students home.
Keep parents and the media informed.

Food Poisoning

If students and/or staff appear to be sick from food poisoning, notify the appropriate health authorities and discontinue food services, isolate food service areas, don't allow the use of drinking fountains or vending machines until they have been checked.

Keep any left-over food from the meals served, as well as packages, in a refrigerated area so that they can be tested.

Document as much information as you can—when people started to get sick, symptoms, who was affected, and so on.

Discontinue all food services until the cause of the sickness has been identified.

Gas Leak

A gas leak refers to a leak of natural gas from a pipe or other containment. Because natural gas explodes when exposed to flame or sparks, a gas leak can be dangerous. In the event of a gas leak, the person identifying the gas leak should notify the principal/provost. School evacuation routes should avoid gas tanks on the school site.

If a serious gas leak is found:

- Evacuate the building in the same manner as in a fire.
- Notify the maintenance department immediately.
- Notify the fire department immediately.
- The custodian should shut off the gas at the meter.

Toxic Spill

In the event of a toxic spill:

- From a considerable distance, if possible, attempt to read the hazard warning and numbers on the diamond-shaped placards placed outside of vehicle or tank.
- Call 911 or appropriate emergency number.
- Relay placard information.
- Indicate wind direction in case upwind evacuation is directed by emergency personnel.
- Gather students and staff from outside areas.

Shelter in area not affected by spill or fumes:

- Close all doors and windows.
- Shut down the air conditioning and ventilation systems until it is safe to keep them on.
- Only trained rescue personnel should assist in the cleanup and rescue.
- Attempt to keep unauthorized persons away from the scene.

Decontamination Procedures

Decontamination requirements for facilities are handled by the local county health department and implemented by the local fire department. If your facility has a decontamination need, the county health department will inform you or the district administration regarding the proper procedures and any necessary equipment.

Elevator Emergencies

The mechanical, electrical, and operational designs of elevators are dictated by various elevator codes, which may be international, national, or state. Because an elevator is part of a building, it must also comply with fire standards, electrical wiring rules, and so on. The US National Elevator Standard Code is ASME A17.1-1990 *Safety Code for Elevators and Escalators*. Additional requirements relating to access by disabled persons may be mandated by laws or regulations, such as the Americans with Disabilities Act.

In the event that a person is trapped in an elevator, he or she should pick up or press the emergency communications unit to connect with persons outside the elevator. The other end of the communication unit should contact either the elevator servicing company or the fire department. Constant communication must be maintained with the elevator occupants. Consideration should be given to the occupant(s) possible medical needs or conditions. If any indication of panic is noted either by emergency phone contact with the trapped passengers, or by communicating with the passengers from outside the trapped elevator, the responsible party should immediately notify the local fire department.

If response by the elevator mechanics is going to take more than 30 minutes, the responsible party should notify the fire department for faster response to the elevator emergency.

Nuclear Accident

A nuclear power plant (NPP) is a thermal power station for which the heat source is one or more nuclear reactors generating nuclear power. In the event of a nuclear plant accident:

- Evacuate a zone within a 10-mile radius of the NPP
- Warn personnel of emergency by alarms, emergency broadcast system, NOAA weather radio, door-to-door notification, Channel 16 VHF
- Instruct school personnel to stay indoors or evacuate
- Make arrangements for persons with special needs
- Follow school evacuation procedures

(See also "Chemical, Biological, and Radiological Threats")

Epidemic and Pandemic

An epidemic is a disease that appears as new cases at the local level in a human population, during a given time period, at a rate that substantially exceeds what is expected or normal. A pandemic is the same as an epidemic except that it happens on a global scale. In the event of an epidemic or pandemic, consult your local county health department for guidelines and policies.

Criminal

A crime is an act that violates a political or moral rule. Crimes in schools include:

Bomb Threats

Treat all bomb threats seriously. The person receiving the bomb threat shall:

- Attempt to have the caller indicate where bomb is located and when it will go off.
- Keep caller talking as long as possible and try to identify the caller's
  - age
  - gender
  - type of voice and/or accent
  - background noises
  - other information that may identify caller
- Immediately notify the principal/dean
- Evacuate building

If students are to be relocated to another facility:

- Notify transportation department of immediate need to relocate staff and students
- Notify parents of relocation of their children
- Secure building as much as possible
- Close all windows; lock all doors

When the police or fire department is called:

- Provide it with all necessary information
- Provide it with an Inventory of School Houses and keys for facility for a bomb search
- Bomb squad will remove the bomb
- Appropriate authorities will investigate

Device Found

- Call 911
- Leave the device alone and isolate the immediate vicinity
- Evacuate the premises

Explosions

In the event of an explosion, evacuate from the danger area in a manner similar to a fire drill. Assess any medical needs and contact the police and fire departments. Evaluate the condition of the building prior to re-occupancy.

Explosions can be the result of the following:

- Arson
- Vandalism
- Terrorism

Suspicious packages

Any suspicious letter or package should not be opened and procedures should be in place for such items to be isolated until they can be inspected.

According to the US Postal Service, a letter or parcel that ought to trigger suspicion includes one or more of the following characteristics:

- It has any powdery substance on the outside
- It is unexpected or from someone unfamiliar
- It has excessive postage, handwritten or poorly typed address, incorrect titles or titles with no name, or misspellings of common words
- It is addressed to someone no longer with your organization or are otherwise outdated
- It has no return address or have one that can't be verified as legitimate
- It is of unusual weight, given their size, or are lopsided or oddly shaped
- It has an unusual amount of tape
- It is marked with restrictive endorsements, such as "Personal" or "Confidential"
- It has a strange odor or stain.

If a suspicious item is found, call 911. Isolate the area and make sure that no one else enters. People who were in the vicinity at the time should stay there in case the item was contaminated. If they leave they will spread the decontamination.

Turn off the air conditioning and heating units to prevent further potential dispersal.

Follow instructions from the police and HazMat teams.

Guns

Between 1994 and 1999, nationwide there were 220 school-associated violent events, resulting in 253 deaths (74.5 percent of these involved firearms). Handguns caused almost 60 percent of these deaths (*Journal of American Medical Association*, December 2001). Almost all schools address the presence of firearms in their School Emergency Plans.

Facility managers should be aware of their board's policies and procedures. If equipments, such as metal detectors or metal detecting wands, are used in the school or community college to prevent gun violence, they should be aware of their role, if any, in maintaining these equipments.

School Shootings

- Notify the police.
- First determine if the shooter(s) is still on the premises.
- If the location of the shooter(s) is known, try to move people away into safe locations and secure the doors to those locations. People should wait there until told by the police that it is safe to come out. Lockdown the rest of the school.
- If the location of the shooter(s) is not known, broadcast a public address system message urging everyone to stay where they are, to get down on the floor, remain quiet and await further instructions. Stress that help is on the way.
- Exit doors should be opened to facilitate an orderly exit.

Hostages

- Call 911 immediately and institute a lockdown.
- Ensure all students and staffs are in as safe and secure area as possible.
- Notify transportation that buses should be on standby to take students to a safer location or their homes.
- Do not allow people to wander round the building and do not allow anyone to leave the building until the police say it is safe to do so.
- Have someone manning the telephone to talk to anxious parents.

Angry Parents

- Stay calm and reasonable.
- Listen to what they are saying.
- Invite another member of staff to join you.
- Try to find a solution that satisfies all.
- If the situation deteriorates, call for help and call 911 if necessary.

Intruders

- Notify the school's resource officer/custodians or call 911.

- Try to establish the location of the intruder, and if possible, his/her intent.
- If the intruder does not appear to pose a threat, approach cautiously and ask to leave.
- Do not try to apprehend the intruder(s) if they appear dangerous.
- Move others away from the threat.
- If the threat remains, institute a lockdown.
- Keep everyone informed.
- Only allow students and staff to leave when the police say it is safe.
- Be prepared to speak to the media afterward and talk to anxious parents.

Missing Child

- Take all reports of missing children seriously, especially if the student disappeared while on the way to or from school or while at school.
- Notify the parents immediately and open lines of communication with them and the police.
- Conduct a search of the school and immediate area and speak to the missing student's friends. Try to establish who was the last person to see the missing student, when and where.
- Notify the district.
- If the child has been kidnapped and the person was seen taking him or her, try to ascertain as much information as possible about the snatcher—personal and physical information, as well as details of any vehicle involved. Cooperate with the police and stay in close touch with the parents. Consider bringing in counselors for the missing student's friends.

Vandalism

- Vandals are not usually caught, but their actions can cause major damage and expense. Report major damage to the police especially if it involves slogans, emblems, and so on, indicating a hate crime or gang activity. Photograph all such damage both for law enforcement and insurance purposes.
- Clean up quickly in order to get back to normal as soon as possible and, as a deterrent to others, prosecute anyone caught. Also seek restitution from them.

Fights Between Students

- Summon assistance if necessary and separate those fighting, keep them apart and disperse those watching.
- Call 911 if necessary—that is, the situation is escalating, there is a serious risk of physical harm, or weapons are involved.
- Institute a lockdown if several students are fighting and there is a danger that more will join in.
- The principal should interview those fighting and others who witness it to try to establish the reasons for it and whether it is likely to happen again. That will determine whether those responsible can be allowed back into the classroom or should be suspended.
- Ensure the incident is fully reported.

Bullying—Physical and Cyber

According to www.stopbullyingnow.com there is one incident of bullying every 7 minutes. Adults intervene in only 4 percent of the incidents and peer intervention occurs in only 11 percent of incidents. Some generalizations about bullying:

- It is done by someone with more power or social support to someone with less power or social support.
- It often includes the abuser blaming the target for the abuse.
- It often it leads to the target blaming him or herself for the abuse.
- In most bullying situations, the target cannot stop the bullying by his or her own actions.

Adults can intervene effectively to reduce bullying. The first step is to identify unacceptable peer actions. These include:

- Physical aggression: hitting, kicking, pushing, choking, punching
- Verbal aggression: threatening, taunting, teasing, starting rumors, hate speech
- Exclusion from activities: This does not mean that a student should have to be friends with every other student; it does mean that children should not be allowed to systematically exclude others: "No one play with Mary"; "No one wants to play with him"; "Don't be her friend"

The components of a successful comprehensive intervention are:

- Consensus-development among staff (with input from students and community members) so people enforce the same expectations consistently. What is your school's definition of unacceptable peer behavior? What behaviors will you commit to changing no matter what it takes? These might include:
    - *Physical aggression: hitting, kicking, pushing, choking, punching*
    - Verbal aggression: threatening, taunting, teasing, hate speech
    - Social exclusion
- Development and consistent enforcement of effective consequences for verbal and physical aggression are predictable, inevitable, immediate, and escalating, and based on uniform expectations for all. Consistent use of consequences will reduce these behaviors and are a necessary component of effective prevention. Inconsistent enforcement makes the problem worse. Effective consequences are small (so they can be used consistently), escalate with repeated aggression, and often involve loss of unstructured times like recess, lunch with peers, or extracurricular activities.
- Building positive connections between staff and students and positive tone in staff-student interactions.
- Monitoring to make sure that consequences and education are effective.
- Effective counseling for youth who bully after enforcement of consequences has generated some anxiety.
- Effective support for targeted youth, including protection from repeated mistreatment.
- Empowering bystanders to tell adults, support targets, and discourage unacceptable behavior.

From a 2008 study of bullying prevention in Colorado schools:

- "A reduction in bullying occurred in schools where teachers and students are willing to intervene, treat each other fairly and demonstrate that they care."
- "Adult and student intervention in bullying is critical from elementary through high school."
- "A positive relationship with adults and students at school and a school culture of trust and fairness are key to reducing bullying."
- "Schools with lower levels of bullying report higher scores on statewide tests."

Web Bullying

The typical bully used to torment unlucky victims out in the open—in school hallways, for instance—spitting out insults and disparaging remarks for anyone to hear and see. But times have changed. Bullying has gone high-tech and anonymous, although it's just as humiliating.

The Internet and cell-phone text messages are the new mediums for "cyber-bullies." They post nasty pictures or messages about others in blogs and on websites or exploit another person's online username to spread rumors about others. In Massachusetts, six students face felony charges in the death of a fifteen year-old girl who hanged herself in January 2010 after being subjected to electronic verbal assaults and threats of physical harm. Many people, it is claimed, were aware of the cyber bullying but no one stepped in to help.

I-Safe America, an organization promoting Web safety, surveyed 1,500 students in forth through eighth grade, and found that 42 percent have been bullied online. And 53 percent admitted to saying something mean or hurtful to someone else online.

The Pew Internet & American Life Project reports that approximately seventeen million kids around twelve to seventeen years of age use the Internet. Teens have embraced Instant Messaging (IM)—74 percent of teens IM, compared to only 44 percent of adults.

Many schools have introduced policies to try to prevent cyber bullying. The state of Florida, for instance, has directed each school board to create a policy regarding cyber bullying and electronic harassment.

ADVICE TO THE VICTIM

Don't reply to messages from cyber-bullies—even though you may really want to, this is exactly what cyber-bullies want. They want to know that they've got you worried and upset. They are trying to mess with your mind and control you, to put fear into you. Don't give them that pleasure.

Do not keep this to yourself! You are NOT alone and you did NOT do anything to deserve this! Tell an adult you know and trust!

Inform your Internet Service Provider (ISP) or cell phone/pager service provider.

Do not erase or delete messages from cyber-bullies—You don't have to read it, but keep it, it is your evidence. You may unfortunately get similar messages again, perhaps from other accounts. The police and your ISP, and/or your telephone company can use these messages to help you. You might notice certain words or phrases that are also used by people you know. These messages may reveal certain clues as to whom is doing this to you, but don't try and solve this on your own, remember, tell an adult you know and trust. GET HELP!

Protect yourself—Never arrange to meet with someone you met online unless your parents go with you. If you are meeting them make sure it is in a public place. You may need to delete your current e-mail accounts, cell-phone/pager accounts and set up new ones. If your cyber-bullying problems persist, do this as soon as possible, unless you are working with the police and your Telecommunications Provider to keep the account(s) active to try and catch the cyber-bully.

If you are more technically inclined, you can do a little cyber-sleuthing of your own to provide the police and your Telecommunications Provider with more information, but NEVER try to meet someone personally whom you suspect might be the cyber-bully. This is best left in the hands of the legal authorities.

If you receive an unsolicited harassing e-mail message from a cyber-bully, you can often use your mouse to right-click on the header of the offending message and choose the "Options" section of the menu. This will often reveal greater details about the message. You can then do a "WHOIS" search with Internic.com. This Web site was been established to provide the public information regarding Internet domain name registration services and is updated frequently. The "WHOIS" search will often provide information as to who owns the domain name and their contact information. Share this information with your local police and your telecommunications or Internet service provider. You can also use software to help protect and/or find out who is sending you harassing messages.

If you are receiving harassing messages from cyber-bullies through Web-based mail services, such as Hotmail, Yahoo Mail, etc., it becomes very difficult to trace such accounts as cyber-bully@hotmail.com. You may have to delete your current e-mail account and start a new one. Tell only a select few people you trust about your new e-mail account when, and if, you choose to reestablish one. If you are receiving harassing messages from cyber-bullies through Instant Messaging (IM), software, such as ICQ, MSN Messenger, AOL Instant Messenger, etc., such programs usually have a "Block," "Ignore," or "Ban" feature. Use this feature to try to "Block," "Ignore," or "Ban" the cyber-bullies.

If you are the victim of a cyber-bully who has set up a Web site that is defaming, or mocking you, contact the Internet Service Provider and inform them about what is happening, also inform the police. Use the "WHOIS" search tool to help you in the manner described above.

If this is a large Web hosting company, it may take a long time to get a response and a promise of action. ISPs are often very reluctant to act in such cases. Unfortunately, some people have only received a response, or seen such cyber-bullying websites taken down after the threat of legal action. For more information go to www.cyberbullying.org.

Student or Staff Deaths (suicide, homicide, unintentional, or natural)

- Hold a faculty meeting as soon as possible to discuss the circumstances and the response.
- Contact family members of the deceased and offer assistance.
- Decide whether it is in the best interests to make a public announcement to the school.
- Be on the lookout for students showing emotional distress and provide counseling if needed.

Sexual Harassment/Assault

- Whenever a sexual assault takes place at school or on the way to and from school, the school must do everything in its power to protect the identity and right of privacy of the victim and alleged perpetrator. Services, such as counseling, must be kept confidential.
- Call 911.
- Isolate the area where the alleged incident took place—it should be treated as a crime scene.
- Instruct staff not to give out any information and to quash any rumors.
- Provide support services if necessary for the victim.
- Cooperate fully with the police.
- Keep full records of the incident.
- If the incident becomes public knowledge, provide counseling for students that request it.

Suicide

Any suicide threat must be taken seriously. If a member of staff or a student thinks that someone is contemplating suicide or has talked about it to close friends, they should report it immediately. The student should be kept under adult supervision and not be allowed to leave the school until interviewed by a qualified school counselor or psychologist who will evaluate the situation. The parents or guardian should also be notified. The best course of action will then be determined. This could be releasing the student into the care of the parents provided they understand the seriousness of the situation and are willing to follow an agreed course of action, including counseling. In extreme cases, the decision may be jointly taken to admit the student to hospital for observation.

If a suicide attempt is made which results in no injury, call 911 if a weapon was involved, do not leave the person alone, call the parent or guardian and prepare a report.

If a suicide attempt results in injury, administer appropriate first aid, call 911, make sure someone stays with the person, call the parent or guardian and prepare a report.

If the person dies, call 911 immediately, isolate the area, and call the parents or guardian saying that there has been an attempted suicide. Do not say that the child is dead—that is not your responsibility.

Hold a faculty meeting to discuss the incident and advise staff on what they should and should not say to students. Arrange to have counselors available once students hear the news.

Make sure that only authorized persons speak to the media.

Accidents

Accidents do happen and procedures should be in place to deal with them.

At School

- Minor Accidents: Treat at school.
- Serious Accidents: Keep injured person still and quite.
- Summon a member of staff if not already present.
- Check for breathing and bleeding and administer immediate first aid, if necessary and trained to do so.
- Call 911, if appropriate.
- Move students away from the area and into classrooms.
- Notify parents and/or guardian and write up an accident report.

Bus accident

- Assuming the driver has called 911, send someone to the scene to provide emotional support and feedback news so that parents, staff, and students can be kept informed.
- In the event of serious injuries, arrange alternate transport for students not injured, organize counselors, contact parents, and speak to the media.

Individual on the way to or from school

- Notify someone in charge.
- Call 911 if not already summoned.
- Notify the principal and contact parents or next of kin in the case of a teacher or adult employee.

Allergic Reaction

- An allergic reaction could be as a result of food, insect bite, exposure to allergen, medication, and so on.
- Send for immediate help. Don't move the person unless necessary for safety reasons.
- Check the student's medical history and administer any prescribed medication.
- Call 911 if necessary.
- Notify a parent or guardian.

Africanized Bees (a growing threat in southeastern states)

In 1956, researchers imported honeybees from Africa into Brazil in an effort to improve beekeeping in the New World tropics. These African bees were well suited to conditions in Brazil, and they began colonizing South America, hybridizing with European honeybees (hence the name "Africanized" honeybees) and displacing the European bees. Compared to docile European bees, Africanized honeybees are extremely defensive. Large numbers of them may sting people and livestock with little provocation. They also take over European colonies by entering them and killing the resident queen.

Africanized bees are sometimes called "Killers Bees," not because of their stings but because they attack in massive swarms and can cause death. They eventually migrated to the United States in 1990 and are now found in much of the south.

Precautions to take:

- Use caution with power mowers—bees are sensitive to low-frequency vibrations; an engine-driven lawn mower may elicit a defense response from a hive or swarm.
- Walk with care in clover and near picnic sites.
- Avoid perfume, cologne, and scented cosmetics; foraging bees will seek out floral scents.
- Beware of head-butting bees. Honeybees usually have a small number of sentry bees patrolling the perimeter of the hive's territory. In some if not most cases, these sentry bees will initially butt (not sting) any animal that enters the hive's territory. If the animal continues closer to the hive, stinging will ensue.
- If bees start butting you, retrace your steps and walk away from the hive. Choosing any other path could lead you deeper into the hive's territory.
- Wear light colors; bees react to dark colors.
- Water meter chambers: Africanized bees are known to frequently nest in ground cavities. In suburban areas, a favored place is an in-ground concrete chamber that contains a water meter. The small lifting hole in the cover becomes the entrance to their hive. If the outside air temperature is over 60/F (15/C), you will probably see bees using this entrance if they are present. Otherwise, use caution if you are in an area known to have Africanized bees.
- Open-air hives: While all bees may build open-air hives in mild climates, this practice is much more prevalent with Africanized bees (as well as certain "rafter bees" native to Southeast Asia).
- Contact a local exterminator for removal of bees.

What to do in case of an attack:

- Run away as fast as possible. Get into a building or vehicle if you can.
- Try to cover your face and head as you run.
- Call 911.
- Start removing stingers from your skin once you are away from the bees. You can remove them by scraping, pulling, or using sticky tape. Do not leave the stingers in any longer than necessary, as they will continue to pump venom.
- Don't stop to remove stingers until you are safely away from the attacking bees.
- Don't jump into water. The bees will see you and wait for you to come up for air.
- Don't panic.

Terrorism

With any manmade threat, it is important to determine who has the intent to cause harm. The aggressors seek publicity for their cause, monetary gain (in some instances), or political gain through their actions. These actions can include injuring or killing people; destroying or damaging facilities, property, equipment, or resources; or stealing equipment, material, or information.

Aggressor tactics run the gamut: moving vehicle bombs; stationary vehicle bombs; bombs delivered by persons (suicide bombers); exterior attacks (thrown objects, such as rocks, Molotov cocktails, hand grenades, or hand-placed bombs); stand-off weapons attacks (rocket-propelled grenades, light antitank weapons, etc.); ballistic attacks (small arms and high-power rifles); covert entries (gaining entry by false credentials or circumventing security with or without weapons); mail bombs (delivered to individuals); supply bombs (larger bombs processed through shipping departments); airborne contamination (CBR agents used to contaminate the air supply of a building); and waterborne contamination (CBR agents injected into the water supply).

Thorough crisis planning will carry the school and district a long way in responding to a terrorist incident. While the risk of a terrorist attack on a school is much lower than the risk of being impacted by many local hazards, it is very important to be prepared. As with other incidents, a terrorist attack may result in the following:

- Damage beyond school boundaries (as with a hurricane)
- Victims who are contaminated (as with a hazardous materials spill)
- A crime scene to protect (as with arson) or
- Widespread fear and panic (as with a school shooting)
- The response will need to involve securing student and staff safety and supporting long-term recovery, just as with any other incident
- As the Department of Homeland Security (DHS), Advisory System signals increased threat, additional protective measures are needed. Several districts have incorporated the DHS system into their crisis planning
- The Red Cross has also issued some general guidance on how schools may adapt these codes. It is useful to consult with local emergency management offices and state or county emergency terrorism task forces. Each state also has a DHS liaison. Check with the Governor's office to identify the contact
- The sample school advisory system is a useful tool to adapt and incorporate into crisis planning. As the risk of attack increases, consider action items under both current and lower threat levels. It is important to assess local conditions and implement actions accordingly

Explosives and Blast Effects

An explosion is an extremely rapid release of energy in the form of light, heat, sound, and a shock wave. A shock wave consists of highly compressed air traveling outward from the source at supersonic velocities. As the shock wave expands, pressures decrease rapidly (with the cube of the distance) and, when it meets a surface that is in line-of-sight of the explosion, it is reflected and amplified by a factor of up to thirteen. Pressures also decay rapidly over time (i.e., exponentially) and have a very brief span of existence, measured typically in thousandths of a second or milliseconds. Diffraction effects, caused by corners of a building, may act to confine the air-blast, prolonging its duration. Late in the explosive event, the shock wave becomes negative, creating suction. Behind the shock wave, where a vacuum has been created, air rushes in, creating a powerful wind or drag pressure on all surfaces of the building.

This wind picks up and carries flying debris in the vicinity of the detonation. In an external explosion, a portion of the energy is also imparted to the ground, creating a crater and generating a ground shock wave analogous to a high-intensity, short-duration earthquake.

In the context of other hazards (e.g., earthquakes, winds, or floods), an explosive attack has the following distinguishing features:

- The intensity of the pressures acting on a targeted building can be several orders of magnitude greater than these other hazards. It is not uncommon for the peak pressure to be in excess of 100 pounds per square inch (psi) on a building in an urban setting for a vehicle weapon parked along the curb. At these pressure levels, major damages and failure are expected.
- Explosive pressures decay extremely rapidly with distance from the source. Therefore, the damages on the side of the building facing the explosion may be significantly more severe than on the opposite side. As a consequence, direct air-blast damages tend to cause more localized damage.
- The duration of the event is very short, measured in thousandths of a second or milliseconds. This differs from earthquakes and wind gusts, which are measured in seconds, or sustained wind or flood situations, which may be measured in hours. Because of this, the mass of the structure has a strong mitigating effect on the response because it takes time to mobilize the mass of the structure. By the time the mass is mobilized, the loading is gone, thus mitigating the response. This is the opposite of earthquakes, whose imparted forces are roughly in the same timeframe as the response of the building mass, causing a resonance effect that can worsen the damage.

Building Damage

The extent and severity of damage and injuries in an explosive event cannot be predicted with perfect certainty. Past events show that the unique specifics of the failure sequence for a building significantly affect the level of damage. Despite these uncertainties, it is possible to give some general indications of the overall level of damage and injuries to be expected in an explosive event, based on the size of the explosion, distance from the event, and assumptions about the construction of the building.

Damage due to the air-blast shock wave may be divided into direct air-blast effects and progressive collapse. Direct air-blast effects are damage caused by the high-intensity pressures of the air-blast close in to the explosion and may induce the localized failure of exterior walls, windows, floor systems, columns, and girders.

The air-blast shock wave is the primary damage mechanism in an explosion. The pressures it exerts on building surfaces may be several orders of magnitude greater than the loads for which the building is designed. The shock wave also acts in directions that the building may not have been designed for, such as upward on the floor system. In terms of sequence of response, the air-blast first impinges on the weakest point in the vicinity of the device closest to the explosion, typically the exterior envelope of the building. The explosion pushes on the exterior walls at the lower stories and may cause wall failure and window breakage. As the shock wave continues to expand, it enters the structure, pushing both upward and downward on the floors.

Floor failure is common in large-scale vehicle-delivered explosive attacks, because floor slabs typically have a large surface area for the pressure to act on and a comparably small thickness. In terms of the timing of events, the building is engulfed by the shockwave and direct air-blast damage occurs within tens to hundreds of milliseconds from the time of detonation. If progressive collapse is initiated, it typically occurs within seconds.

Glass is often the weakest part of a building, breaking at low pressures compared to other components, such as the floors, walls, or columns. Past incidents have shown that glass breakage may extend for miles in large external explosions. High-velocity glass fragments have been shown to be a major contributor to injuries in such incidents. For incidents within downtown city areas, falling glass poses a major hazard to passersby on the sidewalks below and prolongs post-incident rescue and cleanup efforts by leaving tons of glass debris on the street.

Casualties and Injuries

Blast can cause significant casualties. During the bombing of the Murrah Federal Building, 168 people were killed. Severity and type of injury patterns incurred in explosive events may be related to the level of structural damage. The high pressure of the air-blast that enters through broken windows can cause eardrum damage and lung collapse. As the air-blast damages the building components in its path, missiles are generated that cause impact injuries. Airborne glass fragments typically cause penetration or laceration-type injuries. Larger fragments may cause non-penetrating, or blunt trauma, injuries. Finally, the air-blast pressures can cause occupants to be bodily thrown against objects or to fall. Lacerations due to high-velocity flying glass fragments have been responsible for a significant portion of the injuries received in explosion incidents. In the bombing of the Murrah Federal Building in Oklahoma City, for instance, 40 percent of the survivors in the building cited glass as contributing to their injuries. Within nearby buildings, laceration estimates ranged from 25 percent to 30 percent.

Levels of Protection

The amount of explosive and the resulting blast dictate the level of protection required to prevent a building from collapsing or minimize injuries and deaths. The DoD correlates levels of protection with potential damage and expected injuries. The GSA and the Interagency Security Committee (ISC) also use the level-of-protection concept. However, wherein DoD has five levels, they have established four levels of protection. The GSA and ISC levels of protection can be found in GSA PBS-P100, Facilities Standards for the Public Buildings Service, November 2000, Section 8.6.

Large-scale truck bombs typically contain 10,000 pounds or more of TNT equivalent, depending on the size and capacity of the vehicle used to deliver the weapon. Vehicle bombs that utilize vans, down to small sedans, typically contain 500 to 4,000 pounds of TNT equivalent, respectively. A briefcase bomb is approximately 50 pounds, and a pipe bomb is generally in the range of 5 pounds of TNT equivalent. Research performed as part of the threat assessment process should identify bomb sizes used in the locality or region.

In the case of a stationary vehicle bomb, knowing the size of the bomb (TNT equivalent in weight), its distance from the structure, how the structure is put together, and the materials used for walls, framing, and glazing allows the designer to determine the level of damage that will occur and the level of protection achieved. Whether an existing building or a new construction, the designer can then select mitigation measures to achieve the level of protection desired.

Stand-off Distance and the Effects of Blast

Energy from a blast decreases rapidly over distance. In general, the cost to provide asset protection will decrease as the distance between an asset and a threat increases. However, increasing stand-off also requires more land and more perimeter to secure with barriers, resulting in an increased cost.

The critical location of the weapon is a function of the site, the building layout, and the security measures in place. For vehicle bombs, the critical locations are considered to be at the closest point that a vehicle can approach on each side, assuming that all security measures are in place. Typically, this is a vehicle parked along the curb directly outside the building, or at the entry control point where inspection takes place. For internal weapons, location is dictated by the areas of the building that are publicly accessible (e.g., lobbies, corridors, auditoriums, cafeterias, or gymnasiums). Range or stand-off is measured from the center of gravity of the charge located in the vehicle or other container to the building component under consideration.

Defining appropriate stand-off distance for a given building component to resist explosive blast effects is difficult. Often in urban settings, it is either not possible or practical to obtain appropriate stand-off distance. Adding to the difficulty is the fact that defining appropriate stand-off distance requires a prediction of the explosive weight of the weapon. In the case of terrorism, this is tenuous at best.

The DoD prescribes minimum stand-off distances based on the required level of protection. Where minimum stand-off distances are met, conventional construction techniques can be used with some modifications. In cases where the minimum stand-off cannot be achieved, the building must be hardened to achieve the required level of protection (see the DoD UFC—DoD Minimum Antiterrorism Standards for Buildings, UFC 4-010-01, July 31, 2002).

The first step in predicting blast effects on a building is to predict blast loads on the structure. Because blast pressure pulse varies based on stand-off distance, angle of incidence, and reflected pressure over the exterior of the building, the blast load predictions can be very complex. Consultants may use sophisticated methods, such as CFD computer programs, to predict blast loads. These complex programs require special equipment and training to run.

Many shelters can be part of low-rise buildings. Although small weights of explosives are not likely to produce significant blast loads on the roof, low-rise buildings may be vulnerable to blast loadings resulting from large weights of explosives at large stand-off distances that may sweep over the top of the building. The blast pressures that may be applied to these roofs are likely to far exceed the conventional design loads, and unless the roof is a concrete deck or concrete slab structure, it may fail. There is little that can be done to increase the roof's resistance to blast loading that doesn't require extensive renovation of the building structure.

In addition to the hazard of impact by building envelope debris propelled into the building or roof damage that may rain down, the occupants may also be vulnerable to much heavier debris resulting from structural damage.

Safe Rooms in Response to the Domestic Explosive Threat

The concept of safe rooms has been around for quite some time. Bomb shelters were used in the United Kingdom (UK) during World War II to protect the civilian populations against aerial attack, and fall-out shelters were established in cities in the United States during the Cold War to protect against the lingering effects of a feared nuclear attack. More recently, the IDF requires APSs or FPSs to be constructed in every new building or to be added to existing buildings according to engineering specifications. In buildings in which no shelters exist, interior rooms may be converted to shelters by following IDF instructions. In all cases, the shelters must be accessible within two minutes of a warning siren. The protected spaces are intended to serve as a refuge when an attack is suspected, either through early warning or remote detection; however, the protected space is much less effective when the event takes place without warning. Two minutes and 11 seconds elapsed between the time the Ryder truck stopped in front of the Murrah Federal Building in Oklahoma City and the detonation of its explosives, but no one was alerted to the danger until the explosion occurred. At Khobar Towers in Dhahran, Saudi Arabia, US Air Force Security Police observers on the roof of the building overlooking the perimeter identified an attack in progress and alerted many occupants to the threat; however, evacuation was incomplete, and 500 people were wounded and nineteen people were killed by the explosion.

The effectiveness of the safe room in protecting occupants from the effects of an explosive detonation is, therefore, highly dependent on early detection and warning. Unless the attacker notifies authorities of a bomb threat, as often occurred in the terrorist activities in Northern Ireland, the safe room can best be used after an explosion occurs in anticipation of a second attack. The 1998 bombing of the US Embassy in Kenya was preceded by a small explosion that drew curious embassy employees to the windows; such a tactic, if repeated in the United States, would justify the relocation of school occupants to a safe room until school officials are able to determined that it is safe to disperse the students. To these limited objectives, the establishment of a safe room in schools may serve a useful purpose. Given the nature of the explosive threat, however, it may be more effective to provide debris mitigating protective measures for all exterior facade elements.

It is important to understand the nature of the domestic explosive threat in order to effectively plan for the protection of different types of facilities and particularly for the establishment of safe rooms in schools. Although the patterns of past events may not predict the future, they give valuable insight to the protection against a very low-probability, but potentially high-consequence, event. As previously discussed, despite a wide range of terrorist events, such as CBR contamination, an explosion remains the most insidious threat, requiring the least sophisticated materials and expertise. The principal components of an explosive device can be obtained at a variety of retail outlets, without arousing suspicion. Every year, over 1,000 intentional explosive detonations are reported by the FBI Bomb Data Center. In 1998, the last year for which the compiled data were published, there were 1,225 actual incidents of unauthorized explosions in the United States. The majority of these explosives were targeted against residential properties and vehicles; however, seventy-six explosive events were detonated at educational facilities, causing a total of $28,500 in property damage. In addition to these actual events, sixty-three incidents involving hoax devices were investigated. By contrast, only one explosive device was detonated at a federal government facility, causing $1.5 million in property damage, and eight were detonated at local/state government facilities, causing $316,000 in property damage. Over 70 percent of the people involved in bombing incidents were "young offenders" and less than 0.5 percent were members of terrorist groups. Vandalism was the motivation in 40 percent of the intentional and accidental bombing incidents. Although two out of three attacks were perpetrated between 6 p.m. and 6 a.m., the incidents against educational facilities were more uniformly distributed throughout the day. Although each successive major domestic terrorist event exceeded the intensity of the predecessor, this is not particularly relevant to the threats to which a school structure might be subjected; if an explosive were to be detonated in or around a school building, it would most likely be a small improvised device assembled by a youth, and vandalism is most likely to be the motive.

The size of the explosive that might be considered for a protective design is limited by the maximum weight that might be transported either by hand or by vehicle. Despite the large weight of explosives that might be transported by vehicle, there have been relatively few large-scale explosive events within the United States. The 1995 explosion that collapsed portions of the Murrah Federal Building in Oklahoma City contained 4,800 pounds of ammonium nitrate and fuel oil (ANFO), and the 1993 explosion within the parking garage beneath the World Trade Center complex contained 1,200 pounds of urea nitrate. As implied by the FBI statistics, the majority of the domestic events contain significantly smaller weights of low-energy explosives. The explosive that was used in the 1996 pipe bomb attack at the Olympics in Atlanta consisted of smokeless powder and was preceded by a warning that was called in 23 minutes before the detonation. Nevertheless, the protective design of structures focuses on the effects of high energy explosives and relates the different mixtures to an equivalent weight of TNT.

Locating Safe Rooms to Mitigate Threats

The building's facade is its first real defense against the effects of a bomb and typically the weakest component that would be subjected to blast pressures. Although the response of specific glazed components is a function of the dimensions, make-up, and construction techniques, the conventionally glazed portions of the facade would shatter and inflict severe wounds when subjected to a 50-pound explosive detonation at a stand-off distance of about 200 feet. If the glazed elements are upgraded with a FRF, the same facade element may be able to withstand a 50-pound explosive detonation at a stand-off distance of about 70 feet. Unreinforced masonry block walls are similarly vulnerable to collapse when subjected to a 50-pound threat at a stand-off distance of 50 feet; however, if these same walls are upgraded with a debris catching system, they may be able to sustain this same intensity explosive detonation at a stand-off distance of 20 feet. If the weight of explosives were increased from 50 pounds to 500 pounds, the required stand-off distances to prevent severe wounds increases to 500 feet for conventional window glazing, 200 feet for window glazing treated with a FRF, 250 feet for unreinforced masonry block walls, and 60 feet for masonry walls upgraded with a debris catching system. Based on these dimensions, it is apparent that substantial stand-off distances are required for the unprotected structure, and these distances may be significantly reduced through the use of debris mitigating retrofit systems. Furthermore, because blast loads diminish with distance and geometry of wave propagation relative to the loaded surface of the building, the larger threats at larger stand-off distances are likely to damage a larger percentage of facade elements than the more localized effects of smaller threats at shorter stand-off distances. Safe rooms that may be located within the school should, therefore, be located in windowless spaces or spaces in which the window glazing was upgraded with a FRF.

Although small weights of explosives are not likely to produce significant blast loads on the roof, low-rise structures may be vulnerable to blast loadings resulting from large weights of explosives at large stand-off distances that may sweep over the top of the building. The blast pressures that may be applied to these roofs are likely to far exceed the conventional design loads and, unless the roof is a concrete deck or concrete slab structure, it may be expected to fail. There is little that can be done to increase the roof's resistance to blast loading that doesn't require extensive renovation of the building structure. Therefore, safe rooms should be located at lower floors, away from the roof debris that may rain down in response to blast loading.

The building's lateral load-resisting system, the structural frame or shear walls that resist wind and seismic loads, will be required to receive the blast loads that are applied to the exterior facade and transfer them to the building's foundation. This load path is typically through the floor slabs that act as diaphragms and interconnect the different lateral resisting elements. The lateral load-resisting system for a school building depends, to a great extent, on the type of construction and region. In many cases, low-rise buildings do not receive substantial wind and seismic forces and, therefore, do not require substantial lateral load-resisting systems. Because blast loads diminish with distance, a package sized explosive threat is likely to locally overwhelm the facade, thereby limiting the force that may be transferred to the lateral load-resisting system. However, the intensity of the blast loads that may be applied to the building could exceed the design limits for most conventional school construction. As a result, the building is likely to be subjected to large inelastic deformations that may produce severe cracks to the structural and nonstructural partitions. There is little that can be done to upgrade the existing school structure to make it more flexible in response to a blast loading that doesn't require extensive renovation of the building. Safe rooms should, therefore, be located close to the interior shear walls or reinforced masonry walls in order to provide maximum structural support.

In addition to the hazard of impact by facade debris propelled into the building or roof damage that may rain down, the occupants may also be vulnerable to much heavier debris resulting from structural damage. Progressive collapse occurs when an initiating localized failure causes adjoining members to be overloaded and fail, resulting in an extent of damage that is disproportionate to the originating region of localized failure. The initiating localized failure may result from a sufficiently sized parcel bomb that is in contact with a critical structural element or from a vehicle-sized bomb that is located at a short distance from the structure. However, a large explosive device at a large stand-off distance is not likely to selectively cause a single structural member to fail; any damage that results from this scenario is more likely to be widespread and the ensuing collapse cannot be considered progressive. Although progressive collapse is not typically an issue for buildings three stories or shorter, transfer girders or precast construction may produce structural systems that are not tolerant of localized damage conditions. The columns that support transfer girders and the transfer girders themselves may be critical to the stability of a large area of floor space. Similarly, precast construction that relies on individual structural panels may not be sufficiently tied together to resist the localized damage or large structural deformations that may result from an explosive detonation. As a result, safe rooms should not be located on a structure that is either supported by or underneath a structure that is supported by transfer girders unless the building is evaluated by a licensed professional engineer. The connection details for multistory precast structures should also be evaluated before the building is used to house a safe room.

Nonstructural building components (e.g., piping, ducts, lighting units, and conduits) that are located within safe rooms must be sufficiently tied back to a competent structure to prevent failure of the services and the hazard of falling debris. To mitigate the effects of in-structure shock that may result from the infilling of blast pressures through damaged windows, the nonstructural systems should be located below the raised floors or tied to the ceiling slabs with seismic restraints.

Fragment Mitigating Upgrades

The conversion of existing construction to provide blast-resistant protection requires upgrades to the most fragile or brittle elements enclosing the safe room. Failure of the glazed portion of the facade represents the greatest hazard to the occupants. Therefore, the exterior glazed elements of the school facade and, in particular, the glazed elements of the designated safe rooms, should be protected with a FRF, also commonly known as anti-shatter film (ASF), "shatter-resistant window film" (SRWF), or "security film." These materials consist of a laminate that will improve post-damage performance of existing windows. Applied to the interior face of glass, ASF holds the fragments of broken glass together in one sheet, thus reducing the projectile hazard of flying glass fragments. See FEMA 426 Reference Manual to Mitigate Potential Terrorist Attacks Against Buildings for more information.

Most ASFs are made from polyester-based materials and coated with adhesives. They are available as clear, with minimal effects to the optical characteristics of the glass, and tinted, which provide a variety of aesthetic and optical enhancements and can increase the effectiveness of existing heating/cooling systems. Most films are designed with solar inhibitors to screen out ultraviolet (UV) rays and are available treated with an abrasion-resistant coating that can prolong the life of tempered glass. However, over time, the UV absorption damages the film and degrades its effectiveness.

According to published reports, testing has shown that a 7-mil thick film, or specially manufactured 4-mil thick film, is the minimum thickness that is required to provide hazard mitigation from blast. Therefore, a 4-mil thick ASF should be utilized only if it has demonstrated, through explosive testing, that it is capable of providing the desired hazard-level response.

The application of security film must, at a minimum, cover the clear area of the window. The clear area is defined as the portion of the glass unobstructed by the frame. This minimum application, termed daylight installation, is commonly used for retrofitting windows. By this method, the film is applied to the exposed glass without any means of attachment or capture within the frame. Application of the film to the edge of the glass panel, thereby extending the film to cover the glass within the bite, is called an edge-to-edge installation and is often used in dry-glazing installations. Other methods of retrofit application may improve the film performance, thereby reducing the hazard; however, these are typically more expensive to install, especially in a retrofit situation.

Although a film may be effective in keeping glass fragments together, it may not be particularly effective in retaining the glass in the frame. ASF is most effective when it is used with a blast-tested anchorage system. Such a system prevents the failed glass from exiting the frame.

The wet-glazed installation, a system where the film is positively attached to the frame, offers more protection than the daylight installation. This system of attaching the film to the frame reduces glass fragmentation entering the building. The wet-glazing system utilizes a high-strength liquid sealant, such as silicone, to attach the glazing system to the frame. This method is more costly than the daylight installation.

Securing the film to the frame with a mechanically connected anchorage system further reduces the likelihood of the glazing system exiting the frame. Mechanical attachment includes anchoring methods that employ screws and/or batten strips that anchor the film to the frame along two or four sides. The mechanical attachment method can be less aesthetically pleasing when compared to wet glazing because additional framework is necessary and is more expensive than the wet-glazed installation.

Window framing systems and their anchorage must be capable of transferring the blast loads to the surrounding walls. Unless the frames and anchorages are competent, the effectiveness of the attached films will be limited. Similarly, the walls must be able to withstand the blast loads that are directly applied to them and accept the blast loads that are transferred by the windows. The strength of these walls may limit the effectiveness of the glazing upgrades.

If a major rehabilitation of the facade is required to improve the mechanical characteristics of the building envelope, a laminated glazing replacement is recommended. Laminated glass consists of two or more pieces of glass permanently bonded together by a tough plastic interlayer made of PVB resin. After being sealed together, the glass "sandwich" behaves as a single unit. Annealed, heat-strengthened, tempered glass, or polycarbonate glazing can be mixed and matched between layers of laminated glass in order to design the most effective light for a given application. When fractured, fragments of laminated glass tend to adhere to the PVB interlayer, rather than falling free and potentially causing injury.

Laminated glass can be expected to last as long as ordinary glass provided it is not broken or damaged in any way. It is very important that laminated glass is correctly installed in order to ensure long life. Regardless of the degree of protection required from the window, laminated glass needs to be installed with adequate sealant to prevent water from coming in contact with the edges of the glass. A structural sealant will adhere the glazing to the frame and allow the PVB interlayer to develop its full membrane capacity. Similar to attached film upgrades, the window frames and anchorages must be capable of transferring the blast loads to the surrounding walls.

Blast curtains are made from a variety of materials, including a warp knit fabric or a polyethylene fiber. The fiber can be woven into a panel as thin as 0.029 inch that weighs less than 1.5 ounces per square foot. This fact dispels the myth that blast curtains are heavy sheets of lead that completely obstruct a window opening and eliminate all natural light from the interior of a protected building. The blast curtains are affixed to the interior frame of a window opening and essentially catch the glass fragments produced by a blast wave. The debris is then deposited on the floor at the base of the window. Therefore, the use of these curtains does not eliminate the possibility of glass fragments penetrating the interior of the occupied space, but instead limits the travel distance of the airborne debris. Overall, the hazard level to occupants is significantly reduced by the implementation of the blast curtains. However, a person sitting directly adjacent to a window outfitted with a blast curtain may still be injured by shards of glass in the event of an explosion.

The main components of any blast curtain system are the curtain itself, the attachment mechanism by which the curtain is affixed to the window frame, and either a trough or other retaining mechanism at the base of the window to hold the excess curtain material. The blast curtain with curtain rod attachment and sill trough differ largely from one manufacturer to the next. The curtain fabric, material properties, method of attachment, and manner in which they operate all vary, thereby providing many options within the overall classification of blast curtains. This fact makes blast curtains applicable in many situations.

Blast curtains differ from standard curtains in that they do not open and close in the typical manner. Although blast curtains are intended to remain in a closed position at all times, they may be pulled away from the window to allow for cleaning, blind or shade operation, or occupant egress in the case of fire. However, the curtains can be rendered ineffective if installed such that easy access would provide opportunity for occupants to defeat their operation. The color and openness factor of the fabric contributes to the amount of light that is transmitted through the curtains and the see-through visibility of the curtains. Although the color and weave of these curtains may be varied to suit the aesthetics of the interior dećor, the appearance of the windows is altered by the presence of the curtains.

The curtains may either be anchored at the top and bottom of the window frame or anchored at the top only and outfitted with a weighted hem. The curtain needs to be extra long with the surplus either wound around a dynamic tension retainer or stored in a reservoir housing. When an explosion occurs, the curtain feeds out of the receptacle to absorb the force of the flying glass fragments. The effectiveness of the blast curtains relies on their use, and no protection is provided when these curtains are pulled away from the glazing.

Structural Upgrades

The retrofit of existing structures to protect against a potential progressive collapse resulting from the detonation of a terrorist explosive threat may best be achieved through the localized hardening of vulnerable columns. These columns need only be upgraded to a level of resistance that balances the capacities of all adjacent structural elements. At greater blast intensities, the resulting damage would be extensive and termed global collapse rather than progressive collapse. Attempts to upgrade the structure to conform to the alternate path approach will be invasive and potentially counterproductive. Care must be taken not to weaken a structure in the attempt to make it more robust.

Conventionally designed columns may be vulnerable to the effects of explosives, particularly when placed in contact with their surface. Stand-off elements, in the form of partitions and enclosures, may be designed to guarantee a minimum stand-off distance; however, this alone may not be sufficient. Additional resistance may be provided to reinforced concrete structures by means of a steel jacket or a carbon-fiber wrap that effectively confines the concrete core, thereby increasing the confined strength and shear capacity of the column, and holds the rubble together to permit it to continue carrying the axial loads. The capacity of steel flanged columns may be increased with a reinforced concrete encasement that adds mass to the steel section and protects the relatively thin flange sections. The details for these retrofits must be designed to resist the specific weight of explosives and stand-off distance.

The floor slabs are typically designed to resist downward gravity loading and have limited capacity to resist uplift pressures or the upward deformations experienced during a load reversal. Therefore, floor slabs that may be subjected to significant uplift pressures, such that they overcome the gravity loads and subject the slabs to reversals in curvature, require additional reinforcement. If the slab does not contain this tension reinforcement, it must be supplemented with a lightweight carbon-fiber application that may be bonded to the surface at the critical locations. Carbon-fiber reinforcing mats bonded to the top surface of slabs would strengthen the floors for upward loading and reduce the likelihood of slab collapse from blast infill uplift pressures, as well as internal explosions in mailrooms or other susceptible regions. This lightweight high-tensile strength material will supplement the limited capacity of the concrete to resist these unnatural loading conditions. An alternative approach may be to notch grooves in the top of concrete slabs and epoxy carbon-fiber rods into grooves; although this approach may offer a greater capacity, it is much more invasive and has not been evaluated with explosive testing.

CHEMICAL, BIOLOGICAL, RADIOLOGICAL, AND EXPLOSIVE THREATS

Although the likelihood of a direct attack against a school is very low, recent terrorist events have increased interest in the vulnerability of all types of buildings to CBR threats. Of particular concern are building HVAC systems, because they can become an entry point and distribution system for airborne hazardous contaminants. Even without special protective systems, buildings can provide protection in varying degrees against airborne hazards that originate outdoors. Conversely, the hazards produced by a release inside a building can be much more severe than a similar release outdoors. Because buildings allow only a limited exchange of air between indoors and outdoors, not only can higher concentrations occur when there is a release inside, but hazards may persist longer indoors.

Airborne hazardous contaminants can be gases, vapors, or aerosols (small solid and liquid particles). Most biological and radiological agents are aerosols, whereas most chemical warfare agents (CWAs) are gaseous.

The three basic groups of biological agents that would likely be used as weapons are bacteria, viruses, and toxins:

Bacteria are small free-living organisms that reproduce by simple division and are easy to grow. The diseases they produce often respond to treatment with antibiotics.

Viruses are organisms that require living cells in which to reproduce and are intimately dependent upon the body they infect. The diseases they produce generally do not respond to antibiotics; however, antiviral drugs are sometimes effective.

Toxins are poisonous substances found in, and extracted from, living plants, animals, or microorganisms; some toxins can be produced or altered by chemical means. Some toxins can be treated with specific antitoxins and selected drugs.

Most biological agents are difficult to grow and maintain. Many break down quickly when exposed to sunlight and other environmental factors, while others, such as anthrax spores, are very long lived. They can be dispersed by spraying them in the air or by infected animals that carry the disease, as well as through food and water contamination:

Aerosols—Biological agents are dispersed into the air, forming a fine mist that may drift for miles. Inhaling the agent may cause disease in people or animals.

Animals—Some diseases are spread by insects and animals, such as fleas, flies, mosquitoes, and mice. Deliberately spreading diseases through livestock is also referred to as agroterrorism.

Food and water contamination—Some pathogenic organisms and toxins may persist in food and water supplies. Most microbes can be killed, and toxins deactivated, by cooking food and boiling water.

Person-to-person spread of a few infectious agents is also possible. Humans have been the source of infection for smallpox, plague, and the Lassa viruses.

After the presence of an airborne hazard is detected, there are five possible protective actions for a building and its occupants. In increasing order of complexity and cost, these actions are:

1. Evacuation
2. Sheltering in Place
3. Personal Protective Equipment
4. Filtering and Pressurization
5. Exhausting and Purging

These actions are implemented, singly or in combination, when a hazard is present or known to be imminent. To ensure these actions will be effective, a school safety emergency plan specific to each school, as well as training and familiarization for occupants, is required. Exhausting and purging is listed last because it is usually the final action after any airborne hazard incident.

Evacuation

Evacuation is the most common protective action taken when an airborne hazard, such as smoke or an unusual odor, is perceived in a building. In most cases, existing plans for fire evacuation apply. Orderly evacuation is the simplest and most reliable action for an internal airborne hazard, but may not be the best action in all situations, especially in the case of an external CBR release or plume, particularly one that is widespread. If the area covered by the plume is too large to rapidly and safely exit, sheltering in place should be considered. If a CBR agent has infiltrated the building and evacuation is deemed not to be safe, the use of protective hoods may be appropriate. Two considerations in non-fire evacuation are: (1) to determine if the source of the airborne hazard is internal or external to the building; and (2) to determine if evacuation may lead to other risks. Also, evacuation and assembly of occupants should be on the upwind side of the building and at least 100 feet away, because any airborne hazard escaping the structure will be carried downwind.

Sheltering in Place

Typically, buildings offer little protection to occupants from airborne hazards outside the structure because outdoor air must be continuously introduced to provide a comfortable, healthy indoor environment. However, a school can provide substantial protection against agents released outdoors if the flow of fresh air is filtered/cleaned, or temporarily interrupted or reduced. Interrupting the flow of fresh air is the principle applied in the protective action known as sheltering in place.

The advantage of sheltering in place is that it can be implemented rapidly. The disadvantage is that its protection is variable and diminishes with the duration of the hazard. Sheltering requires that two distinct actions be taken without delay to maximize the passive protection a school building which provides the following:

- First, reduce the indoor-outdoor air exchange rate before the hazardous plume arrives. This is achieved by closing all windows and doors, and turning off all fans, air conditioners, and combustion heaters.
- Second, increase the indoor-outdoor air exchange rate as soon as the hazardous plume has passed. This is achieved by opening all windows and doors, and turning on all fans to ventilate the building.

The level of protection that can be attained by sheltering in place is substantial, but it is less than can be provided by high-efficiency filtration of the fresh air introduced into the building. The amount of protection varies with:

- **The building's air exchange rate.** The tighter the school building (i.e., the lower the air exchange rate), the greater the protection it provides. In most cases, air conditioners and combustion heaters cannot be operated while sheltering in place because operating them increases the indoor-outdoor exchange of air.

- **The duration of exposure**. Protection varies with time, diminishing as the time of exposure increases. Sheltering in place is, therefore, suitable only for exposures of short duration, roughly two hours or less, depending on conditions.
- **Purging or period of occupancy.** How long students, faculty, and staff remain in the building after the hazardous plume has passed also affects the level of protection. Because the school building slowly purges contaminants that have entered it, at some point during plume passage, the concentration inside exceeds the concentration outside. Maximum protection is attained by increasing the air exchange rate after plume passage or by exiting into clean air.
- **Natural filtering.** Some filtering occurs when the agent is deposited in the school shell or upon interior surfaces as air passes into and out of the building. The tighter the school building, the greater the effect of this natural filtering.

In a home, taking the actions required for sheltering (i.e., closing windows and doors, and turning off all air conditioners, fans, and combustion heaters) is relatively simple. Doing so in a school may require more time and planning. All air-handling units must be turned off and any dampers for outside air must be closed. Procedures for a protective action plan, therefore, should include:

- Identifying all air-handling units, fans, and the switches needed to deactivate them.
- Identifying cracks, seams, joints, and pores in the building shell to be temporarily sealed to further reduce outside air infiltration. Keeping emergency supplies, such as duct tape and polyethylene sheeting, on hand.
- Identifying procedures for purging after an internal release (i.e., opening windows and doors, turning on smoke fans, air handlers, and fans that were turned off) to exhaust and purge the building.
- Identifying school safe rooms (i.e., interior rooms having a lower air exchange rate) that may provide a higher level of passive protection. It may be desirable to go to a predetermined sheltering room (or rooms) and
  - shut and lock all windows and doors;
  - seal any windows and vents with plastic sheeting and duct tape;
  - seal the door(s) with duct tape around the top, bottom, and sides;
  - firmly pack dampened towels along the bottom of each door;
  - turn on a TV or radio that can be heard within the shelter and listen for further instructions;
  - when the "all clear" is announced, open windows and doors.

Although sheltering is for protection against an external release, it is possible, but more complex, to shelter-in-place on one or more floors of a multistory school building after an internal release has occurred on a single floor. Important considerations for use of sheltering in place are that stairwells must be isolated by closed fire doors, elevators must not be used, and clear evacuation routes must remain open if evacuation is required. Escape hoods may be needed if the only evacuation routes are through contaminated areas.

One final consideration for sheltering in place is that students, faculty, and staff cannot be forced to participate. During an event, some building emergency plans call for making a concise information announcement, and then giving occupants three to five minutes to proceed to the sheltering area or evacuate the building before it is sealed. It is important to develop a plan in cooperation with likely participants and awareness training programs that include discussions of sheltering in place and events (CBR attacks, hazardous material releases, or natural disasters) that might make sheltering preferable to evacuation. Training programs and information announcements during an event should be tailored to help students, faculty, and staff to make informed decisions.

Safe Room Criteria

Any type of room can be used as a safe room if it meets the criteria listed below.

Accessibility—The safe room must be rapidly accessible to all people who are to be sheltered. It should be located so that it can be reached in minimum time with minimum outdoor travel. There are no specific requirements for the time to reach a safe room; however, moving to the safe room from the most distant point in the building should take less than two minutes. For maximum accessibility, the ideal safe room is one in which one spends a substantial portion of time during a normal day. The safe room should be accessible to persons with mobility, cognitive, or other disabilities. Appropriate use of stairs or ramps when shelters are located above or below grade must account for such occupants.

Size—The size criterion for the toxic-agent safe room is the same as tornado shelters. Per FEMA 361, the room should provide five square feet per standing adult, 6 square feet per seated adult, and 10 square feet per wheelchair user for occupancy of up to two hours.

Tightness—There is no specific criterion for air tightness. With doors closed, the safe room must have a low rate of air exchange between it and the outdoors or the adjacent indoor spaces. Rooms with few or no windows are preferable if the windows are of a type and condition that do not seal tightly (e.g., older sliders). The room must not have lay-in ceilings (suspended tile ceilings) unless there is a hard ceiling above. The room should have a minimum number of doors, and the doors should not have louvers unless they can be sealed quickly. The door undercut must be small enough to allow sealing with a door-sweep weather strip or expediently with duct tape.

HVAC system—The safe room must be isolated or capable of being isolated quickly from the HVAC system of the building. If the selected room is served by supply and return ducts, modifications or preparations must include a me*ans of temporarily closing the ducts to the safe room. In the simplest form, this involves placing duct tape or contact paper over the supply, return, and exhaust grilles and turning off fans and air-handling units. If there is a window-type or through-the-wall air conditioner in the selected room, plastic sheeting and tape must be available to place over the inside of the window and/or air conditioner, which must be turned off when sheltering in the safe room.

Ventilation—For Class 1 Safe Rooms, 15 cfm per person is the desired ventilation rate; however, the minimum ventilation rate is 5 cfm per person if that rate is adequate for pressurization. Class 3 and unventilated Class 2 Safe Rooms are suitable only for short-duration use, not only because the low ventilation rate when occupied can cause carbon dioxide levels to rise, but also because protection diminishes as the time of exposure to the hazard increases.

Location—For unventilated shelters (Class 3 and some Class 2), there are three considerations for location within a building. First, relative to the prevailing wind, the safe room should be on the leeward side of a building. Second, if there is a toxic material storage or processing plant in the community, the safe room should be on the side opposite the plant. Third, an interior room is preferable to a room with exterior walls, if it meets criteria for size, tightness, and accessibility. For a low-rise building, there is no substantial advantage in a room on the higher floors, and a location should not be selected based on height above ground level if it increases the time required to reach the shelter in an emergency.

Water and toilets—Drinking water and a toilet(s) should be available to occupants of a safe room. This may involve the use of canned/bottled water and portable toilets. Toilet fixture allowance is presented in FEMA 361.

Communications—For sheltering situations initiated by local authorities, the safe room must contain a radio with which to receive emergency instructions for the termination of sheltering. A telephone or cell phone can be used to receive emergency instructions and to communicate with emergency management agencies. Electrical power and lighting are also required.

Personal Protective Equipment

A wide range of individual protection equipment is available, including respirators, protective hoods, protective suits, CBR detectors, and decontamination equipment.

Of particular note, new models of universal-fit escape hoods have been developed for short-duration "escape-only" wear to protect against chemical agents, aerosols (including biological agents), and some TICs. These hoods are compact enough to be stored in desks or to be carried on the belt. They should be stored in their sealed pouches and opened only when needed. Most of these hoods form protective seals at the neck and do not require special fitting techniques or multiple sizes to fit a large portion of the population. Training is required to use the hoods properly. Depending on hood design, the wearer must bite on and breathe through a mouth bit or use straps to tighten a nose cup around the nose and mouth. Escape hoods should be considered, but may not be an effective or efficient proposed solution for use in schools, under current threats.

There are no government standards for hoods intended for protection against the malicious use of chemical or biological agents. In selecting an escape hood, a purchaser should, therefore, require information on laboratory verification testing. Plans should be made for training, fitting, storing, and maintaining records relative to storage life, and there should be procedures for instructing building occupants about when to put on the hoods. Wearing a hood can cause physiological strain and may cause panic or stress that could lead to respiratory problems in some people. Finally, it should be recognized that no single selection of personal protective equipment is effective against every possible threat. Selection must be tied to specific threat/hazard characteristics.

Air Filtration and Pressurization

Among the various protective measures for school buildings, high-efficiency air filtration/cleaning provides the highest level of protection against an outdoor release of hazardous materials. It can also provide continuous protection, unlike other approaches for which protective measures are initiated upon detecting an airborne hazard. *FEMA 426 Reference Manual to Mitigate Potential Terrorist Attacks Against Buildings* discusses air filtration in detail.

Two basic methods of applying air filtration to buildings are external filtration and internal filtration. External filtration involves drawing air from outside, filtering and/or cleaning it, and discharging the air inside the building or protected zone. This provides a higher level of protection, but involves substantially higher costs. Internal filtration involves drawing air from inside the building, filtering and/or cleaning it, and discharging the air back inside the building.

The relative levels of protection of the two methods can be illustrated in terms of protection factor, and the ratio of external dose and internal dose (concentration integrated over time). External filtration systems with high-efficiency filters can yield protection factors greater than 100,000. For internal filtration, the protection factors are likely to be less and are highly variable. The protection of internal filtration varies with a number of factors, including the efficiency of the filter, flow rate of the filter unit, and size of the room or building in which the filter unit operates.

Exhausting and Purging

Turning on building ventilation fans and smoke purge fans is a protective action for purging airborne hazards from the building and reducing the hazard to which school occupants are exposed, but it is mainly useful when the source of the hazard is indoors.

Purging must be carefully applied with regard to the location of the source and the time of the release. It must be clear that the source of the hazard is inside the school building and, if not, purging should not be attempted. If the hazardous material has been identified before release or immediately upon release, purging should not be employed, because it may spread the hazardous material throughout the school building or HVAC zone. In this case, all air-handling units should be turned off to isolate the hazard while evacuating or temporarily sheltering in place.

Additionally, the ventilation system and smoke purge fans can be used to purge the building following an external release after the hazard outdoors has dissipated, and it has been confirmed that the agent is no longer present near the school building.

CBR Detection

Most strategies for protecting students, faculty, staff, and visitors from airborne hazards will require a means of detection (i.e., determining that a hazard exists). Although effective and inexpensive devices are widely available to detect, for example, smoke and carbon monoxide, there are no detectors that can rapidly alert occupants to a broad range of chemical and biological hazards.

Chemical detection technology has improved vastly since Operation Desert Storm, where many military detection systems experienced high false alarm rates, but biological detection technology has not matured as fast. Biological signatures are not as distinctive as chemical signatures and can take 30 minutes or more to detect. Biological detection systems are expensive and generally require trained specialists to operate. Current chemical detectors work in approximately 10 seconds; furthermore, wide varieties of efficient radiological detectors have been developed for the nuclear industry and are commercially available.

Chemical Detectors—Driven largely by a desire to protect workers from toxic vapors in industrial environments, considerable information is known on the toxicity of CWAs, which often have dual uses in industry. A variety of detection technologies exist, ranging from inexpensive manual point detection devices (e.g., paper strips and calorimetric tubes) utilizing basic chemical reactions to trigger color changes, to sophisticated detection systems utilizing advanced technologies.

Chemical agents do not possess universal properties that permit detection by any single method. Therefore, most chemical detectors are designed to detect specific agents or a group of related agents. Most broad range detection systems actually combine several different sensors utilizing different technologies and can be very expensive and complex. Nevertheless, today there are numerous commercially available chemical detectors. The most capable detectors utilize ion IMS, SAW, or GC/MS technologies to detect chemical agents and TIMs.

Today, there are commercially available IMS detection systems that will detect most chemical agents and many TIMs. They are suitable for integration into a building ventilation system, can interface with HVAC control systems, have reasonable maintenance requirements (every three months), low false alarm rates, and can be programmed to detect specific chemical agents.

Biological Detectors—The current state of biological detection technology is very different from that of chemical agent detection technology. In general, most biological detection systems are currently in the research and early development stages. There are some commercially available devices that have limited utility (responding only to a small number of agents) and are generally high-cost items. Because commercially available biological warfare (BW) detection systems and/or components exhibit limited utility in detecting and identifying BW agents and are also costly, it is strongly recommended that purchasers be very careful when considering any device that claims to detect BW agents.

Indications of CBR Contamination

Most hazardous chemicals have warning properties that provide a practical means for detecting a hazard and initiating protective actions. Such warning properties make chemicals perceptible; for example, vapors or gases can be perceived by the human senses (i.e., smell, sight, taste, or irritation of the eyes, skin, or respiratory tract) before serious effects occur. The distinction between perceptible and imperceptible agents is not an exact one. The concentrations at which a person can detect an odor vary from person to person, and these thresholds also vary relative to the concentration that can produce immediate, injurious effects.

Most of the industrial chemicals and chemical warfare agents are readily detectable by smell. Soldiers in World Wars I and II were taught to identify, by smell, such agents as mustard, phosgene, and chlorine, and this detection method proved effective for determining when to put on and take off a gas mask. An exception is the chemical warfare agent sarin, which is odorless and colorless in its pure form and, therefore, imperceptible. Among the most common TICs, carbon monoxide is one of the few that is imperceptible. Because it is odorless and colorless, it causes many deaths in buildings each year.

Biological agents are also imperceptible and there are no detection devices that can determine their presence in the air in real time. Current methods for detecting bacterial spores, such as anthrax, require a trained operator and expensive equipment. It is not currently possible to base protective responses to biological agents on detection.

Researchers are working on a prototype device to automatically and continuously monitor the air for the presence of bacterial spores. The device would continuously sample the air and use microwaves to trigger a chemical reaction, the intensity of which would correspond to the concentration of bacterial spores in the sample. If an increase in spore concentration is detected, an alarm similar to a smoke detector would sound and a technician would respond and use traditional sampling and analysis to confirm the presence of anthrax spores. Researchers hope the device response time will be fast enough to help prevent widespread contamination.

In the absence of a warning, people can be alerted to some airborne hazards by observing symptoms or effects in others. This provides a practical means for initiating emergency plans, because the susceptibility to hazardous materials varies from person to person. The concentrations of airborne materials may also vary substantially within a given building or room, producing a hazard that may be greater to some occupants than to others.

Other warning signs of a hazard may involve seeing and hearing something out of the ordinary, such as the hiss of a rapid release from a pressurized cylinder. Awareness of warning properties, signs, and symptoms in other people is the basis of an emergency plan. Such a plan should apply four possible protective actions: evacuating, sheltering in place, using protective masks, and exhausting and purging, as already discussed in this chapter.

For protection against imperceptible agents, the only practical protective measures are those that are continuously in place, such as filtering all air brought into the building on a continuous basis and using automatic, real-time sensors that are capable of detecting the imperceptible agents.

CBR materials, as well as industrial agents, may travel in the air as a gas or on surfaces we physically contact. Dispersion methods may be as simple as placing a container in a heavily used area, opening a container, or using conventional (garden)/commercial spray devices, or as elaborate as detonating an aerosol.

Most chemical warfare agents are gaseous, and biological and radiological agents are largely aerosols.

Chemical incidents are characterized by the rapid onset (minutes to hours) of medical symptoms and easily observed indicators (e.g., colored residue, dead foliage, pungent odor, and dead animals, birds, fish, or insects).

In the case of a biological incident, the onset of symptoms takes days to weeks and, typically, there will be no characteristic indicators. Because of the delayed onset of symptoms in a biological incident, the area affected may be greater due to the migration of infected individuals.

In the case of a radiological incident, the onset of symptoms also takes days to weeks to occur and typically there will be no characteristic indicators. Radiological materials are not recognizable by the senses because they are colorless and odorless.

Specialized equipment is required to determine the size of the affected area and if the level of radioactivity presents an immediate or long-term health hazard. Because of the delayed onset of symptoms in a radiological incident, the affected area may be greater due to the migration of contaminated individuals.

Safe Rooms within Schools

It is important to note that the probability of either a CBR or terrorist explosive event occurring in the United States is small. This is evidenced by the relatively few domestic buildings that have been targeted by intentional CBR or explosive events compared to the vast number of buildings that might be considered vulnerable. To date, two incidents of biological terrorism have been recorded and acknowledged to have had significant impacts on coincident populations in the United States: the 2001 anthrax mailings and the 1984 contamination of restaurants with Salmonella bacteria. If a localized CBR event were to occur, the potential for contamination to spread and cause collateral illnesses and fatalities up to four or more kilometers (approximately 21⁄2 miles) from the target site would be likely. Unpredictable meteorological conditions would play a key role in the spread of such CBR contamination. Similarly, if an explosive event were to occur, there would be a significant potential for injuries resulting from debris impact and structural collapse. Therefore, in consideration of the proximity of some schools with respect to higher profile potential targets in the United States, school board members and administrators may determine that their select facilities require the design and construction of safe rooms. Because there are so many different types of school buildings, with so many different types of construction and materials, it is not possible to relate all the following issues to specific building types; nevertheless, an attempt was made to relate the relevant threats and the general principles of protective design to the development of safe rooms within schools.

The needs of persons with disabilities requiring shelter space should be considered. The appropriate access for persons with disabilities must be provided in accordance with all federal, state, and local ADA requirements and ordinances. If the minimum requirements dictate only one ADA-compliant access point for the shelter, the design professional should consider providing a second ADA-compliant access point for use in the event that the primary access point is blocked or inoperable.

Types of CBR Hazards

Chemical contaminants of concern are the CWAs and TICs. Key attributes of CWAs and TICs are their toxicity, volatility, and availability. The most toxic CWAs are the nerve agent liquids, which include VX with high toxicity and low volatility, and sarin with high toxicity and moderate volatility. The measured volatility of a chemical represents the ease with which the quantity of liquid chemical leaves the liquid state and becomes a gas in equilibrium with its volumetric surroundings. So, an occupant of a room where sarin liquid is naturally evaporating is at much greater risk than the same individual being in the same room with the same or (to a degree greater) quantity of a naturally evaporating VX agent. The lethality of VX exceeds the lethality of sarin by dose, but sarin is much more volatile than VX.

Toxic Industrial Chemicals

Though of lower toxicity than nerve agents, TICs are widely available, and some can be easily obtained or produced without sophisticated equipment. Among the hundreds of TICs produced worldwide are several that have been used as CWAs (e.g., arsine, chlorine, hydrogen cyanide, phosgene, hydrogen sulfide, acrolein, and cyanogen chloride). Those that have been used in warfare are considered second-rate CWAs because their toxicity and vapor pressure make them less effective than other agents for open-air battlefield use.

Incapacitating and Tear-producing Agents

Although incapacitating and tear-producing agents are considered non-lethal, indoor releases can, under certain conditions, produce lethal concentrations. In addition to the tear-producing agents, there are commercially available agents containing oleoresin capsicum (OC), the natural oil of chili peppers. The malicious or accidental release of pepper spray has caused many disruptive incidents in recent years. In contact with the eyes, nose, or mouth, OC causes immediate pain and inflammation. Inhaled, its aerosol causes choking and gasping for breath. Of low vapor pressure, OC is easily filtered.

Biological Agents

Biological agents include bacteria, viruses, and rickettsia. Toxins, which are poisons of biological origin and not living organisms, are sometimes grouped with biological agents and sometimes with chemical agents. Although there are hundreds of micro-organisms that could be used as biological agents, the likely number is much smaller when the agents' effectiveness, reliability, availability, ease of manufacture, and stability in storage and dissemination are considered. When disseminated as aerosols, biological agents are most effective in the size range of one to five microns, because they can remain suspended for long periods. Smaller particles are less likely to survive as aerosols, and larger particles settle rapidly, making them less likely to enter the lungs. The settling time in still air for an anthrax spore (one micron by 0.7 micron in size) is approximately 1/2 foot per hour. Particles of this size are readily filtered from an air stream with HEPA filters. Toxins, which may be in crystalline or liquid form, are also filterable with HEPA when disseminated as aerosols.

Radiological Agents

Radiological agents are radioactive materials. Explosive release is the most likely means of disseminating such agents in a terrorist attack (e.g., a "dirty bomb" consisting of radioactive material packaged with a conventional explosive). The likely radioactive ingredients are those used for industrial and medical purposes (e.g., isotopes of caesium, cobalt, and iridium). They are commonly found in hospitals and labs, often with few safeguards. Given the availability of nuclear reactors for research or energy production by universities, research facilities, or private industries, the threat associated with radiological materials is significant. Radiological contaminants are very persistent, in that their decay rate is extremely slow. Unlike chemical or biological agents, decontamination involves only removal, not neutralization. Radiological aerosols present a health hazard if ingested or inhaled, but are easily filtered from an air stream with HEPA filters.

Most Likely Delivery Methods for CBR Agents

For purpose of vulnerability assessments, delivery methods are divided into four types of releases: internal, external proximate, remote, and remote with forewarning.

Internal Release

This involves transporting a container of agent into a building and releasing the contents manually, automatically, or remotely. Such a device may rely simply upon natural evaporation (as in the Tokyo subway sarin attack), with the rate of evaporation proportional to the surface area that develops as liquid agent spills from its container. Aerosolization may occur with movement of an open package or letter containing a biological agent. A sprayer powered by batteries or compressed air can produce an effective dose of an agent quite rapidly. An agent can be released in any area served by return ducts/plenums or in a mechanical room, with dissemination through an air-handling unit. Biological agents can also be placed into certain types of humidifier systems.

External Proximate Release

This involves introducing an agent or a dissemination device from outside the building directly through a penetration in the building shell, such as a fresh-air intake. Vulnerability to this type of release is highest when air intakes are at accessible, unsecured locations at ground level. Agents can also be delivered through other penetrations, but potential effectiveness is less in the absence of a driving force (a fan) to introduce and distribute air within the building. A documented example of an attack through a ground-level penetration is the release of a toxic industrial gas from a pressurized cylinder through a dryer vent. External proximate release also includes forcing open or breaking windows and doors to introduce agents from pressurized cylinders or tossing a grenade or container of an agent into the building.

Remote Release

If directed at a specific facility, this type of attack involves a plume, puff, or line source generated so that the wind carries the agent to the target building; the facility may be the target or collaterally in the direction of the attack. The most efficient type of remote attack is a directed-plume attack with a ground-level source placed upwind of a building's fresh-air intakes or open windows. A ground-level, directed-plume attack was conducted with the nerve agent sarin from a distance of 60 yards in Matsumoto, Japan, in 1994, killing seven and injuring 264 in a zone 500 yards deep and 100 yards wide. A remote attack can also involve an aerial release. Release from an aircraft is much less likely to affect a specific, targeted building, however, because the vertical rate of transport, governed by settling time and atmospheric stability, is extremely difficult to judge.

Remote Release with Forewarning

This type of attack differs from other remote releases because protective actions other than those for no-warning attacks can be applied. This type involves warning in the form of an explosion or an event, such as an accidental or intentional release of an agent from a chemical transport or storage tank. Scenarios involving forewarning include sabotage of toxic industrial storage tanks/trucks, transport accidents, fires, or the impending release of a chemical agent from a point upwind of the building. Quantities of agent that could be released from a single 3,000-gallon tanker truck are approximately 34,000 pounds for phosgene, 35,000 pounds for chlorine, and 17,000 pounds for hydrogen cyanide.

Vulnerability to Remote CBR Release

In the absence of a secure perimeter around the building and a real-time detection system, vulnerability to a remote release is determined by (1) the efficiency of the school building's filtration system in removing aerosols and gases; (2) the unfiltered component of air exchange; and (3) the configuration of the school building and elevation of air intakes. These vulnerabilities can be characterized as follows.

- **Efficiency of gas filtration.** Generally, if adsorbers are found in buildings, they are for the purpose of improving indoor air quality by removing both outdoor and indoor air pollutants, particularly corrosive gases, such as sulfur dioxide, nitrogen dioxide, and ozone, and are used where appropriate for protecting against the deleterious effects of these gases. Although not intended for protecting people from toxic chemical agents, these gas adsorbers do reduce the vulnerability to an attack with certain chemical agents.

With a one-inch bed thickness of coarse (4x6 mesh) sorbent granules and a short residence time of these indoor air quality filters, the efficiency is about 99 percent initially, and it diminishes with time in service, typically to about 25 percent in a year. There is also an initial bypass of roughly one percent through the bed and additional bypasses among filter modules' holding frames. The bypasses may increase with time in service, dropping the net efficiency below the initial level. This compares with an efficiency greater than 99.999 percent for gas adsorbers designed for protection of people in military applications. Removal efficiency for these indoor air quality adsorbers is relatively low and uncertain for some of the threat agents (the capacity for arsine is low, for example). Thus, the efficiency and capacity are highly variable. Manufacturers provide surveillance testing to determine when to change filters and recommend that they be changed when reactive capacity has dropped to 25 percent of the initial value. The typical service life for single stage, one-inch beds is approximately one year.

- **Particulate filtration.** Significant particulate filtration can be accomplished by using a 35-percent pre-filter and a 95-percent filter in series. The efficiency of this filter train is in the range of 95 percent to 99 percent for 1-micron particles in the new condition, and this efficiency increases as the filters load.
- **Unfiltered air exchange.** Typical of schools, a substantial portion of the air exchanged between indoors and outdoors may not pass through the filters of any air-handling units. When this occurs, the level of protection the building structure provides is, therefore, governed not by the efficiency of the filters but rather by that portion of makeup air bypassing the filters. There are several paths by which air exchange is driven by fans, buoyancy, and/or wind pressures. They include operable windows; doorways with flows driven by buoyancy, particularly in summer and winter when indoor-outdoor temperature differentials are highest; and unintentional openings in the building shell. When internal resistance is minimal, less dense (buoyant) warm air rises and flows out of a school building near its top in winter, drawing in cool air at the lower levels. Conversely, in summer, cool air falls and flows out of the building's base. The buoyancy effect tends to be less pronounced in spring and fall because of smaller indoor and outdoor temperature differences. With the standard draw-through configuration of the air-handling units, leakage paths at the access doors and panels are subject to inward pressure; these leakage paths increase as the gaskets age. Typically, access doors also fail to seal well with filter frames, allowing bypass that increases with age.
- **Typical protection factors achievable**. In terms of protection factor, protection against aerosols (biological/radiological agents and others, such as tear gas) provided by the best (standard) filtration systems available in air-handling units is substantial, but relatively low (in the range of approximately five to fifty). Protection factor is a ratio of dose (concentration integrated over time) of an agent outside divided by the resulting dose of agent inside a building). This level of protection is comparable to that achievable with sheltering in place or an active detection-based system that responds by de-energizing fans and closing dampers. The higher value in this range is estimated by taking the inverse of the penetration factor for 1-micron particles through the filters, including an initial bypass of approximately one percent. The lower value in the protection-factor range is estimated for both particulate and gas filters by using 20 percent as an estimate of the unfiltered air exchange; the inverse of the penetration factor (0.2) is five.

With an indoor air quality filter unit having a gas adsorber, the protection factor for gases can be as high as fifty, but only when filters are new and only with gas adsorbers. The protection drops to a low value as the filter efficiency decreases with time in service. This may be less than two if the efficiency drops below 50 percent after one year in service. The protection factor against gases is also reduced by the portion of outside air (which could be at least 80%) that does not flow through gas adsorbers. With a penetration of 80 percent, the protection factor for gases for the whole building is less than two.

Vulnerability to Remote CBR Release with Forewarning

This type of attack involves release of agent by explosion or rapid release from a tanker truck, rail car, or fixed storage tank. The potential for this type of attack is higher when the facility is near rail lines, public roads with truck traffic, or storage tanks of toxic chemicals. This type of attack may also involve an explosive release of a radiological agent (i.e., a "dirty bomb" attack at a distance from the building great enough to allow for protective actions to be taken before wind carries the agent to the building).

The criterion for this aspect of vulnerability is the ability to rapidly assume a sheltering-in-place posture. The main requirements are plans/procedures for sheltering, controls to rapidly turn off all fans, and a communications or public address system to facilitate closing of doors and keeping them closed while an outdoor hazard is present or imminent. Protection factors vary, diminishing with time of exposure; however, scenarios of explosive release under most conditions would present a relatively short exposure to the school building.

Vulnerability to Internal CBR Release

This is a remote possibility owing to the nature and likelihood of other vandalism and that actual targeting of a school has no historical precedence. Nevertheless, internal releases involve covert entry or covert introduction of agents in containers. Vulnerability to internal release is, therefore, determined principally by physical security measures in place. Containers of agents may be hand-carried or delivered in mail, supplies, or equipment. Other factors affecting this vulnerability are internal (recirculation) filtration and how well entry zones where any screening takes place are isolated architecturally and mechanically.

The basis for preventing covert introduction of agents is access control and entry screening. Use of the X-ray machine for hand-carried items facilitates the detection of containers large enough to hold hazardous quantities of chemical agents; however, it requires specific operating procedures for doing so, and it may not be effective in detecting containers of hazardous quantities of biological agents. Obviously, such procedures are not recommended for schools without provocation (i.e., an actual threat) because of cost.

Vulnerability to External Proximate CBR Release

Vulnerability to external proximate release is determined mainly by the accessibility of outside air intakes to covert introduction of agent or agent-dissemination device. Unless air intakes are relocated at a higher elevation, this vulnerability would remain high.

The three strategies for protecting a school building from airborne hazards originating outdoors are air filtration, controlling air exchange, and exclusion by physical security. Options presented here focus on air filtration; however, enhanced filtration techniques discussed earlier would be applied most economically in schools to a selected safe room, such as a school gymnasium or auditorium. Without a secure exclusion zone around the school building, physical security measures are limited to those described above for external proximate release and internal release.

Controlling air exchange is most commonly employed with human detection and warning (i.e., sheltering in place). It can also be applied with automatic, real-time detection equipment, but with very limited effectiveness. Few agents among the full spectrum of threat agents can be detected with accuracy in real time. Protection factors vary with response time and, even with instantaneous response, protection factors are no greater than the maximum protection factors achievable with sheltering in place. As the response time increases, protection factors diminish. With current technology, response times are longest for biological agent detection. The response time for presumptive identification by a detector, such as the Joint Biological Point Detection System, is approximately 30 minutes and far exceeds the response time needed for effective use of sheltering in place. A biological detection system would not, therefore, prevent the contamination of a particular building.

- **Criteria for protective performance.** All of the following discussion represents extreme measures applied to high-risk, high-security assets or, in general, to lesser degrees, safe rooms, and perhaps to would-be safe rooms for schools. There is no standard requirement for protection factors. The US military systems are designed to achieve protection factors greater than 10,000. The criterion applied to military masks and collective protection shelters is 6,667, which is based upon specific levels for chemical agents on the battlefield and for threshold effects of the chemical agents on soldiers. There is no criterion for biological or radiological agents based upon concentrations and doses likely to be developed in an attack on a school building; however, it would be 10,000 or greater.

The selection of a CBR safe room in a school building requires an assessment of factors contributing to infiltration (or wind penetration from the outside). To prevent infiltration through the protected envelope requires an internal pressure of approximately 50 Pascals [0.2 inch, water gauge (iwg)]. This pressure does not prevent infiltration driven by buoyancy and wind pressures under all possible conditions, but it does so under wind conditions most conducive to a (stand-off) plume attack. The level of safe room pressurization should exceed 95–99 percent of the meteorological conditions for the given school location. Note that wind does not exert a uniform pressure on a building face; the pressure varies by location on the building face and the angle of incidence. A 20 mile-per-hour (mph) wind velocity is not uncommon in the United States and, thus, a safe room pressurized to 50 Pascals would prevent infiltration from time-averaged 20 mph winds.

There are several options for improving protection factors with filtration; they involve both the type/configuration of the filter system and the extent of the protective envelope.

- **Options for type of filter system.** Four options for a dedicated type of filtration system for a safe room include:
  - **Improving mixed-air particulate filtration of air-handling units.** Particulate filters may be upgraded to 95-percent filters, providing the potential for substantial improvement in protection against biological agent aerosols. The limit of protection factor against one-micron particles, however, is approximately 100 with pressurization of the protective envelope and reduction of bypass at the filter frames. Reduction of bypass requires sealing and gasketing existing retainers, slide-in tracks, and access doors, and adding gaskets between filter frames in slide-in tracks. Pressurization can be achieved by rebalancing the air-handling units and controlling the flows through open doorways and windows. Among the options for improvement are to upgrade the filters to HEPA with leak-tight holding frames; with pressurization, this would increase the potential protection factor to about 10,000 for biological agents, but not for chemical agents. This option requires special holding fixtures for the filters and may require replacement of supply fans to accommodate higher pressure drop.
  - **Improving mixed-air gas filtration of the air-handling units.** An option to increase protection factors of a school building for chemical agents is to install gas adsorbers in air-handling units. This would involve adding the indoor air quality (IAQ) type adsorbers to existing air-handling units, at a cost of $0.50 per cfm. With a one-year service life, the filter replacement costs would be $0.25 per cfm. Additional energy related operating costs would be incurred due to the pressure drop of the adsorbers (0.75 iwg). This option does not provide high efficiency against all chemical agents.
  - **Installing makeup air units with HEGA and HEPA filters.** A makeup air unit for both gases and aerosols includes the following components in series: pre-filter, fan, HEPA filter, HEGA filter, and heating and cooling coils. The makeup air unit provides filtered outside air to pressurize the protective envelope. It eliminates recirculation and the internally induced infiltration associated with applying a single fan for both makeup air and recirculated air. The most cost-effective HEGA filter units currently available for protection from chemical agents employ the military-standard 200-cfm radial-flow filters per MIL-PRF-51527A, "Filter Set, Gas-particulate, 200 cfm," Type II. These contain ASZM-TEDA carbon of 12x30 mesh size in two-inch-deep beds, which removes all chemical warfare agents and a substantial number of toxic industrial chemicals. These provide removal efficiency greater than 99.999 percent throughout their service life. HEPA filters are standard components with HEGA systems. These military adsorbers cost approximately $4.50 per cfm, and their expected service life is three years, although service life varies with the air quality of the region and the moisture to which the filters are exposed over time. Maintenance costs run approximately $2 per year per cfm. Maintenance also includes changing HEPA filters annually and pre-filters every ninety days. With total pressure drop of six iwg across the adsorber and HEPA filter, energy costs for the high-efficiency filtration run approximately $0.50 per cfm per year.
  - **Installing makeup air units with HEPA only.** Use of only HEPA in a makeup air unit would provide a high level of protection from biological agents, radiological

agents, solid aerosols, such as tear gas, and liquid aerosols of low vapor pressure. High-level protection against biological aerosols is particularly beneficial because there is no capability for real-time detection of biological agents (all strategies that require biological detection are mitigation strategies involving decontamination and medical treatment). Use of HEPA only in a makeup air unit would substantially reduce hardware costs, maintenance costs, and electrical costs of ventilation, as well as the space requirements for the units. Protection at a lower level would still be provided by filtration of recirculated air with gas adsorbers in air-handling units.

Recommendations for CBR Protection

The following actions are recommended for CBR protection:

- To provide a substantial level of protection against an external release of CBR agents, apply any one of the filtration options summarized previously to a renovated school gymnasium or auditorium safe room.
- To protect against a remote attack with a chemical or radiological agent, plans, procedures, and training for sheltering in place should be developed. To support this protective measure, a rapid notification system (public address system) and controls for rapid deactivation of fans and closing of dampers should be defined. A guide for developing protective action plans is available in the Army Corps of Engineers draft Technical Instruction TI 853-01 Protecting Buildings and Their Occupants from Airborne Hazards, dated October 2001.
- To reduce vulnerability to internal release, implement security procedures specific to entry screening for containers of unknown liquids or gases being carried into the secure area. Provide training to employees on awareness of the CBR threat and the protective action plan.

Now we know what to do, we have to incorporate this into a strategic safety and security plan, work out how much it will cost and get senior management buy-in to implement it.

For more detailed responses to specific emergencies see the emergency response flipcharts in appendix 3.

TRAINING AND EXERCISES

No Crisis Management or Emergency Operations Plan can be properly implemented and executed during a crisis or emergency unless all the expected participants are aware that the plan exists, what emergencies are covered, and what is expected of each (administrators, faculty, staff, and students) during an actual emergency. All participants must be adequately trained on those portions of the plan that affect them and they must know that other participants understand and are able to perform their respective tasks safely and effectively.

No Emergency Operations Plan can be considered adequate and complete until it has been tested and exercised, and participants cannot be considered to understand the plan and feel comfortable in their respective roles and assignments until they have participated in drills and exercises. All participants must have a level of trust in the plans and their personal ability to accomplish their assigned tasks in a stressful environment; this is only achieved through training and exercising.

Further, all Schools employees must have at least a minimal understanding of the ICS utilized by all Public Safety First Responders in the country to any emergency, regardless of its size.

Administrators, teachers, and some staff will have specific ICS positions based on the nature and scope of the emergency. Upper-level administrators will have more expanded roles as the emergency becomes greater or encompasses more than one school or facility. By law, all schools are required to utilize an ICS that is compliant with the National Incident Management System.

Goals

The goal sought in testing the Emergency Operations Plan is to prepare Schools personnel and students for a real emergency; the goal is to save lives and limit property damage. It is important to remember that testing is a part of the training process—we are testing the plan, not the personnel. We can discover weaknesses in the plan, identify resource needs, improve coordination, and improve individual performance and readiness for a real emergency.

Training

All Schools personnel will be provided with basic awareness-level training on the plan upon implementation; this training will be presented as basic orientation and will include the general concepts of the plan, specific assignments, and general emergency procedures. Those individuals requiring hands-on training in specialized skills (e.g., CPR-AED) will be provided this training based on the needs of the Schools during the initial training phase and yearly thereafter as required (new hires, policy changes, new buildings, etc.)

National Incident Management System-compliant ICS training will be provided to all personnel based on their anticipated assignments, duties, and responsibilities during an emergency. The ICS training will be provided on three levels, as follows:

- *Level 1*—Provided to all teachers and staff and will consist of Basic ICS understanding at the orientation level. Trainees will have an understanding of ICS, how it works, and why it is used. They will understand the relationship between ICS used by the Schools and its interface with the Public Safety First Responders ICS. (2 hours)
- *Level 2*—Provided to principals, assistant principals, secretaries and custodial staff at each school (personnel who would have a position on the Crisis Response Team at the school.) Level 2 includes all Level 1 training plus ICS at the operational level. Trainees will understand the ICS utilized by the Schools, will be able to implement ICS and execute the Crisis Management Plan at their respective school, and will be

able to staff those positions that would be required during the various emergencies addressed in the plan. (4 hours)
- *Level 3*—Provided to district managers and administrators who would have a position on the Crisis Management Team and/or make policy-level decisions during a major emergency. Level 3 is similar to Level 2 but will focus on the ICS at the policy level. Trainees will have the same ICS knowledge base as the Level 2 group but will be prepared to implement the ICS during a crisis at the district level, regardless of the nature and scope of the emergency. Trainees will understand the ICS and its ability to aid in the management of a crisis and will be able to interact with the Public Safety First Responder ICS at the state and federal level, if warranted by the emergency. (4 hours)

DRILLS AND EXERCISES

Exercises are the preferred method of testing the Crisis Management Plan on both a school and district level. Exercises fall into two general categories, the tabletop and the functional; both are used to test the plan and the Schools will do at least one tabletop and one functional exercise during the school year.

A tabletop exercise is the simulation of an emergency during which a scenario is presented to all participants and each participant explains what they or their operational unit would do to respond to the scenario as presented. The timeline is artificial and is controlled by the facilitator. There is no physical deployment of assets, no operations are actually conducted, and the Crisis Management Plan and response procedures are tested in a safe, controlled environment. All four phases of emergency management are tested, and gaps in the Crisis Management Plan are identified. Outside agencies are generally invited to participate and provide a heightened sense of realism. Tabletop exercises are the preferred method of testing a new or revised Crisis Management Plan, due to their low cost and minimal resource requirements. They usually run for about 3–4 hours.

Drills are utilized to practice and perfect a single emergency response, such as fire drills, bomb threats, tornadoes, and so on, and will continue on the current schedule used by the schools. Drills are not a good method of testing a Crisis Management Plan.

A functional exercise is the simulation of a real emergency under high-stress conditions involving multiple responders and multiple agencies. An actual emergency situation is physically created as a simulation and all participants and responders deploy and perform the same functions using the same equipment as they would in a real incident. Communications equipment is used, command posts are physically set up and staffed, EOC's are opened, and a real time line is used, compromised only by the pre-staging of equipment and responders to save time during the exercise. A functional exercise tests the Crisis Management Plan, the ICS, the response procedures, and the public safety responders in a stressful environment. Role players are used, victims are presented, and all equipment deployed. Functional exercises are a superb test of the entire Crisis Management Plan but are time consuming and fairly expensive. They usually run about 6–8 hours.

Section III

Response and Recovery

> Knowing how to respond quickly and efficiently in a crisis is critical to ensuring the safety of our schools and students. The midst of a crisis is not the time to start figuring out who ought to do what. At that moment, everyone involved—from top to bottom—should know the drill and know each other.
>
> —Margaret Spellings, Secretary, US Department of Education, 2004–2009

OVERVIEW

Here are four phases of crisis management:

1. **Preparedness** focuses on the process of planning for the worst-case scenario.
2. **Response** is devoted to the steps to take during a crisis.
3. **Recovery** deals with how to restore the learning and teaching environment after a crisis.
4. **Mitigation/Prevention** addresses what schools and districts can do to reduce or eliminate risk to life and property.

PREPAREDNESS

Action Checklist

- Determine what crisis plans exist in the district, school, and community.
- Identify all stakeholders involved in crisis planning.
- Develop procedures for communicating with staff, students, families, and the media.
- Establish procedures to account for students during a crisis.
- Gather information that exists about the school facility, such as maps and the location of utility shutoffs.
- Identify the necessary equipment that needs to be assembled to assist staff in a crisis.

Action Steps

Start by identifying who should be involved in developing the crisis plan. Include training and drills. Delegating responsibilities and breaking the process down into manageable steps will help planners develop the plan.

Please refer to "Planning for the Worst" in section two on how to

- identify and involve stakeholders;
- consider existing efforts;
- determine what crises the plan will address;
- define roles and responsibilities;
- develop methods for communicating with the staff, students, families, and the media;
- obtain necessary equipment and supplies;
- prepare for immediate response;
- create maps and facilities information;
- develop accountability and student release procedures;
- practice; and
- address liability issues.

RESPONSE

A crisis is the time **to follow the crisis plan,** not to make a plan from scratch. This section summarizes some of the major recommendations gathered from experienced practitioners and other experts about points to remember when called on to implement your crisis plan.

Action Checklist

- Determine if a crisis is occurring.
- Identify the type of crisis that is occurring and determine the appropriate response
- Activate the incident management system.
- Ascertain whether an evacuation, reverse evacuation, lockdown, or shelter-in-place needs to be implemented.
- Maintain communication among all relevant staff at officially designated locations.
- Establish what information needs to be communicated to staff, students, families, and the community.
- Monitor how emergency first aid is being administered to the injured.
- Decide if more equipment and supplies are needed.

As a rule, the Crisis Management Team, sometimes called the ICS will be activated when an incident has the potential to escalate and/or the following has occurred or is likely to occur:

- Fatality(ies) and/or serious personal injury(ies).
- Incidents requiring partial and/or full evacuation.
- Major threat to school/school system, for example, terrorism alert, direct threat to school/district, threat to adjacent institutions that could result in school/district impact, severe weather threat, IT systems disruption/threat, power outage, transportation system impact, other natural disasters.
- Adverse/media attention that could affect district image and reputation.
- Significant or catastrophic incident occurring at a school, facility, or school function.
- Hostage/kidnap/shooting/criminal act/violence-based incident/and so on.
- Adverse government or political scrutiny.

Table 3.1 Crisis Contacts

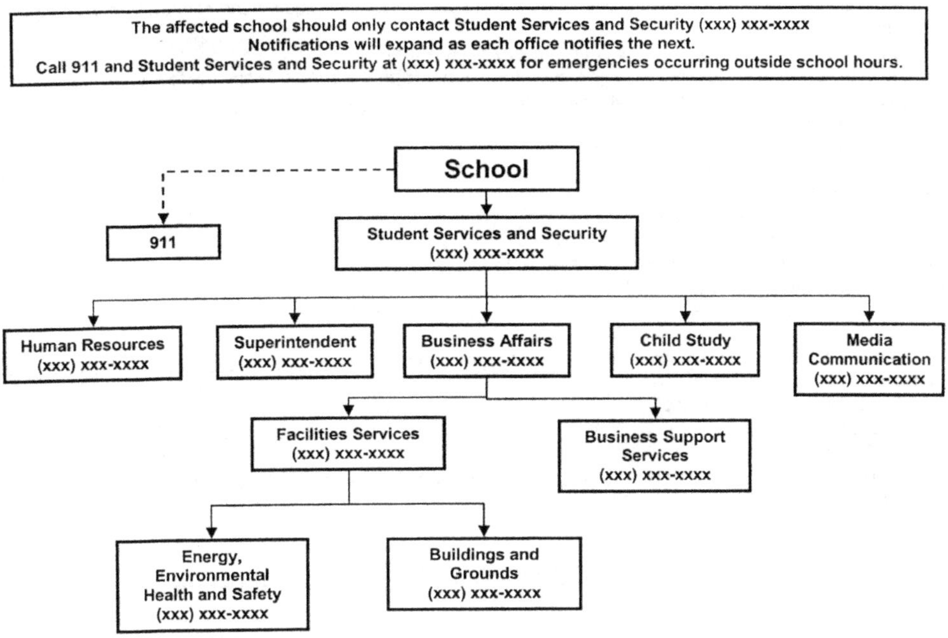

Table 3.2 Initial Call Assessment

Used initially by SS incident commander (principal/delegated representative) and district Crisis Management Team representative (security, executive director of Business Affairs) to decide initial response actions
Incident description • What has happened? • Where has it happened? • When did it happen? • Is everybody accounted for? Casualties? • How did we find out about it? • Who is at risk? • How did it happen?
Incident status • Contained or escalating? Potential to escalate—what/who is potentially at risk? • What safety and security measures are being implemented? • Has a 911 call been made? • Is the on-site SS officer involved? • What are your/our response objectives? (What are you trying to make happen/prevent from happening)? • Have fire/police departments provided any orders or recommendations? • What actions are being taken? Is area secured from unauthorized access? • Who is taking these actions? Who is responding? What resources (equipment/manpower) are being used? • Are exterior evacuation routes or staging areas accessible? • Are the effects of wind direction and slope of the grounds on the spread of smoke, fire, toxic gases, or liquids being considered in response/evacuation planning?
Incident assessment • What are the health and safety issues? • Do we need to evacuate? • Is a lockdown required? • Do we need to shelter-in-place? • What immediate messaging do we want to provide to the staff? • Do we need to isolate building ventilation systems? • What community impact? • What contacts have been made with community, government, and media? • What advice has been received from safety officials? • What is the operational impact—short, medium, and long term?
Immediate response decisions • Evacuation, shelter-in-place? • Lockdown? • Staff notification? • CMT activation and staffing (emergency operations center and on-site)? • Emergency operations center activation? • Crisis Management Plan activation? • Support school site/facility? • Interact with the Safety/Incident Command Organization? Where? How?

Table 3.3 Discussion Checklist

For executive director of Business Affairs, superintendent *(and Crisis Management Team members)*
Situation status: • What is status of safety and security of students, staff members, and others? • What happened, when, and what do we know about how and why? • What is the current status of the incident/site? • What is the magnitude and probable duration of the incident? (one day—local coverage, or multiple days—national coverage) • What has been or potentially can be the human impact on employees, community, contractors—casualties, evacuations? • Has there been any media interest? Have we issued any press releases? • Should the Crisis Management Team come to the remote scene? Who? When? Is there a plan to get them in/out? • What other organizations are involved? Law enforcement, fire department, government/political entities, and so forth. • What is the impact on the school/facility? And on other SS interests? Are there secondary impacts on other areas that need consideration? • Are there national implications to this incident? • Have all necessary notifications been made? • Has the emergency operations center been notified? • Who is the incident commander? Crisis manager? Relations leader? • What role is the superintendent going to assume and where will he/she be located?
Strategy for handling incident: • What are the current objectives, priorities, and plans to deal with the situation? • What message(s) are we giving/proposing to give to our employees/students/families? • Who should make the contacts with government—federal /state/local? Messages? • What community/special interest groups are being contacted? By whom? Messages? • Have we sent any messages updating employees? • What should we be telling the community? • What community contacts should be made? By whom? Messages? • Who is going to be the SS spokesperson(s)? • What resources are needed? What help is needed? • Who is staffing the key positions at the site and at the emergency operations center?
Communications: • Let's agree on schedule of when next to talk, how often to talk, and where each of us will be. • How are we going to deal with off-hours coverage?
Special agreed actions:

Table 3.4 Contacts List

Police	911
Fire	911
Emergency Medical Service	911
Health Department	911
State Highway Patrol	911
County sheriff	911

Emergency Contacts

Crisis Management Team Position	Name	CMT Position	Work	Pager
Superintendent				
Executive director of Business Affairs				
Coordinator, Facility Services				
Coordinator, Business Support Services				
Director, Information Services				
Assistant superintendent, Curriculum and Instruction				
Manager, Energy, Environmental, Health, and Safety				
General Counsel				
Executive director, Communications				
Executive director, Human Resources				
Student Services and Security				
Director, Community Partnership and Customer Services				
Treasurer				

Contacts-School System Crisis Teams

This Table Can Be Used to Identify Team Members during an Actual Incident/Mobilization						
Crisis Management Team Position	Contact/Location	Work	Home	Mobile	Fax	E-mail
Incident commander						
Safety officer						
Information officer						
Liaison officer						
Operations Section						
• Chief						
• Student Care						
• Search and Rescue						
• Student Release						
• Site Facility/Security						
Planning Section						
• Chief						
• Documentation						
• Situation Analysis						
Logistics Section						
• Chief						
• Supplies/facilities						
• Staffing						
• Communications						
Finance/Administration Section						
• Chief						
• Timekeeping						
• Purchasing						

Contacts: Utilities

Utility Type	Company Name	Daytime #	24-Hour #
Electric Power			
Natural Gas			
Water/Sewer			
Telephone			
Steam			

Other Key Contacts

Entity	Contact
Federal Regulatory Agencies • State Environmental Protection Agency • Food and Drug Agency • State Division of Labor and Worker Safety	
City of • Mayor's Office • Safety Director • Fire Marshall • Police Chief • Regional Transit Authority	
Hospitals • City Hospital • General Medical Center • Children's Hospital	
American Red Cross (/Summit County)	
Child Abuse Hotline	
Local TV • Channel 3 • Channel 5 • Channel 8 • Channel 10 • Channel SS 12	
Local Radio • WXYZ - • WXYZ - • WXYZ -	
Newspapers •	

CONCEPT OF OPERATIONS

The School System should have a comprehensive, two-tier, fully integrated ICS-based organization capable of responding to all incident and crisis situations that may arise and affect its operations. The organization should consist of:

- *Tier 1 School Crisis Response Teams (CRTs)*—composed of designated principals, teachers, and other trained staff located at individual schools who can respond safely and rapidly to an incident at their respective school/site and mount and sustain initial tactical response operations. The on-scene SS Initial Incident Commander (principal/delegate) will assume initial command and control of the incident response until arrival of a safety official (fire/police), whereupon command and control will transition to fire/police officials.
- *Tier 2 Crisis Management Team (CMT)*—composed of district-based administrative staff and other personnel. The CMT provides strategic and onsite tactical assistance to the CRTs. The CMT interfaces with government agencies, the media, and the; analyzes the implications of an incident and response operations on SS's viability, operability, and credibility; and assists CRTs in establishing post-incident recovery programs and operations.

Chain of Command

The ICS provides a unified command in a multi-responder emergency in which all agencies, that have jurisdictional responsibility for multi-jurisdictional incidents, contribute to the process of

- determining the overall objectives to gain control of the incident;
- selecting strategies to achieve those objectives;
- planning jointly for tactical activities;
- integrating appropriate tactical operations; and
- making maximum use of all assigned resources.

Generally, if a crime is involved, the police department will initially take control of the scene. However, under individual State law, for most incidents, the senior fire official on scene will assume the role of Incident Commander until the exact nature of the incident has been determined. In certain crime-related incidents, the Federal Bureau of Investigation may assume control. Other non-criminal events (e.g., fire, evacuations, injuries, accidents, infrastructure failures, etc.), will be controlled by the fire department, and/or emergency medical service. Law enforcement, emergency personnel of the fire department, paramedics, and other regulatory agencies adhere to a clear Incident Command System-based chain of command, as defined within the National Incident Management System (NIMS). The SS response organization will be expected to conform and readily integrate into the Unified Incident Command System to ensure an efficient and effective ramp-up of response efforts.

Tier 1 School Crisis Response Teams

Tier 1 involves the maintenance of school Crisis Response Teams (CRT) at each school and major operating facility. The role of the CRT is to immediately assess and respond to incidents and emergencies at their respective locations. Each CRT will be led initially by a senior school/ site representative with delegated authority to direct initial response actions, interface with safety emergency first responders, and interact with school district administration. CRT leaders are referred to as Incident Commanders (IC). The SS IC may be relieved by arriving police and/or fire officials who will assume command of the incident response from the SS initial IC. Command can formally transition back to the SS at an appropriate time once life safety and other agency jurisdictional issues have been resolved. The initial SS Incident Commander will interact with the district as noted below.

CRTs are predominately composed of local site staffing; but are supported by other school/facility as well as district-level staff. Table 3.1 provides a matrix that identifies specific staff positions that can be mobilized to assume various response roles and positions within the CRT. The nature and extent of actual team mobilization and staffing will greatly depend on the type, severity, and magnitude of the incident. Site-specific staffing will be supplemented with additional personnel activated by the CMT from other SS school and facility sites. The CRT organization is illustrated in figure 3.5. CRT position description and action checklists are included in Section 5. Certain CRT functions, included within planning, logistics, and finance, may be staffed at the district CMT level and be housed within the district Emergency Operations Center. In these cases delegates will be deployed to the scene to assist in onsite support coordination between the site and the CMT.

Tier 2 District Crisis Management Team

Tier 2 involves the maintenance of a district-level Crisis Management Team (CMT). The CMT will be activated in situations where the CRT requests support; the incident involves serious injury or death, or serious crime; the incident exceeds resource capabilities of a single facility, the incident has disrupted or has the potential to disrupt school/school district operations; and the incident has resulted in or has the potential to result in significant attention and/or involvement of parents, community officials, media, and/or other special interest groups. The CMT will operate within the District Emergency Operations Center and be composed of district administrative and support staff. The CMT will support initial efforts of the CRT and assume responsibility for sustaining response and restoration operations upon completion of initial emergency response actions.

Table 3.5 Crisis Response Team Organization

Primary and Support Roles											
Staff Position / Function	Command	Communications	Information	Search and Rescue	Safety and Security	First Aid	Mental health	Hazmat	Utilities	Logistics	Family Reunion
Principal	P	P	P						S1		
Assistant principal	S1	S2	S1						S2		P
Resource officer	S2			P	P						
Nurse						P	S1			S2	
Counselor			S2				P				S2
Custodian								P	P	P	
Cafeteria manager						S1				S1	
Secretary		S1			S2	S2					
Teacher											S1

P—Primary S1—Support 1 (First alternate) S2—Support 2 (Second alternate)

When the primary designee is not able to perform the assigned functional leadership role, the first alternate (S1) assumes that responsibility. Likewise, if both the primary and first alternate are unable to perform the leadership responsibility, then the second alternate (S2) assumes the role.

Terminology of the Incident Command System will be utilized to make clear the role and relationships of various school personnel and safety emergency responders.

As soon as the principal or alternate is ready (on-scene and informed of the situation) she/he assumes incident command. The incident commander should appoint a safety coordinator and liaisons (to district office and emergency services as required) to serve the command group. For most emergency operations, the principal will turn over incident command to the safety emergency response professional arriving on the scene.

The district Crisis Management Team will arrange to staff all unfilled positions that may be required in a particular emergency event.

Key First Actions

Initial response actions and decision making will depend upon many factors including:

- Nature of the incident or threat
- Magnitude of the impact or perceived potential impact on people and property
- Particular school, facility, or another related site affected
- Whether the incident occurs during or outside of normal school hours
- Extent of external involvement by governmental agencies, community, and the media

Critical and immediate actions and considerations by staff and administrators must focus on the safety and welfare of people.

It is imperative that designated on-scene staff (principal/delegate/facility manager), the SS Executive Director of Business Affairs (SS Crisis Management Team Manager), and SS security, consult with each other to confirm the **following critical first actions**:

- Confirm actual or potential emergency situation. An emergency is considered to be any sudden event that endangers or threatens to endanger the safety or the health of any person or that destroys or threatens to destroy or damage property or the environment.
- Conduct initial incident assessment.
- Ensure appropriate notifications are completed.
- Determine onsite support needs. Assist on-scene SS IC in deploying and staffing the on-scene School Crisis Team as appropriate.
- Ensure immediate life/safety decisions are addressed:
  - Notification of authorities.
  - Evacuation/shelter-in-place/lockdown.
  - Communication/messaging to staff and students.
  - Activation of immediate emergency response actions.
  - Student/staff/visitor status accounting.
  - Aid to injured/missing persons.
- Determine extent of involvement of safety-first responders.
- Employ Initial Incident Assessment Checklist (by security and the Executive Director of Business Affairs or delegate.
- Hold incident assessment consultation between the Executive Director of Business Affairs, and the district superintendent.
- Agreement/modification of initial actions.
- Initial impact assessment.
  - Need for consultation/conference calls with SS CMT and school/facility staff to discuss school district impact issues.
  - Determine whether and to what extent the SS CMT should be activated to the EOC. If CMT is activated:
    - Establish communication protocol and schedule.
    - Activate the EOC (Located in SS Distribution Center).
    - Convene CMT at designated EOC at earliest practicable time to direct response operations.
  - Determine and implement most appropriate means to interact with onsite safety first responder organizations using the Unified Incident Command System.
  - Agree to staff and student messaging and communication protocols.
  - Engage in incident management planning by the CMT to support initial response operations.
  - Position the CMT to assume direct control and longer term management of response operations.

Incidents and Emergencies...

An emergency is considered to be any sudden event that endangers or threatens to endanger the safety or the health of any person or that destroys or threatens to destroy or damage property or the environment. Examples include but are not limited to the following:	
❖ Fatality	❖ Impact by equipment/machinery/aircraft
❖ Serious injury/serious assault/sexual assault	❖ Earthquake or other natural event
❖ Siege/hostage/firearms	❖ Fire in school/on grounds
❖ Disappearance or removal of student/other	❖ Fumes/spill/leak contamination by hazardous material on-site
❖ Bomb threat	❖ Fumes/spill/leak contamination caused by off-site non-school-related incident
❖ Collapse/major damage to building or equipment	❖ Outbreak of disease
❖ Motor vehicle impact with school	❖ Flood/windstorm/ice/other inclement weather
❖ Suicide/attempted suicide	❖ Violent crime
❖ Gang violence/threats	❖ Perceived threats/rumors
❖ Accidents/incidents associated with school activities and excursions occurring before/after hours; on-site/off-site, and while traveling to and from school.	
When in Doubt—Report it!	

School Crisis Response Team

Minor and other day-to-day incidents will be managed by local schools and facilities employing appropriate members of their respective school Crisis Response Teams. These incidents will be reported to the district via routine written reporting protocols and verbal briefings during staff meetings.

More significant incidents, accidents, injuries, disruptions to school functions, threats, perceived threats, and/or other related concerns will be immediately reported to the district per notification procedures.

The Executive Director of Business Affairs, serving as the District Crisis Management Team Manager, will engage in an incident assessment process and, if appropriate, will confirm that the incident should be handled locally. He or she also may assign staff or other resources to assist the local team resolve the incident.

Depending on the magnitude and severity of the incident, the Executive Director of Business Affairs may activate the district Crisis Management Team to either support or assume control of response efforts, as appropriate.

In situations involving fire and/or law enforcement response, the SS local school/facility will activate its school Crisis Response Team and assimilate team members into the safety agency Incident Command System organization, as required.

The SS CRT Incident Commander (principal or facility manager) will cooperate with the safety Incident Commander(s) (police and fire) to coordinate on-scene response actions.

Partnering With Community Agencies Using Incident Command System

Homeland Security Presidential Directive HSPD-5 establishes a comprehensive National Incident Management System to enhance the abilities of federal, state, and local agencies and response entities to management domestic incidents effectively. The SS will collaborate with community safety agencies to establish an on-scene organization and incident management procedures in conformance with the NIMS. NIMS is the nationally recognized emergency operations system that is adapted for incidents where multi-agency response is required.

NIMS facilitates priority-setting, interagency cooperation, and the efficient flow of resources and information. NIMS establishes common standards of organization, procedures, communications, and terminology among agencies. This unified command structure functions under the direction of the Incident Command System. All agencies and organizations work together under this system to respond more effectively to any major incident.

The basic principle behind the Unified Incident Command System is that all agencies and jurisdictions involved in an incident will work together towards a common set of goals and objectives, with centralized management of operations and resources. It is a team effort and process. All agencies continue with their own authority, responsibility, and accountability, but they all work together and share information.

There are several advantages to using the Unified Incident Command concept:

- One set of objectives is developed for the entire incident.
- A collective approach is made for developing strategies to achieve incident goals.
- Information flow and coordination is improved between all jurisdictions and agencies involved in the incident.
- All agencies with responsibility for the incident have an understanding of one another's priorities and restrictions.
- No agency's authority or legal requirements will be compromised or neglected.
- Each agency is fully aware of the plans, actions, and constraints of all others.
- The combined efforts of all agencies are optimized as they perform their respective assignments under a single Incident Action Plan. A coordinated effort equals a safe response.
- Duplicate efforts are reduced or eliminated, thus reducing costs and chances for frustration and conflict.

By adopting and utilizing this system, the School System will be able to respond to any incident with better communication and coordination with other responding agencies and organizations.

The SS CRT organization is based on the Incident Command System. It consists of five functions:

- Management
- Planning/Intelligence
- Operations
- Logistics
- Finance/Administration

Management: Provides overall policy and operational direction.

Incident Commander: This position assumes sole responsibility for directing emergency response operations. Until the arrival of the safety (police/fire) Incident

Command Officer, the site administrator is in charge. The site administrator is most often the principal or building/site manager, unless circumstances dictate otherwise.

Safety Officer: This position ensures that all activities are conducted in as safe a manner as possible under existing circumstances.

Information Officer: This position acts as the official spokesperson for the school/district in an emergency.

Liaison Officer: This position serves as the point of contact for agency representatives from assisting organizations and agencies outside the school district and assists in coordinating the efforts of these outside agencies by ensuring proper flow of information.

Planning/Intelligence: Collects, evaluates, documents, and uses information about the development of the incident and the status of resources. Maintains accurate records and site map. Provides ongoing analysis of situation and resource status.

Documentation: This position is responsible for collection, evaluation, documentation, and use of information about the development of the incident and status of resources.

Situation Analysis: This position is responsible for collection, evaluation, documentation, and use of information about the development of the incident and status of resources. Maintains accurate site map. Provides ongoing analysis of situation and resources.

Operations: Implements priorities established by the Incident Commander, which can include site facility check/security, search and rescue, medical response, student care, and student release.

Student Care: Ensures the care and safety of all students and staff except those that are in the medical treatment area. This position knows where all first aid supplies are located, oversees first aid prior to paramedics' arrival, coordinates with paramedics.

Search and Rescue: Accounts for everyone onsite as quickly as possible. Position must have access to attendance records, visitor sign-in sheets, emergency data cards of students, and any other information that will assist in accounting for the school population. This position works closely with the student assembly Shelter and Release Coordinator.

Student Release: Assures the reunification of students with their parents or authorized adult using pre-established protocols and documentation procedures.

Site Facility Check/Security: Provides site safety and security support. Knows layout of building and grounds, location of shutoff valves and utility lines, and is familiar with school blueprints.

Medical Response: Provides emergency medical response, first aid, and counseling. Notifies Incident Commander when the situation requires health or medical services that staff cannot provide and ensures that appropriate actions are taken in the event of deaths.

Logistics: Provides facilities, services, personnel, equipment, and materials in support of the incident.

Supplies/Facilities: Provides facilities, equipment, supplies, and materials in support of the incident.

Staffing: Coordinates assignment of personnel (staff, students, disaster volunteers) in support of the incident.

Communications: Establishes, coordinates, and directs verbal and written communications within the school disaster site and with the school district. If the school district cannot be contacted, communications may be made with outside agencies when necessary.

Finance/Administration: Performs financial tracking, procurement, and cost analysis related to the incident. Maintains financial records and tracks and records staff hours.

Timekeeping: Maintains accurate and complete records of staff hours.

Purchasing: Maintains accurate and complete records of purchases.

For more information on the specific roles of the Crisis Management Team members see appendix 2.

CRISIS MANAGEMENT TEAM

The Crisis Management Team determines what, if anything must be done to support CRT response efforts, and to identify, evaluate, and proactively address the implications of the incident and incident response operations on the School System in its entirety.

When activated, the CMT is responsible for strategic planning and control of incident-related activities. The group consists of district and school administrators and support staff with authority to determine short- and long-term strategies and objectives, order evacuation or shutdown of the facilities, commit SS resources, and interface with outside organizations and media.

The School System superintendent serves as the chief district spokesperson to the outside world, supported by relations and communications. Table 3.3 summarizes the specific roles of the CMT. The CMT comprises the following individuals:

- Superintendent
- Executive Director, Business Affairs
- Executive Director, Communications
- General Counsel
- Executive Director, Human Resources
- Director, Community Partnerships and Customer Services
- Coordinator, Business Support Services
- Coordinator, Facility Services
- Manager, Energy, Environmental Health & Safety
- Director, Student Services & Security
- Treasurer

Overview of crisis management team roles

• Ensure response is consistent with SS policy and meets or exceeds stakeholder expectations.
• Consider worst-case scenarios for all situations and issues and identify triggers to assist in proactive recognition of escalation toward a worst-case scenario. Develop an action contingency strategy, as appropriate.
• Ensure all key emergency response and statutory notifications have been made.
• Develop response strategy based on priorities.
• Protect people/environment; implement response and recovery measures.
• Locate and mobilize additional resources requested by CRT members.
• Identify all stakeholders and associated issues and develop a communications strategy to keep all informed.
• Draft and coordinate release of information to media, staff members, and relevant third parties to ensure a common and consistent messaging.
• Consider how the SS makes the best of the situation, and identify the positives for the company.
• Maintain an auditable trail including Events and Issues and Actions Logs.
• Initiate and support the incident investigation.
• Identify strategic issues and response strategies as they might affect other SS interests and/or the community at large; coordinate actions with external response partners.
• Draft and coordinate the release of information to the media liaising with relevant third-party organizations to ensure a common and consistent message is transmitted.
• Notify and liaise with insurers. Track the cost of response to maximize insurance returns.
• Liaise with emergency services at a senior level.
• Forecast costs and provide data for use in discussions with stakeholders.
• Provide finances as necessary.

EMERGENCY OPERATIONS CENTER ACTIVATION

CMT members will consult with each other via telephones, mobile phones, radios, and/or text messaging pagers.

If so decided, the CMT will convene in the Emergency Operations Center to undertake a more thorough incident assessment and to initiate and coordinate sustained response operations.

Emergency Operation Center Locations:

- **Primary:** ______________
- **Alternate:** ______________

The CMT administrator will be responsible for setting up the Emergency Operation Center facilities to support the CMT. The critical first actions include:

- Advise security that the Emergency Operation Center has been activated.
- Set up communications equipment.
- Set up incident information center status boards.
- Distribute office supplies as needed.
- Distribute copies of Crisis Management Plan, other relevant emergency plans, responder checklists, and contact lists.

Emergency Operation Center Activation checklist

____ **Communications Setup**
Remove all telephones from telephone cabinet and place on tables.

____ **Group Page**
Notify the CMT via group phone/page/text message that the EOC has been activated.

____ **Security**
Notify security that the center has been activated. Determine if center needs to be secure. If so, request a security guard be stationed at the center.
Instruct security to forward all calls that they take to the EOC.

____ **Computers**
Notify director of Information Services that the center has been activated. Determine need for additional computers, printers, network connections, and peripheral equipment. Request activation of agreed-to equipment and support personnel.

____ **Facsimile Machines**
Ensure that machines are on and properly loaded with paper. Two machines are immediately available.

____ **Contact Lists**
Distribute latest updated listing of emergency contact listings and SS phone directory.

STATUS BOARDS

____ **Contact Board**
Place all pertinent contact names and numbers on lined contact board in the center.

____ **Incident Board**
Set up board to maintain log of factual information known, and update as facts are received.

____ **Communications Board**
All facsimile numbers should be written on the communications board. Include confirmation name and numbers as well. Place an incoming and outgoing facsimile log form on the board and maintain.

____ **Message Board**
Located at the entrance to the center, assign personnel to a block on the board, and inform responder of where to locate messages.

____ **Response Team Support**
Make available to all responders the following support items: in-/out-box, pen, pencil, paper, blank responder log books, phone instruction handbooks, SS telephone book, and so on.

____ **Organization Chart**
Use poster-size form and slot responder positions held in the center. *Organization chart and phone numbers should be developed for other locations as well.*

____ **Reference Items**
Depending on school/facility involved, pull emergency plans relating to the incident. Question what other reference items you can supply.

____ **Secretarial Support**
Secretarial support personnel should be activated.

____ **Digital Camera**
Plug in and recharge batteries. Shoot pictures for permanent records of wall charts and other key information written on white boards.

____ **Time Clocks**
Set incident time clocks.

____ **Forms**
Refer to the form cabinet in the support area and pull any appropriate forms.

____ **Copy Machine**
Contact **General Services** and request relocation of copy machine to EOC.

____ **Telephone Forwarding**
Forward individual office phones to EOC.

____ **Television/VCR**
Turn TV on to CNN (picture only, no volume). Turn on VCR and make machine ready for taping. Blank videocassette tapes are located in the TV cabinet in the center.

____ **Repdials**
Determine any and all pertinent telephone numbers (site telephone number, etc.), and set up repdials on all phones in the center. Key numbers would also be on the contact board.

____ **Marker Boards**
Erasers and markers should be in both marker board trays as well as the tray. Check supply cabinet in main support area.

____ **Easel Board**
Check pad of paper for easel board and place markers in trays. Additional easels and pads can be obtained from Facility Services.

____ **Food**
Contact Facility Services to notify them of possible rush food requests. Order refreshments (coffee, pop, tea, etc.), and have them replenish food on an ongoing basis until they are further notified. Determine what the needs are for breakfast, lunch, and dinner.

____ **Cash Advance**
Contact Finance for Emergency Cash Procedures. Key contact numbers are located in the SS Crisis Management Plan.

SHUTDOWN PROCEDURES
Return center to original nonactivation status. Refer to this list.
The two most critical items are as follows.

- Deactivate the phone system.
- Notify security that center is no longer in emergency mode.

Collect all incident documentation and consult with legal to determine disposition.

Initial Briefing

This is the recommended format for initial briefing of the CMT. CMT leader and coordinator should ensure briefing is kept short and effective.

- CMT Leader: Introduce self to establish team leadership.
  Conduct roll call.
  Brief the CMT on the incident using the checklist below.
- CMT Coordinator: Capture key information on display boards.
- Documentation Leader: Record time of briefing.
  Capture key issues/actions on display boards.

CMT to brainstorm on the following issues:

- Identify additional resources/support required.
- Identify major stakeholders/issues (see Stakeholder Management Plan template).
- Identify strategies for dealing with stakeholders and issues.
- Identify triggers to assist in proactive recognition of escalation towards worst-case scenario.
- Consider how to make best of situation and determine positives for SS.

CMT Leader to: Summarize issues identified, set priorities, and initial objectives.
Assign actions for each CMT member.
Set time of next meeting.

Incident Briefing Checklist

Incident description:

- **What** has happened?
- **Where** has it happened?
- **When** did it happen?
- Are all **personnel** accounted for? Casualties?
- **Potential to escalate**—what is potentially at risk?
- What is the **impact on the facility**/SS?
- **Who** is supplying information?
- How **accurate** is information?
- What are the **timing issues**—security mandates, weather, daylight, and so on?

Incident Status:

- What are the CMT objectives? (What are we trying to make happen/prevent from happening?)
- What actions are being taken and who is taking them? Responding?
- What resources [equipment/manpower] are being used?
- How effective is the response?

Initial Incident Briefing Checklist – Supplemental

Incident Status: • Are there additional facilities at risk? — Administration Building — Other building/facility — Neighboring properties — Third parties • What is the impact on the facility/district? — Open/lockdown/shelter-in-place/evacuate — Late open/early dismissal — Continuing • Is the area secured from unauthorized access?
Personnel Issues: • Safe and accounted for — Employees, visitors, contractors, others • Casualties — Injured — Dead — Missing/unaccounted for • Interaction with Employers/Affiliations, and so on — Contractors/subcontractors — Neighbors — Visitors — General
Administration Building Status: • Evacuation plans activated? — What floor(s)? — Status/location of people? — Any injuries? — Persons with disabilities taken care of? • Status of involvement of safety within building? (fire/police/medical) • What messaging to the staff regarding building status? — Certain floors OK? — Building safe/all clear? — Building closed? — Other message?

Incident/Threat Assessment Update:

Real incident or hoax:

- Does the safety incident commander agree with initial evacuation or shelter-in-place decision?
- Has a search of the building been conducted and the incident commander advised of results?
- Has there been an actual detonation, fire, shooting, or other incident?
- Has anything unusual been discovered?
- If anything has been found, will fire, police, and so on, conduct a further search?
- Has the area, office, or floor with the suspicious item been closed off; all employees, visitors, and contractors evacuated; and items of obvious evidentiary value left untouched?
- Are there any suspicious unidentified persons in the building? If yes, has another search been ordered? By whom?
- Does the Incident Command safety officer agree with another search?
- Is the CMT prepared for media interest or arrival at the scene?
- Have the cable news, television, and radio monitors discovered any new information regarding the extent of the incident?
- If electrical power is out in the building, is it also out in the surrounding area?
- Have any additional local or federal safety/law enforcement responders arrived on the scene?
- If safety and CMT incident commanders agree that the incident is a hoax or under control, have employees been formally dismissed or encouraged to reenter the building?
- Has a formal "All Clear" order been issued by safety?
- If the incident is a bomb threat and a search was conducted, do CMT and safety incident commander agree that nothing was found and that it is safe to reenter?
- Is the log keeper keeping the log updated and current?

Incident classified as actual:

- Is the incident a natural disaster or a man-made event? By whom?
- If it is a man-made event, has it been classified as a crime or a terrorism incident? By whom?
- Since a terrorism incident is a federal crime and the affected area has been designated as a crime scene, has a decision been made to close the building?
- Has any SS property deemed essential been damaged, contaminated, or removed as evidence?
- Has the CMT communicated with safety incident commander regarding building or area reentry?
- Has a decision been made concerning a joint press release/conference?
- Do the CMT/CRT and the safety incident commander agree on the media statement?
- Has a formal, complete "hands off" of the incident to the incident commander taken place and been so logged by the log keeper?

Stakeholders:

- Which stakeholders do, or potentially will, have interest?

Media: • What is the SS media strategy? • What media coverage has there been so far (radio, TV, print—local, national, international)? • Has it been accurate? • Is it negative toward SS? • Has a statement/release been given to the media so far? • What is the potential for additional media coverage? • Who is the most appropriate SS spokesperson? • Has the safety incident commander designated an information officer? • Have we liaised with him/her? • Has a press conference been called/is one necessary? Does the safety incident commander concur? • Who else is making press releases?
Incident Management Capability: • What cannot be done that should be done? Why? • Has the county Emergency Management Agency (EMA) been notified? • What human/material resources are required? — SS internal/facility/county EMA/other governmental — Specialty consultants and contractors, other specialists?
Incident Assessment: • What are the health, security, and safety issues? • What is the likely environmental impact? • What are the community, government, media, and/or reputation issues? • What are the legal implications? • What is the operational impact—short, medium, and long term?
Set Objectives and Priorities? Objectives must be • Appropriate • Clear • Measurable • Achievable

Once the Crisis Management Team is fully engaged, the CMT leader should advise the onsite Incident Commander (principal) at the incident facility that the CMT is fully functional and is in a position to perform required duties. A communications protocol should be established. At a minimum, the protocol should cover the following topics:

- Names and contact numbers of the CMT and contact number for the Emergency Operations Center.
- CMT and CRT participation in briefing and assessments.
- Mutual expectations regarding information reporting and document sharing and timing.

The purpose of the protocol is two-fold. First, it is designed to ensure that appropriate communications occur in a timely and intelligible fashion. Second, it reinforces the chain of command within the CMT and remote team(s) relationship(s) by ensuring clear guidance and consistency in objectives, priorities, strategies, messaging, and actions.

Emergency Operations Center Information Center

As information is gathered from the incident site(s) and generated in the Emergency Operations Center, it should be prominently displayed for use by all CMT members at the Information Center.

The Information Center should be viewed as the one place in the Emergency Operations Center where anyone can go, at any time, to learn about the nature and status of an incident and incident response operations, and the nature and status of CMT response efforts.

One of the primary purposes of the Information Center is to assist the CMT in its efforts to establish and maintain command and control, and to assist in forward action planning.

The Emergency Operations Center Information Center should include status boards addressing the following issues:

- Incident Facts
- CMT Sign-In Status Board
- CMT Issues/Impacts Status Board
- Help Requested by Affected Asset Status Board
- CMT Action Plan Status Board
- Schedule of CMT Meetings Status Board
- Action Items from CMT Meetings
- Stakeholder Management Plan

The Information Center should be established and maintained by the CMT Coordinator and Log Keeper. It is the obligation of all CMT members to ensure that information pertaining to their respective activities is accurately posted and up to date.

CMT Assessment and Issue Identification Meetings

The duration that the CMT resides within the Emergency Operations Center will be incident specific and depend upon many variables. If the CMT elects to staff the Emergency Operations Center intermittently versus continually, than additional CMT meetings should be held at a frequency to be determined by the CMT leader and the CMT members.

The objectives of the subsequent Assessment and Issue Identification Meetings are the identification of issues and concerns, and the development of actions to address the issues and concerns. The topics to be discussed include:

- Progress being made on work under way
- Problems being encountered that are restraining progress

- Action items to address prior to the next meeting
- Agreed action items for CMT members

To the maximum extent possible, CMT meetings should be scheduled to follow assessment/ incident status updates received from the incident site from either the SS or Safety Incident Commander. This would allow the CMT to benefit from having access to the latest information available on the incident and tactical response operations. The CMT leader may elect to participate in the incident site briefings via conference call. Likewise the CMT Leader may elect to invite remote response team management to participate in select portion of CMT meetings to foster information transfer.

CMT Action Plan

The product of the CMT's work and deliberations should be a written CMT Action Plan. This document should list the issues and concerns raised by CMT members, the agreed-upon actions to be taken to address the issues and concerns, and the name(s) of the CMT member(s) responsible for ensuring that the actions are carried out in a timely fashion. The plan itself should be a stand-alone document that is constantly updated as new issues/actions are identified, and work progresses on the actions. In addition, the Emergency Operations Center Information Center contains a status board that can be used to summarize the Action Plan.

To the extent that the actions listed in the plan involve, or have an impact upon, remote response team staff, the plan should be forwarded to the remote response team manager.

Local, County, State and Federal.....Information

Agency	Role	24-Hour Phone No.
Emergency Medical Service	• Emergency medical response	
Fire Department	• Fire response • Hazardous material spill/release response • Search and rescue • Emergency medical response • Evacuation • Assistance for special needs groups • Damage assessment • Terrorism response	911
Police Department	• Law enforcement • Traffic control • Area control • Evacuation • Search and rescue • Communications	911
Transit System	• Mass transportation	
American Red Cross	• Student shelter operations • Student care and release operations • First aid shelters • Responder food services	
Bureau of Alcohol, Tobacco, Firearms, and Explosives	• Lead agency for arson and nonterrorist bombings • Arson and Bomb Crime Lab	
Centers for Disease Control and Prevention	• Source for general information on disease-related issues • Manages the National Strategic Stockpile • Responds to biological attacks and epidemics • Activated via safety in consultation with health department	
City of Office of the Mayor	• Safety support • Works support	
Federal Bureau of Investigation	• Lead agency for terrorism and weapons of mass destruction incidents • High-level crime scene investigations • Lead agency for most federal crimes • Coordinates requests for federal resources—military and civilian • Activated via Safety	
Federal Emergency Management Agency	• Coordinates federal disaster assistance • Provides disaster-related funding assistance	
Food and Drug Administration	• Source for general information on food safety and related issues • Food contamination • Food borne illnesses and surveillance	

Agency	Role	24-Hour Phone No.
Hospitals — ______ — ______ — ______ — ______	• Medical Support	
Local Radio — ______ — ______ — ______ — ______	• Broadcast emergency information and instructions	
Local TV — ______ — ______ — ______ — ______	• Broadcast emergency information and instructions	
Internet — www.xxx.com	• Broadcast and observe emergency information and news	
National Response Center	• Regulatory reporting of hazardous materials spill in excess of Reportable Quantities (RQs)	
State Emergency Management Agency	• Activation of state and multi-county emergency resources (equipment and personnel)	
State Environmental Protection Agency	• On-scene direction of hazardous materials spills cleanup	
State National Guard	• Disaster Recovery Assistance • Law Enforcement Support • Civil Support Team • Activation via Safety	
State Patrol	• Law enforcement on state property • Traffic control on state highways	
Department of Health	• Health response and coordination • Infectious disease surveillance • Quarantine matters	
County Emergency Management Agency	• Coordination with officials in affected jurisdictions • Activation and coordination of local, county, regional, and state agency, personnel, and equipment resources • Activated via safety agency	
County Sheriff	• Law enforcement	
US Environmental Protection Agency	• Management of hazardous materials spills • Conducts investigations regarding environmental matters	

For additional Crisis Management Team Action Checklists see appendix 1.

INITIAL RESPONSE

Expect to be surprised. Regardless of how much time and effort was spent on crisis planning, the members of the crisis team/ICS should know that there will always be an element of surprise and accompanying confusion when a school is confronted with a crisis.

Assess the situation and choose the appropriate response. Following the plan requires a very quick but careful assessment of the situation. Determine whether a crisis exists and if so, the type of crisis, the location, and the magnitude. Because the team has practiced the plan, leaders are ready to make these decisions. After basic protective steps are in place, more information can be gathered to adjust later responses.

Respond within seconds. When a crisis actually happens, make the basic decisions about what type of action is needed and respond within seconds. An immediate, appropriate response depends on a plan with clearly articulated roles and responsibilities, as well as training and practice. With proper training, district and school staff and students will respond appropriately within seconds.

Notify appropriate emergency responders and the school crisis response team. One common mistake is to delay calling emergency responders, such as the police or fire departments. In the midst of a crisis, people often believe that the situation can be handled in-house. It is better to have emergency responders on the scene as soon as possible, even if the incident has been resolved by the time they arrive, than to delay calling and risk further injury and damage. For instance, it is better to have emergency responders arrive at a school to find a fire put out than to arrive too late to prevent loss of life or serious property damage.

Notifying a district's or school's crisis team allows them to begin the necessary measures to protect the safety of all persons involved. Unless informed otherwise by the incident commander, school crisis team members should proceed with their responsibilities.

Evacuate or lock down the school as appropriate.

This step is crucial and should be one of the first decisions made, regardless of the order in which initial decisions are implemented.

Triage injuries and provide emergency first aid to those who need it. The plan should assign emergency medical services personnel and school staff with relevant qualifications to determine who needs emergency first aid. Designate a location for EMS to treat the seriously injured on the scene.

Keep supplies nearby and organized at all times. If you move to another location, remember to take your supplies with you. Monitor the amount of supplies and replace them as needed.

Trust leadership. Trust the internal crisis team members and external emergency responders who have been trained to deal with crises. Trust will help calm the situation and minimize the chaos that may occur during a crisis.

During a crisis, leaders need to project a calm, confident, and serious attitude to assure people of the seriousness of the situation and the wisdom of the directions being given. This leadership style will help all involved to respond in a similarly calm and confident manner, as well as helping to mitigate the reactions of anyone who might deny that a crisis has occurred.

In certain situations it may be necessary to yield leadership to others in the plan's designated command structure. In some jurisdictions laws state the protocol for the command structure. This structure may vary from state to state and even from community to community within state. For instance, in a fire, the expertise of firefighters should lead the way, with others filling designated roles, such as manager of family-student reunification.

Communicate accurate and appropriate information. During a crisis, districts and schools will communicate with the school community, as well as the community at large. Use the channels of communication identified in the plan. For instance, all information released to the media and public should be funneled through a single public information officer or appointed spokesperson. This will maximize the likelihood of presenting consistent and accurate information to the public.

The crisis team should communicate regularly with staff who are managing students. A school's most important responsibility, the safety of the students entrusted to the school by their families, cannot be fulfilled during a crisis without timely and accurate information to those caring for students.

At a minimum, families need to know that a crisis has occurred and that all possible steps are being taken to see to the safety of their children. Additional details about assembly and shelter procedures may also be provided, as determined by the plan or those managing the crisis. At some point, families will also need to know when and where their children will be released.

Activate the student release system. Always keep in mind that the earliest possible safe release of students is a desired goal. Often student release will be accomplished before complete resolution of a crisis.

Allow for flexibility in implementing the crisis plan. It is impossible for any crisis plan, no matter how complete, to address every situation that may arise during a crisis. With proper training and practice, emergency responders and staff will be able to respond appropriately and to adapt the school crisis plans to the situation.

Documentation. Write down every action taken during the response. This will provide a record of appropriate implementation of the crisis plan. Also necessary is recording damage for insurance purposes and tracking financial expenditures related to the incident. Keep all original notes and records. These are legal documents.

Response to a disaster includes emergency assistance to individuals affected by the disaster. Response activities also include reducing the probability of additional injuries or damage. Response actions should be performed in a way that speeds later recovery operations.

During the initial response to a disaster, approach buildings with caution. Damage and structural integrity cannot always be determined by what is visible outside. Also, wind may be an issue; care should be taken to avoid injury from flying debris that may still be in the area.

Critical Incident Procedures

Critical incident procedures focus narrowly on situations that involve imminent danger to life and limb. Such procedures emphasize a coordinated interagency response to preserve and protect life. Sound policy requires the development, in advance of a critical incident, of interagency agreements that specify channels of communication, types of service, and areas of responsibility. Such interagency agreements typically are established with public safety (i.e., police, sheriff's office, fire department, emergency services) and mental health agencies (i.e., Community Services Board).

The Incident Command System

The ICS is the combination of facilities, equipment, personnel, procedures, and communications operating within a common organizational structure, designed to aid in domestic incident management activities. It is used for a broad spectrum of emergencies, from small to complex incidents, both natural and man-made, including acts of catastrophic terrorism. ICS is used by all levels of government, federal, state, local, as well as by many private-sector and nongovernmental organizations. ICS is usually organized around four functional areas: Operations, Planning, Logistics, and Finance/Administration.

A fifth functional area, Intelligence, may be established if deemed necessary by the Incident Commander, depending on the requirements of the situation at hand.

The ICS provides a unified command structure in an emergency involving multiple responders and jurisdictions by the following:

- Determining the overall objectives to gain control of the incident.
- Selecting strategies to achieve their objectives.
- Planning jointly for tactical activities.
- Integrating appropriate tactical operations.
- Making maximum use of all assigned resources.

As a rule of thumb, if a crime is involved, the police take over the scene; most other incidents are controlled by the fire department or local emergency response system. School and college staff need to be familiar with their local jurisdiction's ICS to avoid any confusion when responding to a critical incident.

It is a good idea to identify in advance a liaison officer for the school or college. There is a clear chain of command among law enforcement personnel, fire department personnel, and paramedics. It is very important for school and college leaders to meet ICS members ahead of time and learn about their ICS so that the school or college can become a functioning member of the ICS team.

The type of event will often determine response and ICS structure. For example, for a health event the county health department is the lead agency for the management of the disaster. For a toxic spill or other hazardous materials event, the lead agency is the Fire Department. For a comprehensive list of agencies and their responsibilities for responding to different disasters, contact your local Emergency Management office.

ICS Functions

The ICS provides overall direction and sets priorities for an emergency. Under each of ICS's four functions, schools and colleges should assign specific individuals prior to an event, based on their job roles and responsibilities, to assist in the management of school or college site operations.

Unified Command

Although a single Incident Commander normally handles the Command function, an Incident Command System organization may be expanded into a UC. The UC is a structure that brings together the IC of all major organizations involved in the incident to coordinate an effective response while carrying out their own jurisdictional responsibilities. The UC links the organizations responding to the incident and provides a forum for these entities to make consensus decisions. Under the UC, the various jurisdictions and/or agencies and nongovernment responders blend together to create an integrated response team.

The UC is responsible for overall management of the incident. The UC directs incident activities, including development and implementation of overall objectives and strategies, and approval of ordering and release of resources. Members of the UC work together to develop a common set of incident objectives and strategies, share information, maximize the use of available resources, and enhance the efficiency of the individual response organizations.

Partnerships with Community Public Safety Agencies

The National Incident Management System (NIMS) is a nationally recognized emergency operations plan that is adapted for large critical incidents for which multi-agency response is required. Schools and colleges can use NIMS and the ICS as the organizing structure for their Critical Incident Response Plan. NIMS facilitates priority-setting, interagency cooperation, and the efficient flow of resources and information. NIMS establishes common standards of organization, procedures, communications, and terminology among agencies. This unified command structure functions under the direction of the ICS, and all agencies and organizations work together under this system to effectively respond to any major critical incident. By adopting this system, schools and colleges can respond to any critical incident with good communications and coordination with other responding agencies and organizations.

Debriefing

At the conclusion of any critical incident, regardless of size, it is important to conduct a debriefing session. This is often referred to as a "critique." All participating agency representatives should meet to discuss what went right and how the system could be improved. It is critical that all agencies participate in this process in order to refine and improve the plan as needed. The recommendations for improvements and identification of success should be recorded and compiled into an After Action Report to be filed with the local Emergency Management office.

Command Center

Establish a command center location where the Incident Command Team can work. This location should have already been identified in your school or community college emergency plan. If that location is not available, you should consider using an alternate location—one outside school district or community college boundaries. The alternate location should also be identified in your plan. One expensive, but reliable, alternative is a mobile center with a portable generator.

Considerations for equipment and structure of your command center should include:

- Emergency power
  - Communication equipment:
    - Communication tie-in with the operation center
    - Direct line telephone(s)
    - Cellular phones
    - Satellite phones
    - Shortwave radio/800 MHZ radio
    - Public address system
    - Office/clerical space for a small core of personnel
    - Large meeting room(s)

- Adequate parking:
  - For staff and volunteers

- Furniture and equipment:
  - Tables
  - Chairs (comfortable)
  - Copier
  - Computer/printer
  - Facsimile machine
  - Easel
  - White board
  - Paper and office supplies
  - Large pads of paper (24 x 36)
  - A/V equipment
  - Refrigerator
  - Inventory, plans, and photos
  - Insurance policies
  - Telephone numbers
  - Vendor contacts
  - Contracts and contact information
  - Food and drink
  - Toilet facilities
  - Prior to the event: Implement plans
  - During the event: Man with a skeleton crew
  - After the initial event: Gather, assimilate information and directives
  - Disaster team organization
  - Development of disaster organization chart
  - Appointment of chairperson
  - Determination of personnel and alternates to fill organization chart

Emergency Operations Center

The EOC is a central command and control facility led by the state coordinating officer and responsible for carrying out the principles of emergency preparedness and emergency management, or disaster management functions at a strategic level in an emergency situation. It also ensures the continuity of operation of the company or political subdivision. The EOC is responsible for the strategic operation, or "big picture," of the disaster and does not normally have direct control of field assets. It makes strategic decisions and leaves tactical decisions to lower commands. The common functions of all EOCs are to collect, gather, and analyze data; make decisions that protect life and property; maintain continuity of the government or corporation, within the scope of applicable laws; and disseminate decisions to all concerned agencies and individuals. In most EOCs, there is one individual in charge—the Emergency Manager.

Shelters

Types of hurricane shelters that facilities could be used for are:

General (Risk) Shelters

These shelter spaces are designated for use during disasters, especially hurricanes. Facilities designated as Risk Shelters lie in the forecast path and associated error cone of an approaching hurricane or severe storm. The designation does not imply that a facility is capable of affording complete protection or is free from hazards, but only that it meets established minimum safety criteria.

Pet-Friendly Shelters

Some General Risk or Host Shelters may be designated as pet-friendly. These shelters sometime require pre-registration either days or months before the shelter opens. Pet-friendly shelters may also have weight, type of animal, quantity or other pet restrictions. Contact your local EOC at the county for a list of pet-friendly shelters in your area, and learn their procedures and policies.

Special Needs Shelters

Special Needs Shelters (SpNS) are structures that have auxiliary power and are capable of providing safe refuge for people who require assistance with the management of a health condition or supervision of that condition by a health care professional during a disaster. The special needs services provided during an emergency are supplied, when practical, in an environment that can help to sustain pre-disaster levels of health.

Host Shelters

Host Shelters are spaces that are available for general use outside of a forecasted hurricane, severe storm, or disaster impact area. These shelters should be facilities that are relatively safe and provide essential support services. Facilities are designated as

Host Shelters when they are located in an area that is outside the projected path of an approaching hurricane, severe storm, or disaster area. Since local conditions are not expected to present hazards, such as surge inundation, rainfall flooding, high winds, or hazardous materials that exceed the building codes of the facilities in use, shelter selection guidelines in ARC 4496 do not have to be considered.

Shelter Operations

The decision to open a shelter is made by the County EOC. Staff members from the EOC communicate with the appropriate Emergency Support Function to activate and set up the necessary shelters. The shelter manager, responsible for opening the actual shelter, is the primary liaison between the shelter and the EOC.

Shelter Staffing

Shelter staff typically includes the following individuals, at a minimum:

- Red Cross Shelter Manager (if used)
- Principal
- Food Manager
- Maintenance Manager
- Law Enforcement
- Service Technicians (electrical, plumbing, heating, ventilation, and air-conditioning (HVAC), etc.)

Ensure that sufficient shelter staff is scheduled to operate the facility for the approved number of evacuees, including adequate relief staff to allow shifts for sleep. The longer the shelter remains open, the more critical this becomes. Shelter environments are stressful and staff will need rest to work effectively and make sound decisions in dealing with evacuees. Develop a staffing schedule, establish teams, and schedule shifts for each team throughout the day.

Shelter Considerations

Short-term planning (36 hours before possible strike):

- Set up command center for the shelter.
- Prepare rooms to be used for housing individuals in the shelter. Remove computers, equipment, furniture, supplies, and so on that are not needed for sheltering activity, and lock/secure areas with valued items.
- Test-run generator and top off fuel tank.
- Use supplies checklist to confirm adequate supplies of inventory, (e.g., paper towels, toilet paper, hand soap, garbage bags, sanitizers).
- Have appropriate shelter forms (e.g., shelter admission form, emergency contact phone numbers).
- Have a plan for deceased persons in the shelter.

- Confirm food supplies with the food manager (e.g., food, water, commodities).
- Store/secure outside items.
- Plan for disposal of human waste.
- Contact waste removal service to have school or college dumpsters emptied when over one-third full.

Shelter Activation

Registration of evacuees. Law enforcement should maintain a presence at registration. Registration should include:

- Insignia to identify shelter staff
- Individual identification tags for evacuees
- Ensure evacuees bring only allowed items into the shelter

Assign evacuees to specific areas of the shelter:

- Families with children in one area
- Elderly in one area
- Pets (if pet shelter) to the appropriate area

Traffic Control

Individual(s) should be assigned to the parking lot, with flashlight(s), bright vest(s), and traffic cones. Duties include:

- Ensuring vehicles are parked in appropriate locations, not blocking bus traffic.
- Ensuring traffic routes remain clear, allowing emergency vehicle entry/exit.
- Ensuring motor homes are directed to remote or off-campus parking, since they take up multiple parking spots due to their size.

Watch for vans or buses arriving from congregate living facilities. Some facilities have been known to drop off evacuees with special needs or conditions. These facilities are required by law to have alternative arrangements. If this should occur, the individuals should be denied entry to the facility (depending on the severity level of the event), and the name and contact number of the facility should be relayed to the County EOC for further instructions.

Administration

The following administrative procedures should be addressed:

- Develop shelter procedures and rules for individuals sheltered.
- Prominently post rules.
- Create procedures and checklist for human waste issues.
- Develop procedures to follow if power goes out (including emergency lighting).

- Develop procedures for closing facility when event is over.
- Develop staff procedures for cleaning shelter and returning facility to normal use.

Communication

The following communication issues should be considered:

- An ability to maintain communication with the County EOC and board administration during and after an event is one of the greatest problems.
- Ensure that backup communication is available at all shelters.

Preparing Students, Staff and Stakeholders to Respond

Experts have noted that when a crisis occurs, individuals involved tend to go on autopilot. Therefore, when a crisis occurs staff immediately need to know how to react. They need to know, for example, the signals for crisis, the protocol for lockdown and evacuation, how to dismiss students, and what to do if staff or students need help. They should know these things ahead of time. There will not be a time during the crisis to think about what to do next. Chances of responding appropriately in a crisis will be much greater if all players have practiced the basic steps they will need to take. Training and drills are crucial.

In the San Diego, California, school district, staff feel that practice and training should constitute the majority of the crisis planning process. In their "formula for success," practice accounts for 50 percent of the process, training for 30 percent, and planning 20 percent. While the percentages are flexible, training and drills are essential. Key components to facilitate training, and thus a successful reaction, are as follows:

Provide regular, comprehensive trainings for teachers and staff. At least once a year, provide crisis response training for teachers and staff. Also provide make-up trainings for those unable to attend the regular training session. Go through the crisis plan and procedures in order to familiarize all school personnel with it. Periodically remind staff of signals and codes.

Visit evacuation sites with staff and stakeholders. Show involved parties not only where evacuation sites are but also where specific areas, such as student reunification areas, media areas, and triage areas will be.

Give all staff, stakeholders, and families literature corresponding to the crisis plan. While all staff should have a copy of the crisis plan, it will also be helpful to provide them with pamphlets reminding them of key principles. Families and community members should also receive literature summarizing crisis procedures and information pertaining to them. Provide each classroom with a copy of the crisis plan and any relevant materials, supplies, and equipment.

Require a specific number of crisis drills every year. Most states require fire drills; the same should be true of crisis drills. This need not be an extra burden; work with state and district laws for possible options. In Arizona, for example, schools are permitted to use some of the mandated fire drills for crisis drills. Also, speak with students about the importance of drills and explain that, while they are serious, students should not be frightened.

Conduct tabletop exercises and scenario-based drills regularly. While actual drills and training are essential, it is also helpful to have group brainstorming activities that can be done informally around a table. These can be held with stakeholders, staff, community members, and first responders. Students can be involved, as well.

Staff Training

School staff need to be trained in how to respond in a crisis.

Tabletop Exercises—Fairfax County, Virginia

Tabletop exercises are "informal and stress-free exercises intended to facilitate the testing, evaluation and practicing of a school facility's crisis response plan and promote group problem solving." While drills and training are essential, it is also helpful to have group brainstorming activities that can be conducted informally. For this reason, many districts are adopting tabletop exercises. Fairfax County, Virginia, has had great success with these exercises. In Fairfax, the exercises consist of complete written scenarios and "injects"—additional pieces of information or circumstances that can be injected to alter the scenario. These injects range from "suspicious person with firearm behind school" to "electrical service to cafeteria interrupted." Injects include a list of possible responses to assist the facilitator.

The exercise begins with the reading of the scenario; scenarios are often tweaked to fit a particular school. A facilitator then distributes injects to individual participants.

Participants may handle the injects and implement an action individually or seek more information and coordination from other group members. Discussion ensues.

In Fairfax, the objectives include the following:

Test the ability of school personnel to identify, allocate, and utilize resources within their school during a critical incident.

Assess the ability of school personnel to implement their critical incident plan.

The director of safety and security for Fairfax County Public Schools commented, "We believe that the best type of training is experience. Fortunately, most of our schools do not have frequent critical events that require these kinds of responses. Therefore, many of our personnel do not have the opportunity to experience the harsh realities of having to manage these issues. The tabletop exercise allows us to provide an environment that can reasonably simulate the topics and some elements of the stress that are inherent in critical events. We have provided tabletop exercises to all 234 of our schools over the past two years. We now have a rotating schedule that provides an exercise facilitated by our office to all high schools and middle schools every other year, and to each elementary school every three years."

General Evacuation Procedures

The following evacuation tips should be followed to ensure the safety and security of all involved:

- Students should stop work immediately.
- Teachers in technical vocation, home economics, etc., shall verify that motors, torches, gas and water outlets, appliances, and so on, are turned off.

- Have assigned students assist the physically disabled.
- Immediately evacuate to prearranged shelter/gathering places.
- Do not pick up papers, textbooks, and so on.
- Do not close windows or do anything that will delay the evacuation.
- Speed is subordinate to control and order; no talking, running, or skipping steps on stairs.
- The last adult to leave the room should check to be sure that everyone is out and then shut the door.
- The first staff member to reach a door shall open it and keep it opened until everyone is out or the staff member is relieved by another staff member.
- Staff should check adjoining toilet rooms.
- Proceed to assigned shelter/gathering places.
- Pupils in lavatories or out of their rooms shall join the nearest line to exit and proceed to a staging location.
- With permission of individuals in charge, students shall join their own class, report to their teachers, and take their places in line.
- Staff members shall evacuate as soon as their assigned duties are completed.
- Take roll and keep students quiet and calm.
- Report missing students to the principal, assistant principal, or the person in charge.

Evacuating People with Special Needs

To meet special evacuation needs, special provisions will be made for students and staff with disabilities, such as learning and developmental disabilities, visual and hearing impairments, and physical limitations.

- Attention will also be provided for students with limited English proficiency to make sure they understand the emergency procedures.
- During emergency evacuations, blind persons should keep physical contact with their aides by placing a hand on the aide's shoulder.

The following procedures will be implemented to ensure that persons with special needs are supported and cared for during any and all types of emergency situations and incidents that may requires actions and physical activities beyond the norm.

- The school nurse or medical service provider will be responsible for maintaining a roster of students and staff who have sight, hearing, language, mobility, and other impairments.
- Persons with special needs should have in their possession, or on file, an individual emergency card describing their special needs. The cards should list information such as; disability, medications and their application frequencies, mobility constraints, attendant needs, allergies, primary physician, and so on.
- The building administrators will assign "buddy rooms" to ensure that a teacher with impairments is partnered with a teacher who can assist under emergency conditions.
- Classroom teachers will be provided the roster of students with impairments prior to the beginning of each semester, along with a copy of any existing individual Emergency Action Plans for assigned students. It is recommended that these rosters be retained within the Emergency Flip Chart binders.

- A "buddy system" will be employed that pairs each student with special needs with an individual who will locate and be responsible for her/him in an emergency. Responsibilities include ensuring their awareness of an emergency and may extend to carrying or leading them down stairs. Pairing a disabled person with two aides is preferable.
- Alternate aides will be designated in the event that there is an emergency and the person(s) responsible for assisting a handicapped individual is not available.
- Rescue teams will be made aware of the best way to rescue special needs students. As an example, mobility-impaired students should be allowed to instruct rescue team members on the best way to move them from the hazardous area. For example, the fireman's method of carrying an impaired individual may be dangerous to someone with respiratory problems.
- Administrators will maintain procedures for rapid access of the following documents and equipment to support onsite initial response actions:
  - A printout listing each student and current significant medical information. This information will remain in the custody of the school nurse or medical service provider in order to ensure confidentiality.
  - Parent permission to provide immediate treatment in the case of a medical emergency.
  - Medications, as necessary.
  - American Sign Language cue cards (as deemed appropriate).
  - Listing of triage area(s) designated for non-ambulatory students or those needing specific medical assistance.
  - Listing of any special power requirements for sustaining specialized medical support equipment.
  - Evacuation chairs capable of going down a stairway to be used with large students who have limited or no head control.

Lockdown

A school lockdown means no one is allowed in and no one is allowed out. Children are not permitted to leave the school during a lockdown and off-duty staff members are not allowed in. Doors are locked and students must follow the directives of school personnel. Procedures include:

- An emergency lockdown (or shelter-in-place) will be announced by intercom or other voice communication.
- If an intercom is not available, a bell should be used.
- If a situation that may require an emergency lockdown is discovered, the individual making the discovery shall immediately contact Police/Security and provide as much information as possible.
- Fire evacuation alarms should not be sounded.

Incident Command System

Response to all crises requires a clear chain of command between all responders. The ICS is based on the premise that every crisis has certain major elements requiring clear lines of command and control. FEMA is a good source of information on the ICS. FEMA has developed a self-study course that anyone can take. The description of the ICS below borrows from that course and from FEMA's multi-hazard training for schools, a program also offered by many state emergency management agencies.

Before developing school and district ICS teams, work with emergency responders to learn how they will respond to different types of crises. Learn which types of crises will result in fire and police departments leading the response. Learn how they will direct their personnel and interface with outsiders. Let these agencies know who at the school will be their liaison during an incident. Designate two backup liaison officers in case the primary liaison is offsite when the incident occurs.

Although emergency responders may be managing the incident, there is still much for school staff to do, including managing the care of students and the supplies and staffing needs of the situation. While the ICS calls for school staff to serve in all of the critical functions, be prepared for the incident commander to designate outside personnel to manage these responsibilities. According to FEMA, the critical functions are as follows:

Incident commander. This person manages the entire incident and will very often be an emergency responder rather than a school administrator.

Public information officer. This person is responsible for releasing information to families, community members, and the media during a crisis. The media can be a tremendous help in getting information to families and community members.

Safety officer. This person is responsible for the safety of the scene and the individuals at the scene. His or her role might include determining whether students have been evacuated far enough from the school. Often this role will be filled by an emergency responder.

Liaison officer. This person is responsible for coordinating with all of the agencies that have responded to the crisis. It is critical that this person be a good communicator and able to convey important information both to responders about the situation or the school facility and to school staff about necessary actions.

Operations officer. This person manages student and staff care during a crisis. This includes physical (food and water), medical (CPR and first aid), and mental needs (psychological services), as well as student release.

Planning and intelligence officer. This person is responsible for documenting the event, analyzing what has transpired thus far, and planning for possible further action.

Logistics officer. This person manages the supply and staffing needs of the situation. The logistics officer focuses on acquiring the supplies needed to assist the emergency responders. The logistics officer's school staff logistics responsibilities will include long-term needs (beyond the first four hours) for things, such as food, water, and bathroom facilities, as well as transportation (if students need to be bussed off campus). The logistics officer is also responsible for locating and assigning staff to fill various tasks for emergency situations. This could include finding staff to carry messages from the operations officer to those staff members directly managing students.

The Media

Though there are not many certainties in school crises, it is guaranteed that the media will be at the scene. Instead of being overwhelmed and threatened by the media, be prepared to work with them. The media can be a valuable asset during a crisis. In the event of a catastrophic event, the media may be your only outlet for communicating with families. However, as with all crisis planning, it is important to be proactive, not reactive.

If members of the media feel that they are not getting a story, they will seek one out.

Work with local media before a crisis occurs to make sure they understand your needs during an incident. The media can even help report on preparedness efforts –families and community members will appreciate knowing about a plan for dealing with the situation should a crisis arise.

Designate one representative within your crisis team to deal with the media.

This should be the PIO. The PIO may be the principal or another team member designated by the principal or the head of the response team. There also may be media specialists at the district level. Investigate this and make sure that the school-level representative immediately contacts the district-level media representative in the event of a crisis.

Emphasize that only the designated representative will give information to the media. In order to be proactive, only one PIO/spokesperson should speak with the media, even if there is nothing yet available to share. It is helpful for the representative to introduce him or herself as the spokesperson and say, "We don't have/aren't able to release any information yet, but we will keep you updated as soon as we are able. We would really appreciate your cooperation with staying in the media staging area. I will be making all announcements from this area and will keep you informed."

Designate a predetermined site for the media to congregate in event of a school crisis. If it is not possible to use the predetermined site that is away from students and staff, the principal or head of the command chain should designate an alternate site.

Prepare staff to deal with the media trying to get live coverage pictures and interviews. Media personnel will often try to get on campus and interview staff and students. Make it clear to staff that they should direct media people to the media area and to the school spokesperson or PIO.

Arrange for a joint press conference with emergency responders or choose one media representative to disseminate information to all other media outlets. This will give you some control over the content, flow, and timing of information that is released.

Work with state and local emergency management agencies to have the Federal Aviation Administration restrict air space over your site. This will prevent helicopters flying over your school at a time of chaos. Media helicopters can be very frightening to children.

Products

A three-ring binder detailing every aspect of response, complete with floor plans, facilities information, and roles and responsibilities is not the only product you'll need. Teachers should have abbreviated guides, principals should have crisis response boxes, and emergency responders should have floor plans and facilities information. Some school districts have found the following products useful.

Teacher Quick Reference Guides

The director of school safety in Bulloch County, Georgia, discovered that teachers found having copies of the district's safety plan inadequate for crisis response. Using the master plan, they were unable to quickly identify their roles and responsibilities in a crisis. Teachers recommended that the district develop something they could hold in their hands and quickly flip through. Staff at one high school, including teachers, nurses, and media center staff, were drafted to develop such a tool.

Their *Quick Guide* was piloted by all teachers at that school for one year. Overall, teachers were happy with the guide but did report some bugs. Over the summer, the district-level team worked to refine the guide to address the bugs and make sure the guide contained all key information from the district-level plan. The guide is a spiral-bound notebook with plastic insert pages. The pages contain district- and school-specific information.

General district procedures are on the front pages and school-specific information, such as evacuation locations for fire drills, are on the back pages. The title of each incident is at the bottom of the page so staff can quickly flip to the procedures for the situation at hand.

The *Quick Guide* has been designed to be a dynamic document that can be updated every year. Now all faculty members, from teachers to cafeteria workers, have a copy of the guide and only principals and members of school safety team have the big book.

Crisis Boxes

The California Safe Schools Task Force realized school administrators should have crisis boxes so that they will immediately have the information essential for effective management of a critical incident. They created a monograph that can be found at http://www.cde.ca.gov/ls/ss/cp/documents/crisisrespbox.pdf.

The monograph contains tips on how to organize the information that should be in the crisis response box, recommendations for who should get copies of the box, and details of what should be in the box and why. Recommended contents include such items as:

- ICS key responders' phone numbers
- Student attendance rosters
- Student disposition forms and emergency data cards
- List of students with special needs
- Teacher/employee roster
- Staff roster
- Keys
- Aerial photos of campus
- Maps of the surrounding neighborhood
- Campus layout

- Evacuation sites
- Designated command post and staging areas
- Fire alarm turn-off procedures
- Sprinkler systems turn-off procedures
- Utility shutoff valves
- Gas line and utility line layout
- The guide also reminds schools of the importance of having first aid supplies easily accessible from multiple locations.

Teacher Crisis Bags

Many experts recommend that each classroom be equipped with a crisis bag. These can take the form of backpacks, tote bags, or even five-gallon buckets. The contents should include the following:

- Current class roster
- Copy of emergency procedures
- First aid supplies
- Flashlight and extra batteries
- Activities for students
- Paper and pens
- Clipboard
- Store teacher crisis bags in easily accessible locations

Family Reunification Plans

Staff in Bibb County School District, Georgia, have put a lot of effort into developing the family reunification procedures that are in the district crisis plan. They have worked with the Red Cross to set up evacuation/reunification sites around the county. Not only does every school have two evacuation kits that include student rosters and emergency notification/contact cards, but the district has a system-wide reunification kit. This kit includes drafts of notices that can be faxed to local media outlets with information necessary to let families know both that an evacuation has occurred and where they can collect their children. Bibb County's crisis preparations included discussions with the media on how media outlets could help distribute information in the event of a crisis. The chief of Bibb County School Police noted that the media has been very cooperative in developing these protocols.

School Site Information

When a crisis occurs, emergency responders will immediately need a great deal of information about your school campus. They will need to know the members of your crisis response team, how various sites can be accessed, and the location of utility shutoff valves. Many schools share this information with local police and rescue agencies during the crisis planning process.

Some schools give these agencies copies of floor plans that indicate shutoff information. Some school districts compile site information for all schools on a CD-ROM and distribute copies to responders; other schools post this information on a secure Web site that responders can access from laptops at the scene. The following are two examples of how this information can be assembled.

Maryland Virtual Emergency Response System, or MVERS. MVERS was developed in partnership with the Maryland State Police, Maryland Institute for Emergency Medical Services Systems, and the Maryland Emergency Management Agency. This system can be used to prepare an electronic plan that allows quick and easy access to information in order to expedite a response to a critical situation. MVERS utilizes digital floor plans with specific icons that link the viewer to photographs, panoramic pictures, or spreadsheets containing essential data. The images can include instructions for disconnecting utilities, gaining access to a certain area, and locations of potential hazards.

The combination of floor plans and associated information provides a virtual tour of the structure's interior and exterior, allowing responders to understand the building layout prior to entering. Schools can also load contact information into MVERS. The Virtual Emergency Response System Construction Kit will provide the user with a description of the MVERS, an appendix of resources, and shareware for completing the plan. The MVERS team estimates it takes about 60 hours to collect and load all information to create the digital floor plan for each school. The bulk of this time will be spent taking and editing pictures of the buildings.

Charlotte-Mecklenburg, North Carolina Police Virtual Tour. After an incident where there were communication glitches between school staff and police, the Charlotte-Mecklenburg Police Department realized it needed to better prepare for school crises. An officer was detailed to create *Virtual Tours* for each school. The *Virtual Tour* is a combination of the school plan and the police plan. School resource officers, or SROs, develop basic crisis plans around the plans their school has already developed. The SROs identify on-scene and off-scene command posts and initial road blocks. They also collect information on crucial players at the school and district (maintenance supervisors), bell schedules, aerial photographs of the school and surrounding community, and extensive photos of the school campus.

For each school, a master Web page contains a picture of the school and links to the crisis plan, the *Virtual Tour*, and aerial photos of the school and surrounding neighborhood. The tour allows emergency responders to move around the school building from the safety of a laptop as they prepare to respond to the crisis. The *Virtual Tour* opens with a map of the school. Users can zoom in on a door or window, click on a door and go through, walk down a hallway, look left, right, up and down, and turn around. Each screen includes an orientation map that shows where you are on the sitemap. The program also flags potential hazards, such as closets, windows in unusual spaces, and crawl spaces. This information is loaded on police laptops and computers and updated monthly. This material is stored on a private Web site and cannot be accessed by the public.

Consideration of Special Need Staff and Students

Be sure to give special consideration to the unique needs of staff and students with disabilities when developing the crisis plan. Evacuation and relocation procedures will need to address mental, physical, motor, developmental, and sensory limitations. For example, individuals who use wheelchairs or other auxiliary aids will not be able to traverse the front steps of a building without substantial assistance.

The following issues should be addressed:

In some cases, individuals with disabilities may have limited mobility. In an evacuation, there may not be enough time to move mobility impaired students and staff to traditional shelters. It is important to identify alternative, accessible, safe shelter locations and to communicate these locations to emergency responders.

Individuals with hearing disabilities may not be able to communicate verbally, to read lips, or to hear fire alarms or other emergency signals.

Consider providing basic sign-language training to designated school staff.

Visual impairments might impede reading signs or traversing unfamiliar or altered terrain—consider whether debris might obstruct the evacuation of such staff and students and necessitate alternative shelter locations.

Debris may obstruct the evacuation of individuals with mobility impairments. Be sure to assign sufficient staff to assist these individuals during a crisis or consider identifying alternative shelter locations.

Are staff trained to assist students with developmental disabilities? These students may become upset if routine patterns of activity are disrupted.

Do any students or staff have special needs for medicines, power supplies, or medical devices that are not likely to be available in emergency shelters? Consider what alternative arrangements can be made to provide these necessities.

In addition to addressing these concerns, find out whether specific crises will require additional considerations for hazards, such as fire, severe weather, or earthquake.

For example, mobility impairments might prevent some staff or students from being able to bend over to assume the protective position recommended during tornadoes.

Also, during a fire, elevators will be unavailable to transport wheelchairs. As noted earlier, it is critical to identify safe and appropriate shelter areas inside school buildings that can be reached quickly and accommodate individuals with disabilities.

Student Release

Student release is a crucial part of crisis planning. In all school crisis planning, the safety of the students is the main priority. During a crisis, traditional student release procedures are frequently unsafe or otherwise inoperable. Accordingly, a comprehensive crisis plan needs to include certain procedures:

Update student rosters. Rosters should be updated at a minimum of twice a year; some recommend updating rosters weekly.

Distribute updated rosters. All teachers need updated rosters of all their classes. This should be stored in their classroom so that a substitute teacher could easily find it. A copy of all rosters should also be placed in the crisis response box, as well as with the principal and any other stakeholder as advisable. It is critical to know which students are present during a crisis.

Create student emergency cards. At the beginning of the school year, make sure the school has an emergency card for each student containing contact information on parents/guardians, as well as several other adults who can be contacted if the parent or guardian is not available. The card should also indicate whether the student is permitted to leave campus with any of the adults listed on the card, if necessary. Some districts recommend authorizing one or more parents of children at your child's school to pick up your child. The card should also include all pertinent medical information, such as allergies, medications, and doctor contact information. These cards should be stored in the front office, both in hard copy and electronically, if possible.

Create student release forms to be used in times of crisis and store them with crisis response materials. Create a back-up plan if forms are not available.

Designate student release areas, as well as back-up options. These areas should be predetermined and communicated to families.

If necessary, changes should be communicated through the designated channels.

Assign roles for staff. For example, a staff member is needed to take the emergency cards from the office to the release area, while several staff members are needed to deal with families and sign out students.

These roles should be assigned before a crisis occurs. If roles change, the principal or designated leader should assign new roles.

Create student release procedures. These procedures should create a flexible, yet simple, system for the release of students. Families will want immediate access to their children; emotions will be running high. Create a system that considers this, and train staff to expect it. Procedures should require proof of identity; if necessary, wait until such proof can be ascertained. It is important not to release a student to a noncustodial guardian if custody is an issue for the family. Do not release students to people not listed on student emergency cards. A well-intentioned friend may offer to take a child home; however, school staff must be certain that students are only released to the appropriate people so students' families will know where they are.

Arrange for transportation for students who are not taken home by a parent or guardian. Also arrange for shelter and provisions, if necessary.

Use all communication outlets to keep families, the media, and community informed during and after the crisis.

Signal the end of the crisis, as well.

Key Responders' Phone Numbers

Names and phone numbers for all team participants involved in coordinating with your local emergency response system should be in the Go Kit. These people include the coordinators for the ICS, Public Information, First Aid, Traffic Safety, Student Assembly and Release, and Grounds and Maintenance. Also, include the names and phone numbers of other key staff members, such as the food, water, and supplies coordinator; the bilingual translator (if appropriate for your school); and any additional positions you have identified. Be sure to place these phone numbers on several cards so that more than one person can begin calling them.

Designated Command Post and Staging Areas

Police chiefs involved in several of the recent school shootings recommend that schools and law enforcement officials plan for three distinct staging areas, in addition to the Command Post for the Incident Command Officer. Among other benefits, separate staging areas will prevent the press from converging upon parents or parents from converging upon police. The areas should include:

A staging area for law enforcement and emergency personnel
A media staging area away from the school, at a location that can accommodate a large number of vehicles
A parent center, located away from the Command Post, where parents can retrieve their children

Maps of all staging areas and command posts, listing main phone numbers for each, should be included in the Go Kit. Be aware that these locations may change based upon the circumstances. It is also recommended that the command post and staging areas have telecommunications capability, whenever possible.

Emergency Resource List

A list of individuals and organizations that assist in an emergency should be prepared on a separate sheet of paper and placed in the Go Kit so that the person assigned can immediately begin to make phone calls. Your local emergency management agency can recommend agencies you should call during an emergency.

Please note that any volunteers you enlist for a critical incident response should be prescreened and that the volunteers on your list must receive training prior to becoming a responsible member of your emergency response team. Contact numbers to include are:

Vendors

- Mutual Aid contacts
- Other board personnel with whom relationships have been built over time
- Parent representatives, who should be trained to answer questions, help fellow parents receive information, and maintain calm at the Parent Center. (Fellow parents can be an excellent source of support.)

Evacuation Sites

Maps with evacuation and alternate evacuation routes should be stored in the Go Kit, as well as posted on classroom doors. Be aware that during a shooting spree, the best-laid plans for evacuation may also be under siege. All classrooms at Columbine, for example, had evacuation plans, but with two students shooting throughout the entire school, evacuating the building was itself a dangerous venture. In the Jonesboro, Arkansas, incident, two boys opened fire after students evacuated the building during a false fire alarm. Other factors may require an Incident Command Officer to alter the usual evacuation route. In a chemical spill, for instance, the wind direction will determine where to evacuate. Thus, it is important to have at least two predetermined evacuation sites identified.

Student Disposition Forms and Emergency Data Cards

Imagine hundreds of parents descending upon a school to retrieve their children while you are trying to account for each student and staff member's whereabouts. Forms will be needed to track who has been released and to whom—parents, relatives, emergency personnel, or the hospital. A set of release forms (enough to cover the entire school census) should be stored in the Go Kit and should be given to the appropriate personnel on your Incident Command team. Additionally, if possible, it is helpful to have a set of your student emergency data cards placed in the box. Having all the data stored on a disk is the most convenient way of storing the information. Optimally, the cards and disks should be updated every three months to remain current. Emergency information can also be stored, updated, and retrieved electronically, either from the school office or a remote site; however, updated hard copies must always be available in the event that power is out.

Inventory of Staff Resources

Survey your staff members to build an inventory of special skills and training they possess. Document your findings and place the list in the Go Kit under the "Incident Command System" heading. For instance, experience can include prior medical and triage experience, bilingual capabilities, grief counseling background, search and rescue training, hostage negotiations, first aid/CPR certification, and volunteer firefighter or reserve police officer/deputy experience. These skills could prove to be very helpful in a critical incident.

First Aid Supplies Location

Sets of first aid supplies should be located throughout the campus. Storage locations should be included in the Go Kit. Include the locations on one of the building layout maps in the Go Kit.

Emergency First Aid Supplies

In the Jonesboro, Arkansas, shooting, large bins of first aid supplies were readily accessible on the school grounds and are credited with saving two children's lives and preventing other individuals from going into shock. These supplies were situated inside and outside of the school building in anticipation of an earthquake because the school property is located on a fault line. The accessibility of these supplies proved to be lifesaving. It will do no one any good if first aid supplies are locked away deep within the confines of the school. Some schools have stationed first aid boxes in every classroom, with basic emergency aid instructions to treat various injuries.

Although not designed for first aid, duct tape is very useful and versatile and should be available in every classroom. Whichever methods you deem will work best for your school, make sure that ample supplies are readily accessible.

Building Supply Kit

It is advisable to have a tool kit with supplies and tools ready for initial needs that could be addressed by the facility manager. Such a kit could include:

- Hammer
- Axe
- Nails
- Duct tape
- Putty
- Other tools that would be helpful in a disaster

Legal Considerations

Boards have the authority to adopt a rule giving the Superintendent/President emergency purchasing authority to restore the school districts and community colleges to full operational condition immediately after a disaster. This emergency authority needs to be in place before disaster strikes.

Boards also have the authority to adopt a rule relating to leave for emergency personnel directly affected by a hurricane. Such a policy allows employees, under specific conditions determined by the Board, to be granted personal leave that is not deducted from sick leave. This can help greatly with the financial burdens suffered by employees as a result of a disaster. If you plan to use this provision, a mutual agreement with local unions prior to a storm would be most helpful. Because of the personnel issues, labor contracts are impacted. It is questionable whether the contracts will require renegotiating, but that will depend on actions of the boards. At the very least, calendars will need to be renegotiated in areas suffering the worst damage. Personnel issues need immediate attention. In areas where campuses remain closed for a week or two, the Superintendent, or President, may have authority to declare that teachers/professors are on duty preparing for schools to open (check your board's policies carefully). Staff will need funds as quickly as possible to get their personal lives in order, as well as prepare for school to reopen.

Mental Health

Some special counseling assistance may be necessary to help deal with the trauma that some staff and youngsters experience during a hurricane.

What to Expect After Trauma:

Possible Reactions of Elementary School Students

1. Feelings of anxiety, fears, and worries about safety of self and others (Students may act more clingy with teacher or parent)
2. Worries about reoccurrence of violence
3. Increased levels of distress (whiny, irritable, moody)
4. Changes in behavior:
   a. Increased activity level
   b. Decreased concentration and/or attention
   c. Withdrawal
   d. Angry outbursts
   e. Aggression
   f. Absenteeism
5. Increased somatic complaints (e.g., headaches, stomachaches, aches and pains)
6. Changes in school performance
7. Recreating event (e.g., talking repeatedly about it, reenacting the event)
8. Increased sensitivity to sounds (e.g., sirens, planes, thunder, backfires, loud noises)
9. Statements and questions about death and dying

At home, parents may see:

1. Changes in sleep habits
2. Changes in appetite
3. Withdrawal
4. Lack of interest in usual activities
5. Increased negative behaviors (e.g., defiance) or emotions (e.g., sadness, fears, anger, worries)
6. Regression in behaviors (e.g., baby talk, bed-wetting, tantrums)
7. Hate or anger statements

Possible Reactions of Middle School Students:

1. Feelings of anxiety, worries, and fears about safety of self and others
2. Worries about reoccurrence or consequences such as war, as well as worries about school violence
3. Changes in behavior:
   a. Decreased attention and/or concentration
   b. Increase in hyperactivity
   c. Changes in academic performance
   d. Irritability with friends, teachers, events
   e. Anger outbursts and/or aggression
   f. Withdrawal
   g. Absenteeism
4. Increased somatic complaints (e.g., headaches, stomachaches, chest pains)
5. Discomfort with feelings, particularly those associated with revenge.

6. Increased likelihood to discuss the gruesome details
7. Repeated discussions of event
8. Increased sensitivity to sounds (e.g., sirens, planes, thunder, backfires, loud noises)
9. Negative impact on issues of trust and perceptions of others, particularly of individuals who are "different"
10. Repetitive thoughts and comments about death and dying

At home, parents may see:

1. Changes in sleep
2. Changes in appetite
3. Withdrawal
4. Lack of interest in usual activities (e.g., after-school activities, time with friends)
5. Increased negative behaviors (e.g., defiance) or emotions (e.g., sadness, fears, anger, worries)
6. Hate or anger statements
7. Denial of impact

Possible Reactions of High School Students

1. Worries, fears, and anxiety about safety of self and others
2. Worries about reoccurrence or repercussions, such as war or school violence
3. Changes in behavior:
   a. Withdrawal
   b. Irritability with friends, teachers, events
   c. Anger outbursts and/or aggression
   d. Changes in academic performance
   e. Decrease in attention and concentration
   f. Increase in hyperactivity
   g. Absenteeism
4. Discomfort with feelings, including revenge and vulnerability
5. Increased risk for substance abuse, including drinking
6. Discussion of events and review of details
7. Negative impact on issues of trust and perceptions of others, particularly individuals who are "different"
8. Increased sensitivity to sounds (sirens, planes, thunder, backfires, or loud noises)
9. Repetitive thoughts and comments about death or dying (including suicidal thoughts)

At home, parents may see:

1. Changes in sleep
2. Changes in appetite
3. Withdrawal
4. Lack of interest in usual activities (after-school activities or time with friends)
5. Increased negative behaviors (e.g., defiance) or emotions (e.g., sadness, fears, or anger, worries)

6. Hate or anger statements
7. Denial of impact

Families

Many facets of school safety planning impact families. Much of the literature on school safety planning provides guidelines for communicating with families and advice for families on how to deal with their children after a crisis. Additionally, verbatim statements from families of children attending school near the World Trade Center on September 11 provide insight into crisis planning. The following addresses the school's role in communicating with families, both before and immediately following a school crisis, and what families can do to facilitate their children's recovery.

Communication Information to Families before a Crisis

Families will appreciate information on crisis preparations. It is especially useful to explain family members' roles before an incident occurs. Some school districts send families letters describing the school's expectations for their response. Other school districts have found it useful to work with local media to disseminate this information.

School and district staff and emergency responders need to be able do their jobs. Families need to know that they should rely on media outlets for information during an incident, rather than telephoning schools. It is very important that families understand that during a crisis, school phones will be needed to manage the situation.

Families should also know that they should wait for instructions on student release rather than rushing to the school. It is helpful to explain to families that emergency responders need the area clear to do their job.

Also explain that only after emergency responders determine that a safe student release is possible will families be reunited with their children. It is also useful to remind families that in many situations, their children will be safer in the school building than outside or in a car, particularly in cases of severe weather.

Communicating Information to Families During a Crisis

Communicating with families. It is important to have a mechanism for communicating with families in the event of a crisis. The mode of communication could be a telephone voice recording with information about welfare of the children, evacuation sites, or information about releasing students. Arrangements could be made with TV and radio stations to release such information.

In the case of an extended crisis, such as the sniper attacks on the Washington, DC, metropolitan area, a school official may want to write a letter to families each day of the crisis to update them on safety measures devoted to the safety of their children. Schools should be sensitive to the communities they draw upon and enlist volunteers to help communicate with families who do not speak English. It is important to acknowledge cultural differences in responding to crises.

Contact information for students. Schools need contact information from families, including numbers where they can be reached during the day. In addition, each child should have several alternative contacts, such as a relative or family friend who would be able to pick up the child in the event of an emergency. One of the backup adults should live outside of the immediate area, if possible.

Guidelines for Families in Dealing with Their Children After a Crisis

Remain calm. It is important to remain calm in the aftermath of a crisis. Children are greatly influenced by their family's sense of well-being, and anything that families can do to reassure students will be helpful. At the same time, families need to be compassionate listeners when their children speak of the crisis.

Attend to children's reactions. Be alert to children's emotional needs. Individuals recover from crisis at their own pace. Many children will benefit from mental health services regardless of whether they were directly or indirectly involved in the incident.

Return children to normal routine as quickly as possible. Families should adhere to the schedule of the school, and if the school remains open immediately after the aftermath of a crisis, it is important to let children return to school. Adhering to a typical routine will help children in the recovery process.

Refer the media to the PIO. Undoubtedly, the media will try to interview families and children during or after a crisis. Families can make a very positive contribution to the school by referring the media to the PIO.

Attend community meetings. Families will receive invaluable information and support by attending community or school meetings. Community meetings often provide information to help dispel rumors and establish mechanisms of communication with parents, the media, and other affected parties.

The following statement, made by a parent of a child in a school near the World Trade Center on September 11, 2001, emphasizes the points made earlier. "Children's reactions are reflections of their parents. Too many parents expose children to their every emotion: fear, anxiety, anger, worry, etc. The fact is that children want parents to be heroes. If parents can be strong, this will benefit their recovery."

Models of Crisis Intervention for Students

There are many approaches to crisis intervention for students. Most experts agree that school- or classroom-based stress management needs to be conducted for all students and that those with more severe reactions need to be referred for evaluation and possibly counseling.

Who provides interventions? Teachers, school counselors, and social workers, as well as community service providers may be involved in conducting interventions following a crisis. Families may also use school personnel as a resource for seeking outside counseling. During the planning phase, districts should identify service providers in the community that have the skills and appropriate credentials and develop a list of referrals.

What are the types of interventions? The following bullets briefly describe several approaches:

- **Group crisis intervention, or GCI,** a school-based intervention, is often defined as "psychological first aid." GCI is an efficient and cost-effective way of helping students cope in the aftermath of a crisis. Basically, GCI is offered to homogeneous groups of students (class membership) and involves guided group discussions in a supportive environment. The agenda for GCI includes an introduction and sessions on providing facts, dispelling rumors, sharing stories, sharing reactions, providing empowerment, and offering a closing. Students with severe reactions to the crisis should receive more intensive interventions (Brock et al., 2002).

- **Acute traumatic stress management** for educators, another school-based intervention, offers a "road map" for educators to deal with the aftermath of a crisis. ATSM takes a practical approach to dealing with the psychological consequences of a traumatic event. The goal is to stimulate adaptive coping mechanisms and to stabilize more severe reactions among students. ATSM has ten stages:
  Assess for danger/safety for self and others.
  Consider the mechanism for injury.
  Evaluate the level of responsiveness.
  Address medical needs.
  Observe and identify.
  Connect with the individual.
  Ground the individual.
  Provide support.
  Normalize the response.
  Prepare for the future.
- **Individual counseling.** Students who experience severe symptoms after a crisis may need individual counseling. It is important for these individuals to be referred for further evaluation by a mental health professional. There are many forms of individual counseling depending on the age of the child and presenting symptoms. Some of the approaches to individual counseling include play therapy, art therapy, talking therapy, drug therapy, and a combination of therapies.
- **Cognitive-behavioral therapy**, among others, has shown to be an effective therapeutic intervention in the literature. Dr. Robert Pynoos, Director of Trauma Psychiatry at the University of California at Los Angeles, developed an interview guide for working with students who have been traumatized.

The interview guide contains the following sections:

- Triage questions
- Individual's reaction to the event/traumatic reminders
- Life changes/changes in behavior
- Grief responses
- Problem solving/taking constructive action, affirmation
- Affirmation and reinforcement of student's strengths and assets

RECOVERY

The goal of recovery is to return to learning and restore the infrastructure of the school as quickly as possible. Focus on students and the physical plant, and take as much time as needed for recovery.

School staff can be trained to deal with the emotional impact of the crisis, as well as to initially assess the emotional needs of students, staff, and responders. One of the major goals of recovery is to provide a caring and supportive school environment.

Action Checklist

- Strive to return to learning as quickly as possible.
- Restore the physical plant, as well as the school community.
- Monitor how staff is assessing students for the emotional impact of the crisis.
- Identify what follow-up interventions are available to students, staff, and first responders.
- Conduct debriefings with staff and first responders.
- Assess curricular activities that address the crisis.
- Allocate appropriate time for recovery.
- Plan how anniversaries of events will be commemorated.
- Capture "lessons learned" and incorporate them into revisions and trainings.

Plan for recovery in the preparedness phase. Determine the roles and responsibilities of staff and others who will assist in recovery during the planning phase. District-level counselors may want to train school staff to assess the emotional needs of students and colleagues to determine intervention needs. Experience shows that after a crisis many unsolicited offers of assistance from outside the school community are made. During planning, you may want to review the credentials of service providers and certify those that will be used during recovery.

Assemble the Crisis Intervention Team. A Crisis Intervention Team, or CIT, is composed of individuals at either the district or school level involved in recovery. A model, there is a centralized CIT at the district level, which serves all schools in that district. In another model, the district trains school-based CITs. Even when crisis intervention teams exist within individual schools, it may be necessary for the superintendent to allocate additional resources on an as-needed basis.

Service providers in the community may want to assist after a crisis. With prior planning, those with appropriate skills and certifications may be tapped to assist in recovery. This will help district and school personnel coordinate activities of the community service providers and see that district procedures and intervention goals are followed.

Return to the "business of learning" as quickly as possible. Experts agree that the first order of business following a crisis is to return students to learning as quickly as possible. This may involve helping students and families cope with separations from one another with the reopening of school after a crisis.

Schools and districts need to keep students, families, and the media informed. Be clear about what steps have been taken to attend to student safety. Let families and other community members know what support services the schools and districts are providing or what other community resources are available. Messages to students should be age appropriate. It may be necessary to translate letters and other forms of communication into languages other than English depending on the composition of the communities feeding the affected school(s). Be sure to consider cultural differences when preparing these materials.

Focus on the building, as well as people, during recovery. Following a crisis, buildings and their grounds may need repairing or repainting/re-landscaping. Conduct safety audits to determine the parts of the building that can be used, and plan for repairing those that are damaged.

Provide assessment of emotional needs of staff, students, families, and responders. Assess the emotional needs of all students and staff, and determine those who need intervention by a school counselor, social worker, school psychologist, or other mental health professional. Arrange for appropriate interventions by school or community-based service providers.

In addition, available services need to be identified for families, who may want to seek treatment for their children or themselves. Appropriate group intervention may be beneficial to students and staff experiencing less severe reactions to the crisis. Group interventions should be age appropriate.

Provide stress management during class time. Trauma experts emphasize the need to create a caring, warm, and trusting environment for students following a crisis. Allow students to talk about what they felt and experienced during the traumatic event. Younger children who may not be able to fully express their feelings verbally will benefit from participating in creative activities, including drawing, painting, or writing stories. Young adolescents benefit from group discussions in which they are encouraged to talk about their feelings, as well as from writing plays or stories about their experiences. Engage older adolescents in group discussions, and address any issues of guilt ("I could have taken some action to change the outcome of the crisis").

Conduct daily debriefings for staff, responders, and others assisting in recovery. Mental health workers who have provided services after crises stress the importance of ensuring that those who are providing "psychological first aid" are supported with daily critical incident stress debriefings. Debriefings help staff cope with their own feelings of vulnerability.

Take as much time as needed for recovery. An individual recovers from a crisis at his or her own pace. Recovery is not linear. After a crisis, healing is a process filled with ups and downs. Depending on the traumatic event and the individual, recovery may take months or even years.

Remember anniversaries of crises. Many occasions will remind staff, students, and families about crises. The anniversary of crises will stimulate memories and feelings about the incident. In addition, other occasions may remind the school community about the crises, including holidays, returning to school after vacations and other breaks, as well as events or occasions that seemingly do not have a connection with the incident. This underscores the notion that recovery may take a longer time than anticipated.

Staff members need to be sensitive to their own, as well as the students' reactions, in such situations and provide support when necessary. School crisis planning guides suggest holding appropriate memorial services or other activities, such as planting a tree in memory of victims of the crises. Trauma experts discourage memorials for suicide victims to avoid glorifying and sensationalizing these deaths.

Evaluate. Evaluating recovery efforts will help prepare for the next crisis. Use several methods to evaluate recovery efforts. Conduct brief interviews with emergency responders, families, teachers, students, and staff. Focus groups may also be helpful in obtaining candid information about recovery efforts. The following are examples of questions to ask:

- Which classroom-based interventions proved most successful and why?
- Which assessment and referral strategies were the most successful and why?
- What were the most positive aspects of staff debriefings and why?
- Which recovery strategies would you change and why?
- Do other professionals need to be tapped to help with future crises?
- What additional training is necessary to enable the school community and the community at large to prepare for future crises?
- What additional equipment is needed to support recovery efforts?
- What other planning actions will facilitate future recovery efforts?

Recovery is the ability to recuperate from the loss or damage to a school facility, whether due to natural disaster or malicious intent. Disaster recovery strategies include rebuilding and implementing procedures to restore a system or community to its pre-disaster condition. Recovery planning includes consideration for assets and services following a disaster. The recovery period also allows for the identification of systems and buildings that performed well in the disaster and others that failed. Learning from the experience is an important part of the recovery process.

Initial Site Inspection and Recording Impacts

After the disaster, be sure to take pictures and document damage before recovery efforts begin. As you document damage, think long-term. The immediate damage, such as damage to roofs and windows, is obvious. The long-term, less detectable damage, such as air quality issues, also needs to be included in your estimates and reimbursement requests.

Both insurers and FEMA will require well-documented evidence of damages claimed. Personnel responsible for taking pictures should be instructed on the type of photos required, the proper labeling of photos, and what to avoid. Several thousand pictures of wet, damaged ceiling tile are worthless unless care has been taken to identify the school, building, and room where the tiles are located.

There is no single assessment process followed by all school systems; each school system has its own particular method. The assessment process used will depend on the severity of the event, number of facilities, staff available to perform damage assessments, distance between the facilities, and the ability to travel between campuses. Assessments vary as to level of detail; some boards with widely disbursed campuses will perform an initial "drive-by" inspection, attempting to obtain cursory information as to the extent of exterior damage, and follow up with a more detailed inspection at a later time. Other schools will begin with a detailed assessment, performing interior and exterior inspections right away. Some school systems are able to complete detailed reviews for all their campuses quickly (i.e., in two days), while others may take several days or weeks.

The Palm Beach Community College (PBCC), Florida, has developed a comprehensive set of processes and procedures for taking pictures and documenting damage.

The system also includes forms used to document damage. PBCC has used this system and has found it to be very effective when dealing with FEMA and insurance providers.

Areas that should be considered when taking photos include the following:

Interior Areas

- Main Office, Lobby, Reception
- Administrative Areas and Staff Offices
- Mailboxes
- Health Clinic
- Guidance Office
- Conference Rooms
- Corridors, Circulation, Lockers
- Stairs, Stairwells, Landings, Steps
- Ramps
- Restrooms
- Classrooms
- Art Rooms
- Music Rooms
- Labs, Shops, Computer Rooms
- Dance Classrooms
- Gymnasiums
- Locker Rooms
- Media Center
- Auditorium and Theaters
- Cafeterias and Student Commons
- Coolers, Freezers
- Vending Machines
- Storage Rooms
- Equipment Rooms
- Elevators
- Portable, Modular, or Temporary Classrooms
- Non-Structural Building Hazards
- Obvious Building Damage
- Entryways
- Interior Walls
- Interior Doors, Windows
- Ceilings
- General Fire Requirements for Existing Buildings
- Utilities
- Air Handling and Filtration
- Fresh Air Intakes
- Gas Tank/Piping
- Interior Water Pipes
- Interior Lighting

- Lighting Fixtures/Poles

Building Access Control

- Building Notification Systems
- CCTV Surveillance Systems
- Telephone Systems
- Public Telephones
- Radio/Wireless Communication Systems

Exterior Areas

- Site Access
- Fencing
- Landscaping, Trees and Shrubs, Land Washout
- School Sign
- Flagpoles
- Playground Equipment
- Walkways
- Canopies, Awnings, Breezeways, Covered Walkways
- Courtyards
- Building Access
- Exterior Walls
- Siding
- Exterior Doors
- Windows
- Skylights
- Roof
- Exterior Roof Damage
- Rooftop Vents
- Gutters, Downspouts
- Rooftop HVAC Units
- Exterior Water Pipe
- Water Fountains
- Water Supply and Storage
- Sewage, Backup, Sewage
- Plants
- Exterior Wiring

Aerial photography works well to document before and after situations affecting both school grounds and structures, and is well received by FEMA and insurance companies.

Infrared photography can also be used; it does an excellent job of identifying where water has penetrated the roof, as well as where it may be leaking down walls. Single ply and built-up roofing that has experienced broken seals can look solid on the exterior when, in fact, its integrity has been compromised.

Purchasing

The School Board has the authority to adopt a rule giving the Superintendent emergency purchasing authority to restore the district to full operational condition immediately after a disaster. This emergency authority needs to be in place before a disaster strikes. Purchasing authority may extend to:

- Purchase cards (P-cards)
- Establish limits for emergency situations before an event occurs
- Identify who can increase those limits
- Open purchase orders
- Establish limits on your open accounts.
- Establish purchase limits.
- Negotiate and establish pre-contracts.
- Maintain purchasing flexibility
- Determine limits with and without approval.

Depending upon the extent of the damage, boards may experience some cash flow issues in reopening schools.

Reopening Schools

Parents attempting to get their lives back in order need someone to supervise young people. Extended school programs and Saturday library programs assist parents and can be grant funded in some cases. With some schools inoperable for a period of time, students may be transported to other schools. Some of these transportation costs may be grant funded, as well.

Local AM and FM radio stations can be helpful in communicating which schools are open after a storm: schedule these announcements to be made at the same time each day. In previous storms, school districts and community colleges used local radio stations to get information to personnel, parents, students, and the community.

Reactions of Staff

The trauma of a disaster can affect everyone. You do not have to be a primary victim to feel the impact or stress of an event. Possible behaviors after a disaster include:

- Increased irritability and impatience with students and staff. Decreased tolerance of minor student infractions.
- Decreased concentration.
- Worries and fears that answers or responses to students could make things worse for them.
- Worries about reoccurrence and repercussions.
- Increased concern or hypersensitivity about school violence.
- Feelings of discomfort, with intense emotions, such as anger and fear.
- Denial that the traumatic event may impact the students.

Closing The Loop

After a disaster, even with rebuilding underway, a community may have changed. Demographic and financial shifts may include:

Shifts in population and students to:

- Non-damaged areas of schools or community colleges
- Other schools, districts, or community colleges
- Loss of FTE
- Loss of students
- Excess number of teachers under contract
- Reallocation of teaching personnel
- Shift of personnel
- Increased travel distances
- Loss of tax revenue due to devaluation of real estate
- Capital programs affected by unknowns
- Construction cost
- Loss of commerce
- Loss of industry
- Bus routes revamped due to shifts in population

It is important to document all activities in dealing with disasters especially that worked and those that didn't and then share these with other emergency preparedness planners.

Recovery may seem like an end, but it is also the beginning. You must close the loop on the circle. A critical step in emergency preparedness planning is to evaluate each incident. What worked? What didn't? How could you improve operations? Take what you have learned and start at the beginning. Update and strengthen the plan so that in a crisis, no child is left behind.

Section IV

Facility Design

FACILITY DESIGN PRINCIPLES—IN THE PAST

Schools are typically in use for long periods of time; as a result, teaching continues to be conducted in facilities that were designed and constructed at the beginning of the twentieth century. Early-twentieth-century school design was based on late-nineteenth-century models and was relatively static until after World War II. Schools ranged from one-room rural schoolhouses to major symbolic civic structures in large cities. Other inner-city schools were more modest, inserted into small sites on busy streets and constrained by budget limits.

The typical city school was one to three stories in height and consisted of rows of classrooms on either side of a wide, noisy corridor lined with metal lockers; there were asphalt play courts and, sometimes, rooftop recreational areas. The larger schools sometimes had a library, special rooms for art and science, a shop, and an auditorium.

The surge to meet the school construction demands of the postwar baby boom was primarily a suburban development. Much larger sites were available, schools were one or two stories in height, auditoriums became multiuse buildings, and large parking lots appeared. However, many rural schools were located far away from towns and their resources, such as fire departments and other services.

But the fundamental school program of classrooms along double-loaded corridors did not change very much. However, in warm climates, the one-story finger plan school, constructed of wood and a small quantity of steel, was both economical and more human, and the noisy tiled double-loaded corridor became a covered walk, open to the air, with the classrooms on one side and a grassed court on the other. Compact versions of these plans appeared as schools became larger and sites smaller.

Inner-city high schools were usually large facilities, housing 2,000–3,000 students (basically small towns with complex social, economic, and class systems). In the 1960s and 1970s, some design experiments were tried, such as team teaching, which spawned large open classrooms with poor acoustics.

Some of the new, large high schools were built as air-conditioned enclosures, with many windowless classrooms, in buildings similar to the shopping malls that replaced the main street retail centers. At the same time, by adding prefabricated classrooms to accommodate a surge in enrollment, many schools were expanded. Although the prefabricated classrooms were originally intended as temporary space, many are now used as permanent classrooms. Schools built in the 1980s and 1990s assumed a wide variety of forms, often combining classrooms into clusters and focusing on providing an attractive learning environment. However, demographic needs, shortage of affordable land, and limited funding has also resulted the adaptation of existing noneducational buildings into schools.

Present School Design

New statements of design principles are beginning to emerge, although some of the following represent perennial concerns:

- The building should provide for health, safety, and security.
- The learning environment should enhance teaching and learning and accommodate the needs of all learners.
- The learning environment should serve as the center of the community.
- The learning environment should result from a planning/design process that involves all stakeholders.
- The learning environment should allow for flexibility and adaptability to changing needs.
- The learning environment should make effective use of all available resources.

These principles lead, in turn, to a number of current design principles, including the following:

- Design for protection against natural hazards;
- Increased design attention to occupant security;
- Careful lighting design and increased use of day lighting and comfort control;
- Design for durability;
- Long-life/loose-fit approach: design for internal change and flexibility; and
- Design for sustainability, including energy efficiency and the use of "green" materials.

Some new schools already respond to these needs, and, indeed, their originators, school districts, communities, and designers are among those defining the schools of the next decade. Some of the changes are the result of ideology and analysis; others are enforced by the effort to provide an improved learning environment and enhanced learning resources in an increasingly financially limited school construction economy. Some school districts will be hard pressed to provide a minimal learning environment with buildings of the utmost simplicity, while meeting the requirements for health, safety, and security.

Future School Design

Schools will continue to vary widely in size; however, even in the suburbs, land has become scarce and expensive. New schools will be more compact and the sprawling one-story campus will become less common. The desire for more supportive environments and the rejection of traditional school plans will result in more imaginative and often more complex layouts. Moreover, the move to repopulate the inner cities will result in the construction of even more dense and compact schools.

However, many educational researchers believe that students improve their learning skills in smaller schools. Although small schools may be economically unrealistic, methods of organization are being explored that provide some of the benefits of small size within a large physical complex. Some schools are organized into "learning academies" for each grade, with classrooms that can expand and contract and other activity rooms of various sizes.

Other researchers believe that the conventional library will disappear. The trend in many new schools is for the library to take the form of a multimedia center and material collections, including laptop computers that are distributed from mobile units to "classroom clusters."

Schools are increasingly seen as community resources that go beyond the educational functions. Adult education and community events now take place on evenings, on weekends, and throughout the traditional vacation periods; therefore, the school day and week have been expanded. These uses are seen as ways of finding affordable methods of enhancing community-service resources by ensuring that a facility's utilization is maximized.

Indications are that the school building will probably increase in importance to the community, as its roles expand beyond that of merely providing a K–12 education for students during a school year. At the same time, modern technology means that today's schools, already far more complex than the relatively simple buildings of a few decades ago, will tend to be more fragile and consequently more vulnerable to nature's and society's threats unless special attention is paid to their design and construction.

The natural hazards will remain: earthquakes and tornadoes will continue to be, for some locations, a source of worry and fear. Besides protecting their occupants, schools in earthquake-prone regions are often used as post-earthquake shelters. In California, this is particularly appropriate because the State's Field Act, enacted in 1933, following the Long Beach earthquake, requires that public schools be designed by a licensed architect or engineer, their plans checked, and the construction on site inspected by staff of the Department of State of Architecture. Elsewhere, floods and high winds are a familiar threat that also must be addressed by knowledgeable design and good construction practice. Schools or designated areas within them, located in hurricane- and tornado-prone areas are increasingly being constructed to provide shelter for the occupants.

BUILDING DESIGN

Every building is unique and there is great variety in school design; however, the purpose of schools, their occupancy, their economic basis, and their role in the social scene mean that there are certain common features of schools that distinguish them from other building types.

The estimated value of the national public school inventory is well over $361.6 billion. Of the almost 15,000 local education agencies found throughout the United States, 41.9 percent are in small towns and rural areas, and enroll 30.4 percent of the students; 25.9 percent are in large towns and cities and enroll 30.7 percent of the students; and 32.2 percent of the education agencies are in suburban areas and enroll 39 percent of the students.

Over half of our school facilities are at least forty years old and, even with minor renovations, have passed their prime in terms of adaptability to modern teaching methods and tools (e.g., computers, in-class electronic information displays, and group learning activities). Almost all states require new construction once replacement costs reach a certain level (usually 60 percent). The most recent studies (completed at the close of the last decade) show a range of $100 billion to over $300 billion would be needed to bring our nation's schools into good teaching condition.

As new schools are built and older ones are remodeled, the following concerns have to be addressed:

- The building should provide for health, safety, and security.
- The learning environment should enhance teaching and learning and accommodate the needs of all learners.
- The learning environment should serve as the center of the community.
- The learning environment should result from a planning/design process that involves all stakeholders.
- The learning environment should allow for flexibility and adaptability to changing needs.
- The learning environment should make effective use of all available resources.

These principles lead, in turn, to a number of current design principles, including the following:

- Design for protection against natural hazards;
- Increased design attention to occupant security;
- Careful lighting design and increased use of day lighting and comfort control;
- Design for durability;
- Long-life/loose-fit approach: design for internal change and flexibility; and
- Design for sustainability, including energy efficiency and the use of "green" materials.

Design Basis Threat Worksheet

Objective: Identify & document who you are protecting your facility against

Purpose: This chart will capture who your potential adversaries are and what are their potential capabilities

Action 1: **Data Gathering**
In order to narrow your threats to the most realistic ones your 1st step would be a Data collection activity there are several sources you will need to draw upon, your local law enforcement (FBI, Sheriff etc), Other similar facility
type facilities, neighboring facilities etc.

Question *Is there historical data that would identify potential threats to your facility*
Question *Do the law enforcement authorities have specific intelligence that could narrow the list of potential threats*
Question *Talk to similar facilities--are they aware of specific risks to your facility Type*
Question *Talk to your local community, neighbors--have there been any recent events that could profile potential adversarial profiles.*

Action 2: Develop your adversary list based on the above Data gathering
A--List your potential adversaries
B--What is there motive to execute their attack
C--What the adversaries profile--how many, armed , on foot, equipped with hand tools
D--Identify what is the realistic probability of an attack

Vulnerability
Low--very slim chances
Medium -- could happen
High--Highly probable

Note : *Action 2D is critical because , the core team will need to establish what adversaries to protect the facility against, a scenario based analysis will be done for each realistic adversary type and actions will be taken to protect the facility against these specific threats.*
As a core team you will need to define the threshold for your DBT (for example ***Medium*** *probability and above)*

Example

	A	B	C	D		
	Adversary List/type	**Motive**	**Profile/ Characteristics**		Probability of Attack	
				Low	**Medium**	**High**
1	*Insider*	Disgruntled Employee	Alone, has access to critical assets			H
2	*Vandal*	Just for Kicks	Adolescent group of 3 to 5		M	
3	*Terrorist*	Media, sent on a mission	alone, suicide bomber	L		

Design Basis Threat Worksheet

Building Considerations

Clustered versus dispersed functions. There is a strong correlation between building functions and building layout and forms. Typically, the former dictates the other two. Depending on the site characteristics, the occupancy requirements, and other factors, school buildings may cluster key functions in one particular area or have these functions designed in a more dispersed manner. Both patterns have compelling strengths and weaknesses in terms of security.

Concentrating key functions in one place may create a target-rich environment and increase the risk of collateral impacts. Additionally, it increases the potential for the establishment of more single-point vulnerabilities.

If these areas become a target, the school may be closed for a substantial period of time, even if the attack is not severe and the rest of the school remains unharmed. However, grouping high-risk activities, concentrations of personnel, and critical functions into a cluster can help maximize standoff from the perimeter and create a "defensible space." This also helps to reduce the number of access and surveillance points, and minimize the size of the perimeter needed to protect the school areas.

In contrast, the dispersal of key functions reduces the risk that an attack on any one part of the site will impact the other parts. However, this could also have an isolating effect and reduce the effectiveness of onsite surveillance, increase the complexity of security systems and emergency response, and create a less defensible space.

Visitor areas and receiving/loading areas constitute a school's innermost line of defense, because they are the first places where people and materials enter the school building. Logically, they should be physically separated from other key functions, such as the main operational areas or where people concentrate.

School building orientation. The orientation of a school building can have significant impact on its performance, not only in terms of energy efficiency, but also the ability to protect occupants.

Vehicular and Pedestrian Circulation

The movement of people and materials into, through, and out of a school facility is determined by the design of its access, circulation, and parking systems. Such systems should be designed to maximize efficiency while minimizing conflicts between vehicle and pedestrian modes. Designers should begin with an understanding of the school's transportation requirements based on an analysis of how the school will be used. This includes studying the number and types of access points that are required, bus requirements, the parking volume needed, where users need to go to and from, and the modes of transportation they will use. Several aspects of transportation planning can impact security and are discussed below.

Roadway network design. Streets are generally designed to minimize travel time and maximize safety, with the end result typically being a straight path between two or more endpoints. Although a straight line may be the most efficient course, designers should use caution when orienting streets relative to school buildings requiring high protection. Designers should design a roadway system to minimize vehicle velocity, thus using the roadway itself as a protective measure. This is accomplished through the use of several strategies.

First, straight-line or perpendicular approaches to school buildings should not be used in a school at high risk, because these give vehicles the opportunity to gather the speed necessary to ram through protective barriers and crash into or penetrate buildings. Instead, approaches should be parallel to the facade, with berms, high curbs, appropriate trees, or other measures used to prevent vehicles from departing the roadway. A related technique for reducing vehicle speeds is the construction of serpentine (curving) roadways with tight-radius corners. Existing streets can be retrofitted with barriers, bollards, swing gates, or other measures to force vehicles to travel in a serpentine path. Again, high curbs and other measures should be installed to keep vehicles from departing the roadway in an effort to avoid these countermeasures. Less radical than these techniques are traffic calming strategies, which seek to use design measures to cue drivers as to the acceptable speed for an area. These include raised crosswalks, speed humps and speed tables, pavement treatments, bulbouts, and traffic circles. In addition to creating a more pedestrian-friendlier environment, which increases "eyes on the street" surveillance, designing roadways to physically limit speeds can have the added benefits of increasing safety and, subsequently, lowering liability. Designers should be aware, however, that many of these techniques can have detrimental effects for emergency response, including slowing response time, interfering with en route emergency medical treatment, and increasing the difficulty of maneuvering fire apparatus. They also may present problems for snow removal, and their outer ends should remain flat so that bicycles can proceed unimpeded.

Parking. Surface lots can be designed and placed to keep vehicles away from school buildings, but they can consume large amounts of land and, if constructed of impervious materials, can contribute greatly to storm water runoff. They can also be hazardous for pedestrians if dedicated pedestrian pathways are not provided.

Landscape design. Many landscape features can be used in school design to enhance security. Landscape design features should be used to create the level of protection without turning the school into a fortress. Elements, such as landforms, water features, and vegetation, are among the building blocks of attractive and welcoming spaces, and they can also be powerful tools for enhancing security. These features can be used not only to define or designate a space, but also to deter or prevent hostile surveillance or unauthorized access. Vegetative groupings and landforms can even provide some level of blast shielding. Stands of trees, earthen berms, and similar countermeasures generally cannot replace setbacks, but they can offer supplementary protection. However, landscaping can also have detrimental impacts for safety and security, and designers should consider the unique requirements of the school project to ensure that the landscape design elements they choose will be appropriate and effective. With careful selection, placement, and maintenance, landscape elements can provide visual screening that protects school gathering areas and other activities from surveillance without creating concealment for covert activity. However, dense vegetation in close proximity to a school building can screen illicit activity and should be avoided. Additionally, thick ground cover, such as English ivy, or vegetation over four inches tall, such as monkey grass, can be used to conceal bombs and other weapons; in setback clear zones, vegetation should be selected and maintained with eliminating concealment opportunities in mind. Similarly, measures to screen visually detractive components, such as transformers, trash compactors, and condensing units, should be designed to minimize concealment opportunities for people and weapons.

Urban design. Numerous urban design elements present opportunities to provide school security. The scale of the streetscape should be appropriate to its primary users, and it can be manipulated to increase the comfort level of desired users while creating a less inviting atmosphere for users with malicious intent. However, even at the pedestrian scale, certain operational requirements must be accommodated. For example, although efficient pedestrian and vehicle circulation systems are important for school functions and operations, they are also critical for emergency response, evacuation, and egress, and must be able to accommodate vehicles up to the largest fire apparatus in the community. Furthermore, despite an emphasis on downsizing the scale of the streetscape, it is critical to maintain the maximum standoff distance possible between vehicles and structures.

At the school perimeter, walls and fences used for space definition may be hardened to resist the impact of a weapon-laden truck; however, planters, bollards, or decorative boulders could accomplish the same objective in a much more aesthetically pleasing manner. Such an approach also creates permeability, which would allow pedestrians and cyclists to more easily move through the space. Landscape and urban design inherently define the "lines of sight" in a space. These techniques seek to deny aggressors a "line of sight" to a potential target, either from on or offsite.

Standoff Distance

The most cost-effective solution for mitigating explosive effects on school buildings is to keep explosives as far away from them as possible. The distance between an asset and a threat is referred to as the standoff distance. There is no ideal standoff distance; it is determined by the type of threat, the type of construction, and desired level of protection. The easiest and least costly opportunity for achieving appropriate levels of protection against terrorist threats is to incorporate sufficient standoff distance into school designs. Maximizing standoff distance also ensures that there is opportunity in the future to upgrade school buildings to meet increased threats or to accommodate higher levels of protection. Standoff distance must be coupled with appropriate building hardening to provide the necessary level of protection to the school.

For schools located in high-risk areas, additional considerations follow:

The first mode of site protection is to create "keep out zones" that can ensure a minimum guaranteed distance between an explosion (i.e., from a vehicle) and the school structure.

The perimeter line is the outermost line that can be protected by the security measures incorporated during the school design process. It is recommended that the perimeter line be located as far as is practical from the building exterior. Many vulnerable school buildings are located in urban areas where only the exterior wall of the building stands between the outside world and the building occupants. In this case, the options are obviously limited. Often, the perimeter line can be pushed out to the edge of the sidewalk by means of bollards, planters, and other obstacles. To push this line even further outward, restricting or eliminating parking along the curb often can be arranged with local authorities. In some extreme cases, elimination of loading zones and the closure of streets are an option.

"Keep out zones" can be achieved with perimeter barriers that cannot be compromised by vehicular ramming. A continuous line of security should be installed along the perimeter of the site to protect it from unscreened vehicles and to keep all vehicles as far away from the school as possible.

The following critical building components should be located away from main entrances, vehicle circulation, parking, and maintenance areas. If this is not possible, harden as appropriate.

- Emergency generator, including fuel systems, day tank, fire sprinkler, and water supply;
- Normal fuel storage;
- Telephone distribution and main switchgear;
- Fire pumps;
- Building control centers;
- Uninterrupted power supply (UPS) systems controlling critical functions;
- Main refrigeration systems if critical to building operation;
- Elevator machinery and controls;
- Shafts for stairs, elevators, and utilities; and
- Critical distribution feeders for emergency power.

Controlled-Access Zones

For a school at high risk, one method to attain the appropriate protection is with the creation of a controlled-access zone. These zones define minimum distances between a school building and potential threats through the installation of barriers (such as bollards, planters, fountains, walls, and fences). The barriers are designed to withstand assaults by terrorist vehicles; however, their placement must be designed to allow for access by fire and rescue vehicles in the event of an emergency. Selection of barriers is based on operational considerations related to vehicle access and parking. Good design principles for high-risk schools endorse the complete surround of a school building with a standoff zone that has perimeters set at distances that consider threat levels, desired level of protection, building construction, and land availability. Entry into the controlled area should only be through an entry control point.

Controlled-access zones may be exclusive or nonexclusive. An exclusive zone is the area surrounding a school building within the exclusive control of the building. Anyone entering an exclusive zone must have a purpose related to the building. A nonexclusive zone is either a public right-of-way or a particular area related to the main school building.

The following are some security considerations applicable to controlled-access zones and enforcement:

- Design and select barriers based on threat capabilities.
- If the limited availability of land precludes the creation of an exclusive zone, the use of screening surrounding the school building is an alternative.
- Design and locate security devices to establish consistent rhythm patterns within the site. Incorporate subtle and aesthetically pleasing security measures to reach the desired level of protection.
- Locate security measures so that they do not impede the free access to school public entrances or internal pedestrian flow. Miscellaneous decorative elements (e.g., flag

poles, fountains, pools, gardens, and similar features) may be located within access ways to slow movement or restrict access.

- Use a combination of barriers. Some barriers are fixed and obvious (fences and gates), while others are passive (sidewalks far away from buildings, curbs with grassy areas, etc.).
- Consider using landscape materials to create barriers that are soft and natural rather than manmade where physical barriers are required.
- Use vehicles as temporary physical barriers by placing them in front of buildings or across access roads.
- Maintain as much standoff distance as possible between potential vehicular bombs and the school building.
  - Provide traffic obstacles near entry control points to slow down traffic.
  - Consider vehicle barriers at building entries and drives.
  - Offset vehicle entrances from the direction of a vehicle's approach to force a reduction in speed.
  - Position gates and perimeter boundary fences outside the blast vulnerability envelope, when possible.
  - Provide a vehicle crash resistance system in the form of a low wall or earth berm, if the threat level warrants it.
- Design entry control points (if provided) to screen the building from vehicles entering it.
- Provide passive vehicle barriers to keep stationary vehicle bombs at a distance from the school building.
  - Use high curbs, low berms, shallow ditches, trees, shrubs, and other physical separations to keep stationary bombs at a distance.
  - Do not allow vehicles to park next to perimeter walls of the secured area. Consider using bollards or other devices to keep vehicles away.
- Provide adequate lighting to aid in threat detection in controlled-access zones.
- Use CCTV to control entry points, the site perimeter, and exclusive and nonexclusive zones.

Entry Control and Vehicular Access

In the case of a school, the objective of the design professional is to save lives by mitigating building damages and reducing the chances of a catastrophic collapse of the building at least until it is fully evacuated. Although there are many forms of attacks against a school, from the standpoint of school structural design, the vehicle bomb governs design because historically it has been used on multiple occasions by terrorists. Where a school perimeter barrier is required for security, it will be necessary to provide points of access through the perimeter for school users (i.e., students, faculty, staff, visitors, and service providers). An entry control point or guard building serves as the designated point of entry for site access. It provides a point for implementation of desired/required levels of screening and access control. The objective of the entry control point is to prevent unauthorized access to school grounds while maximizing the rate of authorized access by foot or vehicle. These measures will not be required for all schools; they may only be appropriate for schools considered at high risk. Designs should be flexible to allow implementation of increased security controls when schools are placed in high alert and easing of controls at lower threat levels. For a school considered to be at high risk, the following should be considered in the design of entry control points:

- Design entry roads to schools so that they do not provide direct or straight-line vehicular access to the main building. Route major corridors away from key school areas and functions.
- Design access points at an angle to oncoming streets so that is difficult for a vehicle to gain enough speed to break through them.
- Minimize the number of access roads and entrances into a school.
- Provide a drop-off/pick-up lane for buses only.
- Minimize the number of driveways or parking lots that students will have to walk across to get to the school building.
- Designate an entry to the school for commercial, service, and delivery vehicles, preferably away from key school areas and functions, whenever possible.
- Design the entry control point and guard building so that the authorization of approaching vehicles and occupants can be adequately assessed, and the safety of both gate guards and approaching vehicles can be maintained when a school is placed at high alert.
- Design (if they are required) traffic calming strategies and barriers (road alignment, retractable bollards, swing gates, or speed bumps) to control vehicle speed and slow incoming vehicles before they reach the gate so that entry control personnel have adequate time to respond to unauthorized activities.
- Provide inspection areas that are not visible to the public. Place appropriate landscape plantings to accomplish screening.
- Provide pullover lanes at site entry gates to check suspect vehicles. Also, provide a visitor/site personnel inspection area to inspect vehicles prior to allowing access to the school site.
- Consider providing a walkway and turnstile for pedestrians and a dedicated bicycle lane.

Signage

Signs are an important element of school security. They are meant to keep intruders out of restricted areas. Confusion over site circulation, parking, and entrance locations can contribute to a loss of site security. Signs should be provided offsite and at school entrances; there should be onsite directional, parking, and cautionary signs for students, faculty, staff, visitors, service vehicles, and pedestrians. Unless required, signs should not identify sensitive areas. A comprehensive signage plan should include the following:

- Prepare entry control procedures signs that explain current entry procedures for drivers and pedestrians.

- Prepare traffic regulatory and directional signs that control traffic flow and direct vehicles to specific appropriate points.
- Consider using street addresses or building numbers instead of detailed descriptive information inside the school grounds.
- Minimize the number of signs identifying high-risk areas; however, a significant number of warning signs should be erected to ensure that possible intruders are aware of entry into restricted areas.
- Minimize signs identifying critical utility complexes (e.g., power stations and significant gas, water, and sewer). Post easily understandable signs to minimize accidental entry by unauthorized visitors into critical areas.
- In areas where English is one of two or more languages commonly spoken, warning signs must contain the other language(s) in addition to English. The signs should be posted at intervals of no more than 100 feet and should not be mounted on fences equipped with intrusion-detection equipment.
- Locate variable message signs, which give information on site/organization special events and visitors, far inside site perimeters.

Parking

Parking restrictions can help to keep potential threats away from a school building. In urban settings, however, curbside or underground parking is often necessary and sometimes difficult to control. Mitigating the risks associated with parking requires creative design measures, including parking restrictions, perimeter buffer zones, barriers, structural hardening, and other architectural and engineering solutions. The following considerations may help designers to implement parking measures for schools that may be at high risk:

- Locate vehicle parking areas away from school buildings to minimize blast effects from potential vehicle bombs.
- Provide separate parking areas for students, faculty, staff, and visitors who may be going in and out during the school day. (This allows the main student parking lot to be closed off during the school day.)
- If possible, locate visitor or general public parking near, but not on, the site itself.
- Locate general parking in areas that provide the fewest security risks to school personnel.
- Consider one-way circulation within a school parking lot to facilitate monitoring for potential aggressors.
- Locate parking within view of occupied school buildings while maintaining standoff.
- Prohibit parking within the standoff zone.
- Request appropriate permits to restrict parking in the curb lane for school vehicles or key employee parking only where distance from the building to the nearest curb provides insufficient setback, and compensating design measures do not sufficiently protect the building from the assessed threat. If necessary, use structural features to prevent parking.
- Provide appropriate setback from parking on adjacent properties, if possible. Structural hardening may be required if the setback is insufficient. In new designs, it

may be possible to adjust the location of the school building on the site to provide adequate setback from adjacent properties.

- When establishing parking areas, provide emergency communications systems (e.g., intercom, and telephones) at readily identified, well-lighted, CCTV-monitored locations to permit direct contact with security personnel.
- Provide parking lots with CCTV cameras connected to the security system and adequate lighting capable of displaying and videotaping lot activity.
- If possible, prohibit parking beneath or within a school building.
- If parking beneath a building is unavoidable, limit access to the parking areas and ensure they are secure, well lighted, and free of places of concealment.
- Apply the following restrictions if parking within a school building is required:
  - Public parking with identification (ID) check;
  - School vehicles and school employees and students only; and
  - Selected school employees only, or those requiring security.

Loading Dock and Service Access

Loading docks and service access areas are commonly required for a school building and should typically be kept as invisible as possible. For this reason, special attention should be devoted to these service areas in order to avoid intruders. Design criteria for school loading docks and service access include the following:

- Separate by at least 50 feet, loading docks and shipping and receiving areas in any direction from utility rooms, utility mains, and service entrances, including electrical, telephone/data, fire detection/alarm systems, fire suppression water mains, cooling and heating mains, and so on.
- Locate loading docks so that vehicles will not be allowed under the building. If this is not possible, the service area should be hardened for blast. Loading dock design should limit damage to adjacent areas and vent explosive forces to the exterior of the building.
- If loading zones or drive-through areas are necessary, monitor them and restrict height to keep out large vehicles.
- Avoid having driveways within or under school buildings.
- Provide adequate design to prevent extreme damage to loading docks. The floor of the loading dock does not need to be designed for blast resistance if the area below is not occupied and or does not contain critical utilities. In certain cases, significant structural damage to the walls and ceiling of the loading dock may be acceptable; however, the areas adjacent to the loading dock should not experience severe structural damage or collapse.
- Provide signage to clearly mark separate entrances for deliveries.

Physical Security Lighting

Security lighting can be provided for overall school ground/building illumination and the perimeter to allow security personnel to maintain visual assessment during darkness. It may provide both a real and psychological deterrent for continuous or periodic observation. Lighting is relatively inexpensive to maintain and may reduce the need for security personnel while enhancing personal protection by reducing opportunities for concealment and surprise by potential attackers.

Provide sufficient lighting at entry control points to ensure adequate lighting for the area. Where practical, place lighting elements as high as possible to give a broader, more natural light distribution. This requires fewer poles (less hazardous to drivers) and is more aesthetically pleasing than standard lighting.

The type of site lighting system used depends on the school's overall security requirements. Four types of lighting are used for security lighting systems:

- **Continuous lighting** is the most common security lighting system. It consists of a series of fixed lights arranged to flood a given area continuously during darkness with overlapping cones of light.
- **Standby lighting** has a layout like continuous lighting; however, the lights are not continuously lit, but are either automatically or manually turned on when suspicious activity is detected or suspected by the security personnel or alarm systems.
- **Movable lighting** consists of manually operated, movable searchlights that may be lit during hours of darkness or only as needed. The system normally is used to supplement continuous or standby lighting.
- **Emergency lighting** is a backup power system of lighting that may duplicate any or all of the above systems. Its use is limited to times of power failure or other emergencies that render the normal system inoperative. It depends on an alternative power source, such as installed or portable generators or batteries. Consider emergency/backup power for security lighting as determined to be appropriate.

Site Utilities

Utility systems can suffer significant damage when subjected to the shock of an explosion. Some of these utilities may be critical for safely evacuating people from the school building. Their destruction could cause damage that is disproportionate to other building damage resulting from an explosion. To minimize the possibility of such hazards, apply the following measures:

- Where possible, provide underground, concealed, and protected utilities.
- Provide redundant utility systems (particularly electrical services) to support school security, life safety, and rescue functions.
- Consider quick connects for portable utility backup systems if redundant sources are not available.
- Prepare vulnerability assessments for all utility services to the school, including all utility lines, storm sewers, gas transmission lines, electricity transmission lines, and other utilities that may cross the site perimeter.
- Protect drinking water supplies from waterborne contaminants by securing access points, such as manholes. If warranted, maintain routine water testing to help detect waterborne contaminants.
- Minimize signs identifying critical utilities. Provide fencing to prevent unauthorized access and use landscape planting to conceal aboveground systems.

- Locate petroleum, oil, and lubricants storage tanks and operations buildings downslope from all other occupied school buildings. Locate fuel storage tanks at least 100 feet from buildings.
- Consider providing utility systems with redundant or loop service, particularly in the case of electrical systems. Where more than one source or service is not currently available, provisions should be made for future connections. In the interim, consider "quick connects" at the building for portable backup systems.
- Decentralize a school's communications resources, when possible; the use of multiple communication networks will strengthen the communications system's ability to withstand the effects of a terrorist attack.
- Place trash receptacles as far away from the building as possible; trash receptacles should not be placed within 30 feet of a building.
- Provide a school-wide public address system that extends from the interior to the exterior of buildings.
- Conceal and harden incoming utility systems within schools to provide blast protection, including burial or proper encasement wherever possible.
- Locate utility systems at least 50 feet from loading docks, front entrances, and parking areas.
- Route critical or fragile utilities so that they are not on exterior walls or on walls shared with mailrooms.
- Ensure that the redundant utilities are not collocated or do not run in the same chases. This minimizes the possibility that both sets of utilities will be adversely affected by a single event.
- Ensure backup systems are located away from the systems components for which they provide backup.
- Mount all overhead utilities and other fixtures weighing 31 pounds (14 kilograms) or more to minimize the likelihood that they will fall and injure school occupants. Design all equipment mountings to resist forces of 0.5 times the equipment weight in any direction and 1.5 times the equipment weight in the downward direction. This standard does not preclude the need to design equipment mountings for forces required by other criteria, such as seismic standards.
- Ensure that access to crawl spaces, utility tunnels, and other means of under school building access is controlled to limit opportunities for aggressors placing explosives underneath buildings.
- Screen, seal, or secure all utility penetrations of the site's perimeter to prevent their use as access points for unauthorized entry into the school site. If access is required for maintenance of utilities, secure all penetrations with screening, grating, latticework, or other similar devices.

Crime Prevention through Environmental Design (CPTED)

CPTED is a crime reduction technique that has several key elements applicable to the analysis of building function and site design against physical attack. It is used by architects, city planners, landscape and interior designers, and law enforcement with the objective of creating a climate of safety in a community by designing a physical environment that positively influences human behavior. Although CPTED principles are not incorporated into the assessment process presented in this primer, it is useful to briefly discuss CPTED because it is often entwined with terrorism protection measures.

CPTED concepts have been successfully applied in a wide variety of applications, including streets, parks, museums, government buildings, houses, and commercial complexes. The approach is particularly applicable to schools, where outdated facilities are common. Most schools in the United States were built thirty to sixty or more years ago. Security issues were almost nonexistent at the time, and technology was dramatically different. As a result, building designs are not always compatible with today's more security-conscious environment.

According to CPTED principles, depending upon purely conventional physical security measures (e.g., security guards and metal detectors) to correct objectionable student behavior may have its limitations. Although employing physical security measures will no doubt increase the level of physical security, in some cases physical security measures employed as stand-alone measures may lead to a more negative environment, thereby enhancing violence. In short, employing stand-alone physical security measures may fail to address the underlying behavioral patterns that adversely affect the school. CPTED analysis focuses on creating changes to the physical and social environment that will reinforce positive behavior.

CPTED builds on the following three strategies:

- Territoriality (using buildings, fences, pavement, sign, and landscaping to express ownership);
- Natural surveillance (placing physical features, activities, and people to maximize visibility); and
- Access control (the judicial placement of entrances, exits, fencing, landscaping, and lighting).

A CPTED analysis of a school evaluates crime rates, office-referral data, and school cohesiveness and stability, as well as core design shortcomings of the physical environment (e.g., blind hallways, uncontrolled entries, or abandoned areas that attract problem behavior). The application of CPTED principles starts with a threat and vulnerability analysis to determine the potential for attack and what needs to be protected. Protecting a school from physical attack by criminal behavior or terrorist activity, in many cases, only reflects a change in the level and types of threats. The CPTED process asks questions about territoriality, natural surveillance, and access control that can

- Increase the effort to combat crime or terrorism;
- Increase the risks associated with crime or terrorism;
- Reduce the rewards associated with crime or terrorism; and
- Remove the excuses as to why people do not comply with the rules and behave inappropriately.

The CPTED process provides direction to solve the challenges of crime and terrorism with organizational (people), mechanical (technology and hardware), and natural design (architecture and circulation flow) methods.

CPTED concepts can be integrated into expansion or reconstruction plans for existing buildings, as well as new buildings. Applying CPTED concepts from the beginning usually has minimal impact on costs, and the result is a safer school. Each school, district, and community should institute measures appropriate for their own circumstances because there is not a single solution that will fit all schools.

Many CPTED crime prevention techniques for a school complement conventional terrorism and physical attack prevention measures. For example, as part of the CPTED strategy of improving territoriality, schools are encouraged to direct all visitors through one entrance that offers contact with a receptionist who can determine the purpose of the visit and the destination, and provide sign-in/sign-out and an ID tag prior to building access. These CPTED measures are similar to and complement physical security entry control point stations.

However, in some cases, CPTED techniques can conflict with basic physical security principles. The CPTED strategy of natural surveillance calls for locating student parking in areas that allow ease of monitoring. A design that locates student parking close to the principal's office also reduces vehicle standoff and could create a vulnerability of the school structure to a vehicle bomb. In cases where CPTED techniques conflict with security principles, designers and school administrators should seek innovative solutions tailored to their unique situation.

Building Exterior

The exterior envelope of the school building is the most vulnerable to an exterior explosive threat because it is the part of the building closest to the weapon. It also is a critical line of defense for protecting the occupants of the school building.

The design philosophy to be used here is that simpler is better. Generally simple geometries, with minimal ornamentation (which may become flying debris during an explosion) are recommended. If ornamentation is used, it is recommended that it consists of a lightweight material, such as timber or plastic, which is less likely to become a projectile in the event of an explosion than, for example, brick, stone, or metal.

Exterior Wall Design

The exterior walls provide the first line of defense to prevent air-blast pressures and hazardous debris from entering the school building. At a minimum, the objective of design is to ensure that these members fail in a flexible mode rather than in a brittle mode, such as shear. The walls also need to be able to resist the loads transmitted by the windows and doors. Beyond ensuring a flexible failure mode, the exterior wall may be designed to resist the actual or reduced pressure levels of the defined threat. Special reinforcing and anchors should be provided around blast-resistant window and door frames.

Poured-in-place reinforced concrete will provide the highest level of protection, but solutions like precast concrete, reinforced CMU block, and metal studs may also be used to achieve lower levels of protection.

When designing schools in areas perceived as high risk, engineers and architects should consider the following recommendations:

- Substitute strengthened building walls and systems when standoff distances cannot be accommodated.
- Use ductile materials capable of very large plastic deformations without complete failure.
- Design exterior walls to resist the actual pressures and impulses acting on the exterior wall surfaces from the threats defined for the school building.
- Design exterior walls to withstand the dynamic reactions from the windows.
- Design exterior shear walls to resist the actual blast loads predicted from the threats specified. Consider shear walls that are essential to the lateral and vertical load-bearing system, and that also function as exterior walls, to be primary structures.
- Consider reinforced concrete wall systems in lieu of masonry or curtain walls to minimize flying debris in a blast.
- Reinforced wall panels can protect columns and assist in preventing progressive collapse, because the wall will assist in carrying the load of a damaged column.
- Give special consideration to construction types that reduce the potential for collapse where exterior walls are not designed for the full design loads.
- Consider use of sacrificial exterior wall panels to absorb blast.

Window Design

Window systems (e.g., glazing, frames, and anchorage to supporting walls) on the exterior facade of a school building should be designed to mitigate the hazardous effects of flying glass during an explosion. In an effort to protect school occupants, designers should integrate the features of the glass, connection of the glass to the frame (bite), and anchoring of the frame to the building structure to achieve a "balanced design." This means all the components should have compatible capacities and theoretically would all fail at the same pressure-pulse levels. In this way, the damage sequence and extent of damage are controlled. Although not all windows in a school can be designed to resist the full forces from very large explosive blasts, hardened window systems can provide significant protection for students, faculty, and staff. Preferred systems include thermally tempered glass with a security film installed on the interior surface and attached to the frame; laminated thermally tempered glass; laminated heat strengthened, or laminated annealed glass; and blast curtains. Glazing systems that do not provide any protection include untreated monolithic annealed or heat-strengthened glass and wire glass.

General guidelines for windows and glazing. General guidelines for windows and glazing include the following:

- Do not place windows adjacent to doors because, if the windows are broken, the door can be unlocked.
- In schools requiring high security, minimize the number and size of windows in a facade. If possible, limit the amount of glazed area in building facades to 15 percent. The amount of blast entering a space is directly proportional to the amount of opening on the facade.

- Consider using burglary- and ballistic-resistant glazing in high-risk school areas.
- Consider using laminated glass in place of conventional glass.
- Consider window safety laminate (such as mylar) or another FRF over glazing (properly installed) to reduce fragmentation.
- Consider placing guards, such as grills, screens, or meshwork, across window openings to protect against covert entry. Affix protective window guards firmly to the structure. Fire egress considerations must be judged against window guards.
- Position the operable section of a sliding window on the inside of the fixed section and secure it with a broomstick, metal rod, or similar device placed at the bottom of the track.
- Provide horizontal windows 6 feet above the finished floor to limit entry.
- Consider using steel window frames securely fastened or cement grouted to the surrounding structure.
- Minimize interior glazing in high-risk areas (e.g., lobbies, loading docks).

Doors

Door assemblies include the door, its frame, and anchorage to the building. As part of a balanced school design approach, exterior doors in high-risk buildings should be designed to withstand the maximum dynamic pressure and duration of the load from the design threat explosive blast. Other general door considerations for these types of buildings are as follows:

- Provide hollow steel doors or steel-clad doors with steel frames. Ensure the strength of the latch and frame anchor equals that of the door and frame.
- Consider blast-resistant doors for schools considered to be at high risk.
- Permit normal entry/egress through a limited number of doors, if possible, while accommodating emergency egress.
- Ensure that exterior doors into inhabited areas are open outward. In addition to facilitating egress, by doing so, the doors will seat into the door frames in response to an explosive blast, increasing the likelihood that the doors will not enter the school building as hazardous debris.
- Replace externally mounted locks and hasps with internally locking devices because the weakest part of most door assemblies is the latching component.
- Locate hinges on interior or use exterior security hinges to reduce their vulnerability.
- Install emergency exit doors so that they facilitate only exiting movement.
- Consider using solid doors or walls as a backup for glass doors in foyers.
- Strengthen and harden the upright surfaces of a door jamb into which the door fits.

Roof System Design

Control access to school roofs to minimize the possibility of aggressors placing explosives or CBR agents there or otherwise threatening school occupants or critical infrastructure. Designers should consider the following:

- For new school buildings, eliminate all external roof access by providing access from internal stairways or ladders, such as in mechanical rooms.

- For existing school buildings, eliminate external access, where possible, or make roof access ladders removable, retractable, or lockable.
- Provide pitched roofs to allow deflection of explosives.
- Make school roof access hatches secure from intruders.
- Consider designing buildings with a sacrificial sloping roof that is above a protected ceiling

INTERIOR SPACES

Systems and Equipment

Mechanical Systems

Mechanical systems design standards address limiting damage to critical infrastructure and protecting school building occupants against CBR threats. The primary goal of a mechanical system after a terrorist attack should be to continue to operate key life safety systems for school occupants. School designers should be aware that during an interior bombing event, smoke removal and control are of paramount importance. They should consider the fact that, if window glazing is hardened, a blast may not blow out windows, and smoke may be trapped in the school building. In the event of a blast, the available smoke removal system may be essential to smoke removal, particularly in large, open spaces. This equipment should be located away from high-risk areas, such as loading docks and garages. The system controls and power wiring to the equipment should be protected. The system should be connected to emergency power to provide smoke removal. Smoke removal equipment should be provided with stand-alone local control panels that can continue to individually function in the event the control wiring is severed from the main control system.

- Avoid mounting plumbing, electrical fixtures, or utility lines on the inside of exterior school walls. When this is unavoidable, mount fixtures on a separate wall at least six inches from the exterior wall face.
- Avoid placing plumbing on the top of the roof deck of the school building.
- Avoid suspending plumbing fixtures and piping from the ceiling.
- Reduce the number of utility openings, manholes, tunnels, air conditioning ducts, filters, and access panels into the school structure.
- Protect school building operational control areas and utility feeds to lessen the negative effects of a blast.
- Design operational redundancies to survive all kinds of attacks.
- Use lockable systems for school utility openings and manholes where appropriate. Infrequently used utility covers/manholes can be tack-welded as an inexpensive alternative to locking tamper-resistant covers.
- Place intakes at the highest practical level on the school building. For protection against malicious acts, the intakes should also be covered by screens so that objects cannot be tossed into the intakes or into air wells from the ground. Such screens should be sloped to allow thrown objects to roll or slide off the screen, away from

the intake. Many existing school buildings have air intakes that are located at or below ground level. For those that have wall-mounted or below-grade intakes close to the building, the intakes can be elevated by constructing a plenum or external shaft over the intake.

- Provide protection for existing school buildings with air intakes below grade, at ground level, or wall-mounted outside secure areas, with physical security measures (e.g., placing fencing, surveillance cameras, and motion detectors around the intakes to facilitate monitoring by security personnel). These measures can help prevent malicious acts, but are less effective than elevating the intakes, because ground level releases under certain conditions can enter the intakes from points outside the area fenced or under surveillance.
- Maintain physical security on mechanical rooms to prevent the direct introduction of hazardous materials into the system of ducts that distributes air to the school building. This includes locking and controlling the access to all mechanical rooms containing HVAC equipment.
- Restrict access to school building operation systems by outside personnel. To deter tampering by outside maintenance personnel, a school staff member should escort these individuals throughout their service visit and should visually inspect their work before final acceptance of the service. Alternatively, schools can ensure the reliability of prescreened service personnel from a trusted contractor.
- Restrict access to school building data. Information on building operations (including mechanical, electrical, vertical transport, fire and life safety, security system plans and schematics, and emergency operations procedures) should be controlled.
- Physically isolate school lobbies, mailrooms, loading docks, and storage areas from the rest of the building. These are areas where bulk quantities of CBR agents are likely to enter.
- Consider "shelter-in-place" rooms or areas in schools, where people can stay in the event of an outdoor release. The goal is to create areas where outdoor-air infiltration is very low. Usually such rooms will be in the inner part of the school in an area with no exterior windows, if possible. The rooms should have doors that are fairly effective at preventing airflow and should contain staging supplies, such as duct tape and plastic, to help further seal the areas from the hallways. Typically, restrooms are a bad choice, because they have exhaust ducts that lead directly to the outside. Opening and closing a conventional hinged door can pump large amounts of air into the room; if practical, replace the door with a code compliant sliding door to reduce this effect. Additionally, it may be possible to provide purified air to the safe area through modifications to the HVAC system.
- Ducted returns offer limited access points to introduce a CBR agent. The return vents can be placed in conspicuous locations throughout a school, reducing the risk of an agent being secretly introduced into the return system. Non-ducted return-air systems commonly use hallways or spaces above suspended ceilings as a return-air path or plenum. CBR agents introduced at any location above the suspended ceiling in a ceiling plenum return system will probably migrate back to the HVAC unit and be redistributed to occupied areas. Schools should be designed to minimize mixing between air-handling zones.

Electrical systems

The major security functions of the electrical system are to maintain power to essential school services, especially those required for life safety and evacuation. When designing a school building, architects and engineers should consider providing lighting and surveillance to deter criminal activities, and provide emergency communications. They should also consider the following recommendations:

- Emergency and normal electric panels, conduits, and switchgear should be installed separately, at different locations, and as far apart as possible. Electric distribution should also run at separate locations.
- Emergency generators should be located away from loading docks, entrances, and parking. More secure locations include the roof, protected grade level, and protected interior areas.
- Main fuel storage for generators should be located away from loading docks, entrances, and parking. Access should be restricted and protected (e.g., locks on caps and seals).
- Fuel tanks should be mounted near the generator, given the same protection as the generator, and sized to store an appropriate amount of fuel. A battery and/or UPS could be a viable alternative for a smaller school.
- Conduits and lines should be installed outside to allow a trailer-mounted generator to connect to the school's electrical system. If tertiary power is required, other methods include generators and feeders from alternative substations.
- Emergency power should be provided for emergency lighting in school restrooms, egress routes, and any meeting room without windows.
- School building access points should be illuminated to aid in threat detection.
- Self-contained battery lighting should be provided in stairwells and for exit signs.
- Suspending electrical conduits from the ceiling should be avoided.
- Adequate lighting of perimeters and parking areas should be provided to aid in visual surveillance and to support the use of CCTV.

Fire protection systems

The fire protection system inside the school building should maintain life safety protection after an incident and allow for safe evacuation of the building when appropriate. Although fire protection systems are designed to perform well during fires, they are not traditionally designed to survive bomb blasts. Fire protection system considerations include the following:

- A school's fire protection water system should be protected from single-point failure in case of a blast event. The incoming line should be encased, buried, or located 50 feet away from high-risk areas. The interior mains should be looped and sectionalized.
- To increase the reliability of the fire protection system, a dual pump arrangement should be considered, with one electric pump and one diesel pump. The pumps should be located away from each other.
- All school security locking arrangements on doors used for egress must comply with requirements of NFPA 101, Life Safety Code.

Communications systems

- **Redundant communications**. The school should have a second telephone service to maintain communications in case of an incident. A base radio communications system with antenna should be installed in the stairwell, and portable sets distributed on floors. This is the preferred alternative.
- **Radio telemetry**. Distributed antennas could be located throughout the school facility if required for emergency communications through wireless transmission of data.
- **Alarm and information systems**. Alarm and information systems should not be collected and mounted in a single conduit, or even collocated. Circuits to various parts of the school building should be installed in at least two directions and/or risers. Low voltage signal and control copper conductors should not share conduits with high voltage power conductors. Fiber-optic conductors are generally preferred over copper.
- **Empty conduits**. Empty conduits and power outlets can be provided for possible future installation of security control equipment.
- **Mass notification**. All inhabited school buildings should have a timely means to notify occupants of threats and instruct them what to do in response to those threats. School buildings should have a capability to provide real-time notification of building occupants and people in the immediate vicinity of the building during emergency situations. The information relayed should be specific enough to determine the appropriate response actions.

Physical security systems

Physical security is defined as that part of security concerned with physical measures designed to safeguard people and to prevent unauthorized access to equipment, certain areas of the school building, and key documents. These days, all security operations face new and complex physical security challenges across the full spectrum of operations.

Although security technologies are not the answer to all school security problems, if applied appropriately, they can enhance security, free up administrators for more appropriate work, and sometimes can save money. At a noneducational building, a typical approach to physical security is

Detect > Delay > Respond

For example, if someone is breaking into a school facility, it is necessary to have a means of detection so that information can be provided to appropriate authorities. Next, the intruder must be delayed as long as possible so that the response force may arrive. Finally, someone, such as the police, must respond to the incident to catch the intruder. For a school, the National Institute for Justice has recommended expanding this model to

Deter > Detect > Delay > Respond/Investigate > Consequences

This educational approach emphasizes deterrence by instituting measures both before and after an event to convince perpetrators not to do whatever they are considering. Although, the efficacy of deterrence for high-risk terrorist tactics is questionable, schools have a broad spectrum of threats to consider with any physical security system. Schools are also unique in that they generally have the authority and opportunity to establish consequences for incidents that involve students and occur on school grounds.

Obviously, when considering any physical security measure for a school, it is important to balance its use with the risk of creating a "bunker" or "prison" atmosphere. Our children must have an environment that is conducive to learning, but above all, they must have one that is safe.

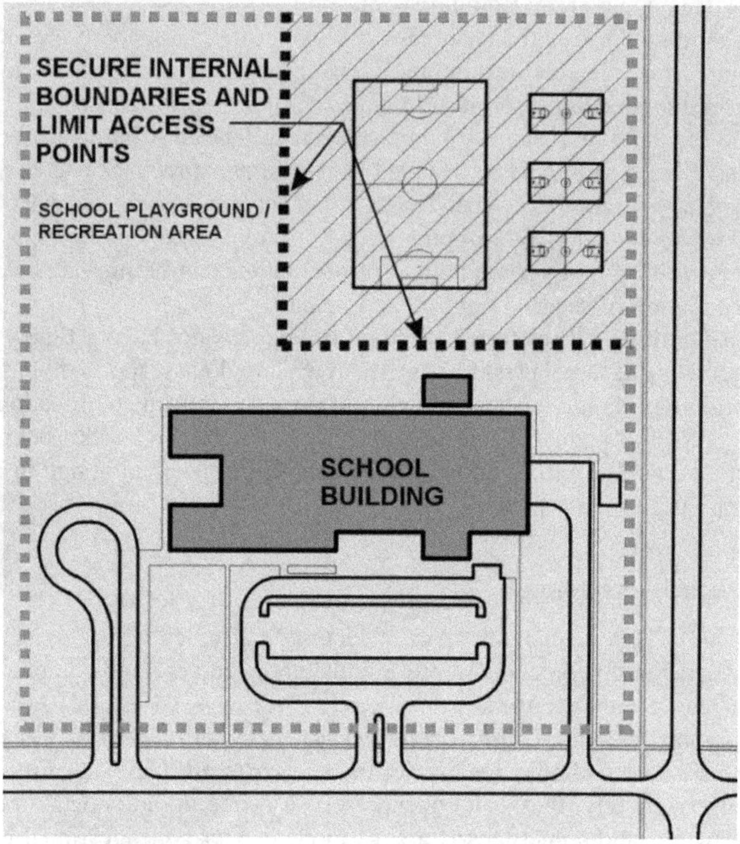

Know the Areas You Need to Protect

Natural Access Control

- Clearly establish and define school property lines.
- Secure the site perimeter and limit access with selected entry points.
- Create boundaries that delineate public, semi-public, semi-private, and private spaces.
- Establish clearly defined and secure boundaries between joint-use facilities and school.
- Where feasible, utilize fencing that does not permit footholds in order to deter unauthorized access.

Natural Surveillance

- Avoid blocking lines of sight with fencing, signage, and landscaping.
- Locate site entry points in areas of high visibility where they can be easily observed and monitored by staff and students in the course of their normal activities.

Territorial Integrity

- Maintain school property to help establish pride of place and a sense of ownership.
- Encourage activities on school grounds that promote community ownership and territorial integrity.

Management

- Utilize fencing materials that resist graffiti.

The location of a school and its relationship to its immediate surroundings is critical in evaluating safety and security concerns. While there is evidence showing that, by and large, schools and community college campuses tend to be safer places than the neighborhoods in which they are located, crime rates and types of crimes in schools nevertheless are affected by their surrounding environment. Despite this, each campus is unique and there are no formulas that can be applied to all. However, there are overarching design principles that are applicable to virtually all locations whether rural, suburban, or inner city urban.

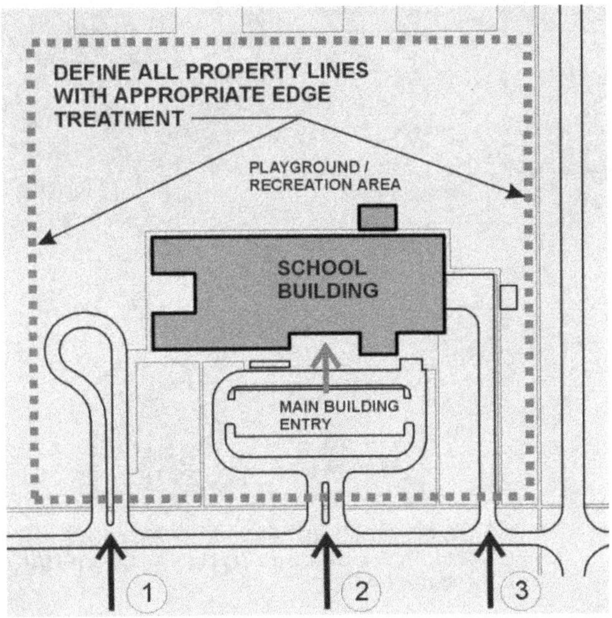

Secure Perimeter with Limited Entry/Access Points
(1) Bus (2) Parent Drop-off (3) Service

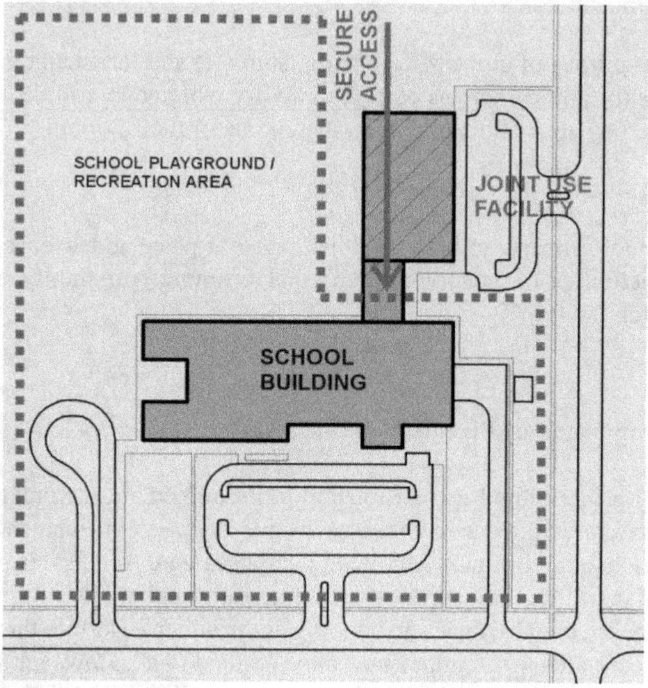

Surveillance

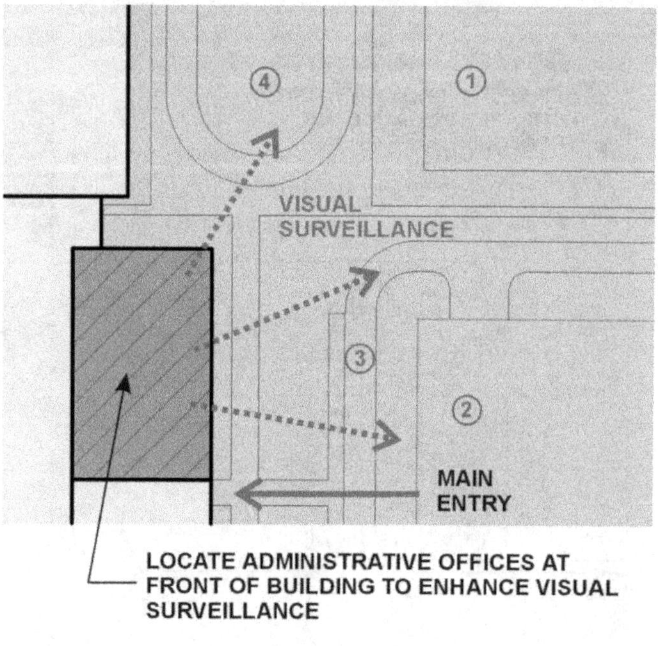

Administration Area with Views to Primary Access Points
(1) Zoned Parking (2) Visitor Parking (3) Parent Drop-off (4) Bus Drop-off

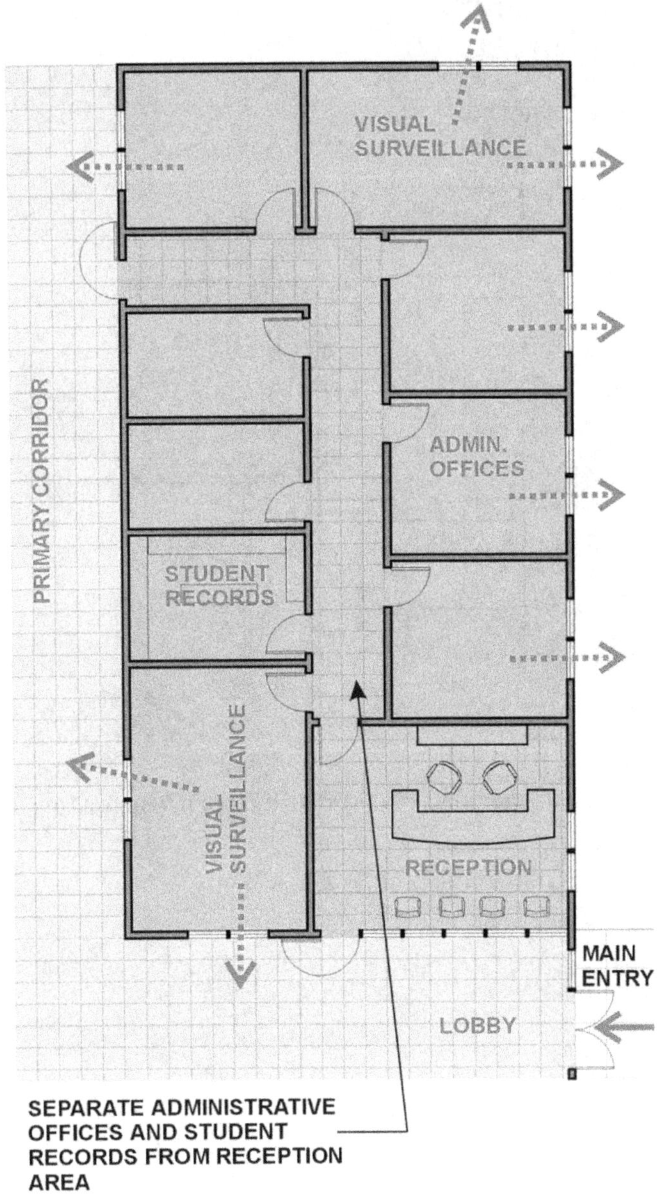
VISUAL
SURVEILLANCE
PRIMARY CORRIDOR
ADMIN.
OFFICES
STUDENT
RECORDS
VISUAL
SURVEILLANCE
RECEPTION
MAIN
ENTRY
LOBBY
SEPARATE ADMINISTRATIVE
OFFICES AND STUDENT
RECORDS FROM RECEPTION
AREA

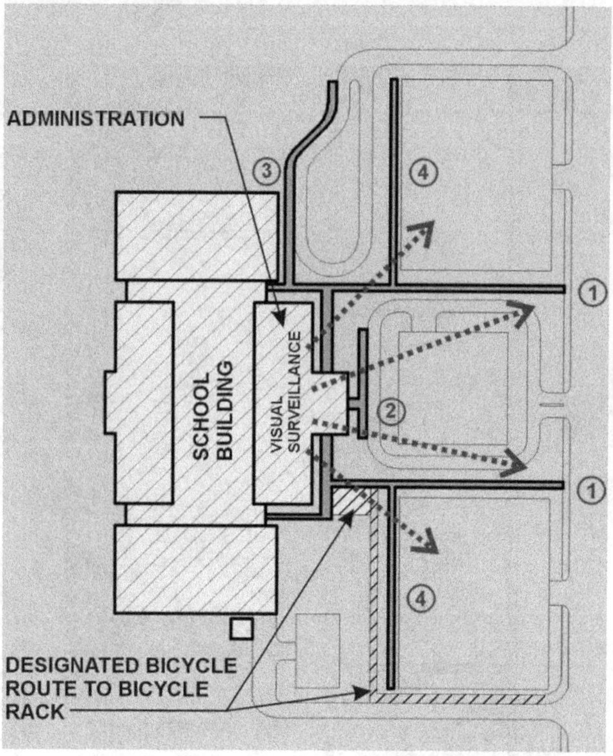

(1) Pedestrian Entry Point (2) Path from Parent Drop-off (3) Path from Bus Drop-off (4) Path from Zoned Parking

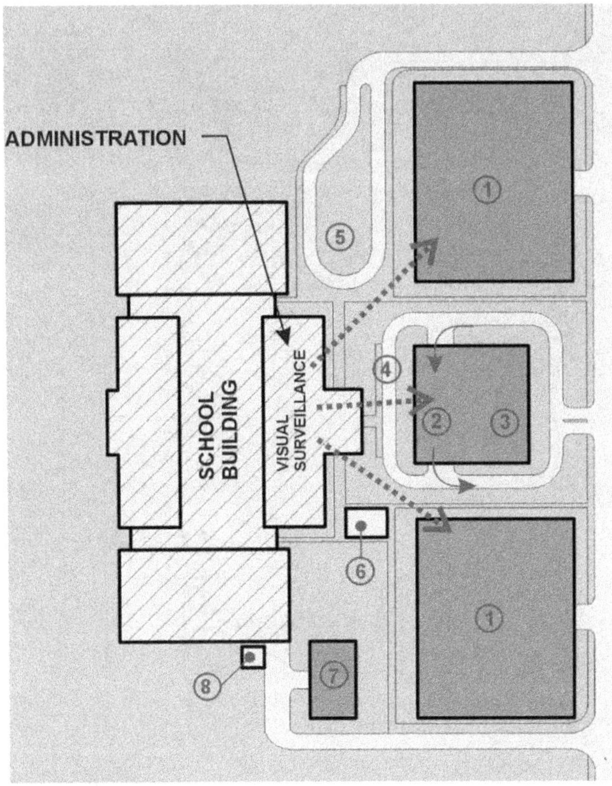

Parking Schematic for a Large School Facility
(1) Zoned Parking (2) Visitor Parking (3) Faculty/Staff Parking (4) Parent Drop-off (5) Bus Drop-off (6) Bicycle Rack (7) Kitchen/Custodial Staff Parking (8) Secure Fenced Area for Off-hour Deliveries

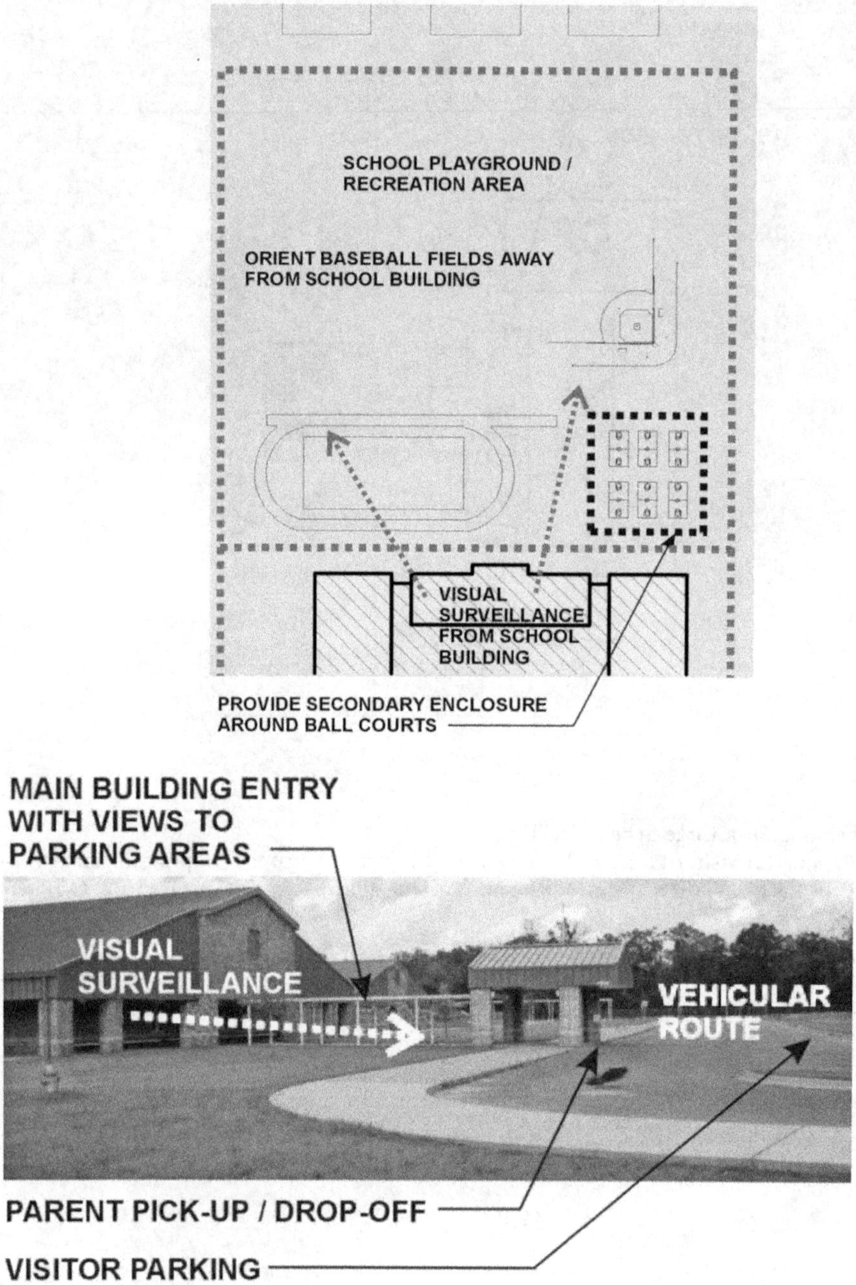

Monitoring the Areas You Need to Protect

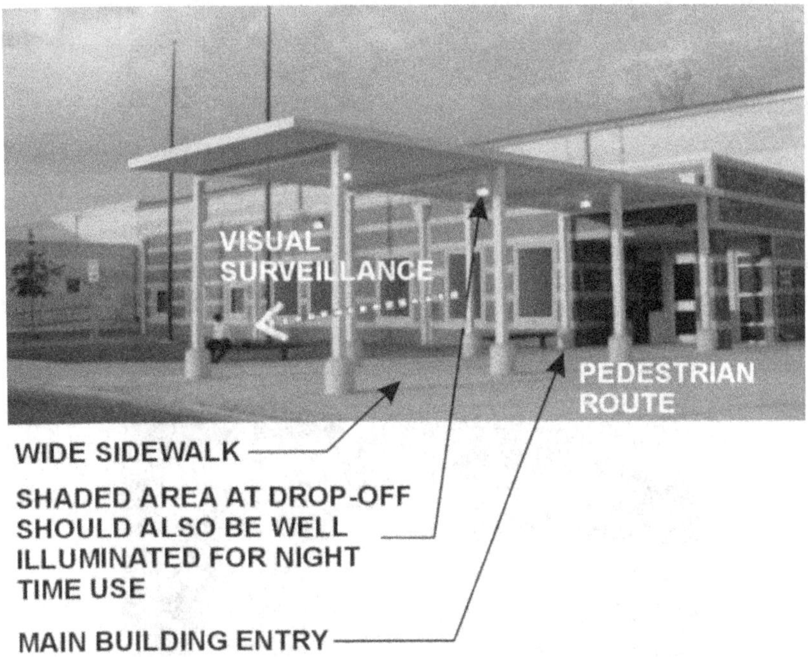
VISUAL
SURVEILLANCE
PEDESTRIAN
ROUTE
WIDE SIDEWALK
SHADED AREA AT DROP-OFF
SHOULD ALSO BE WELL
ILLUMINATED FOR NIGHT
TIME USE
MAIN BUILDING ENTRY

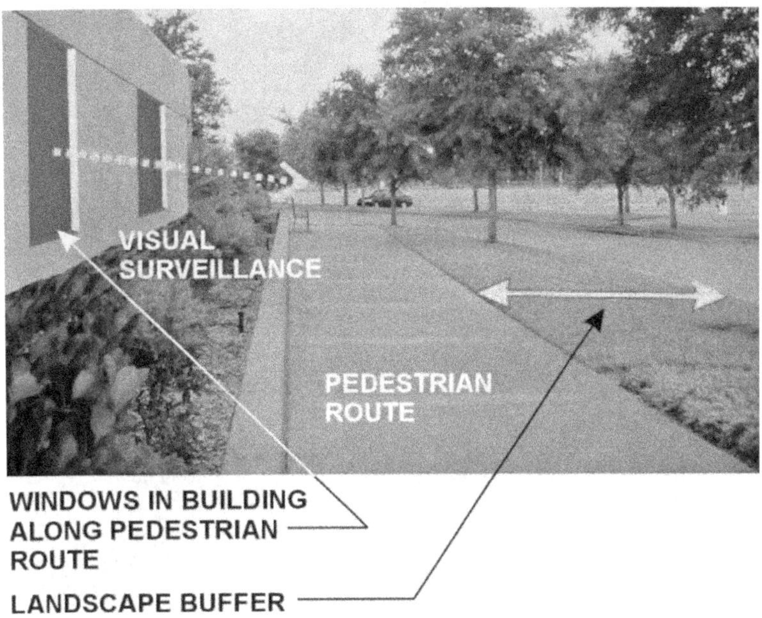
VISUAL
SURVEILLANCE
PEDESTRIAN
ROUTE
WINDOWS IN BUILDING
ALONG PEDESTRIAN
ROUTE
LANDSCAPE BUFFER

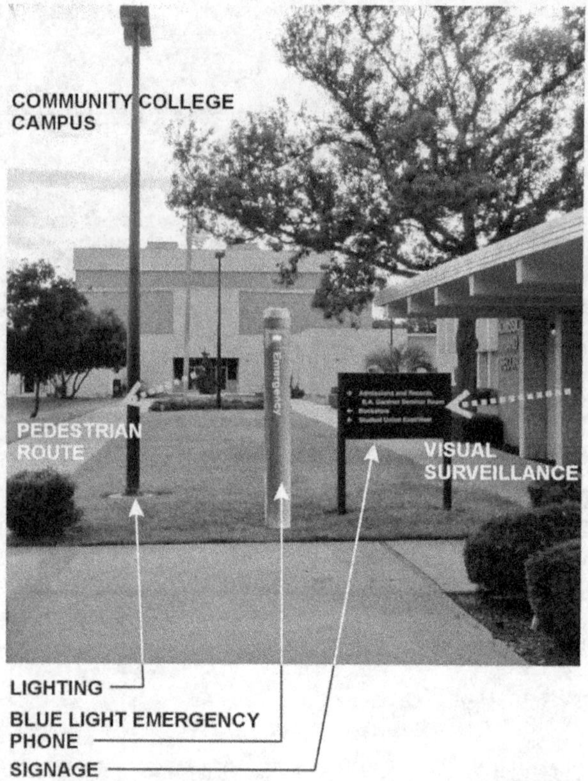
COMMUNITY COLLEGE
CAMPUS
Emergency
PEDESTRIAN
ROUTE
VISUAL
SURVEILLANCE
LIGHTING
BLUE LIGHT EMERGENCY
PHONE
SIGNAGE

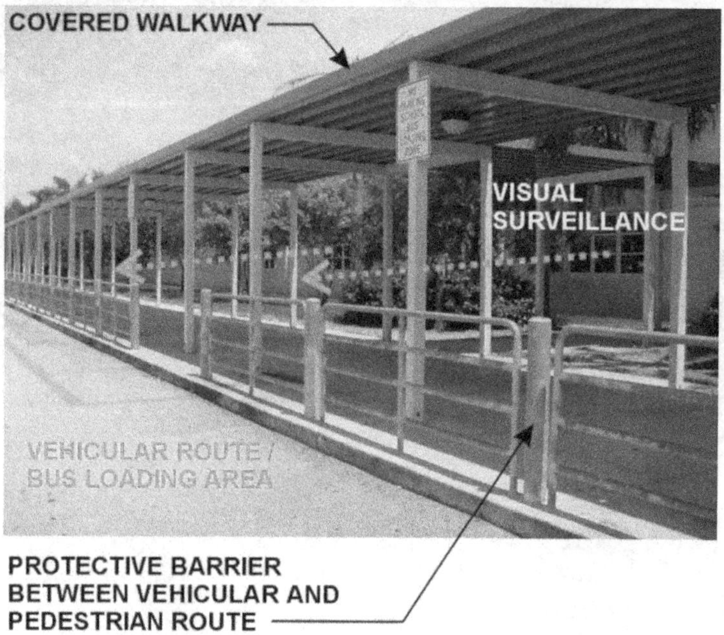
COVERED WALKWAY
VISUAL
SURVEILLANCE
VEHICULAR ROUTE /
BUS LOADING AREA
PROTECTIVE BARRIER
BETWEEN VEHICULAR AND
PEDESTRIAN ROUTE

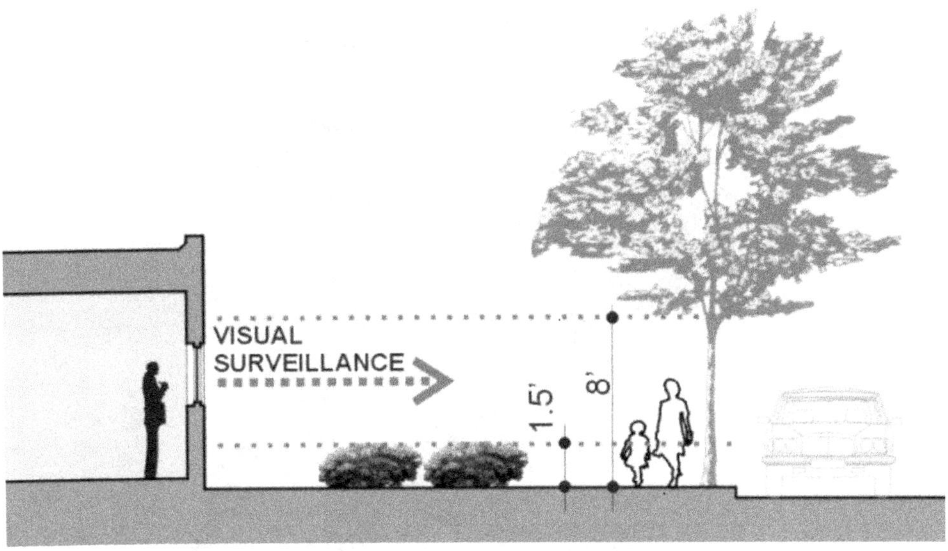

Natural Access Control

- Restrict external access to parking areas to a limited number of controlled entrances.
- Close unsupervised entrances during low-use times to reinforce the idea that access and parking are for school business only.
- Provide clear signage and posted rules as to who is allowed to use parking facilities and when they are allowed to do so.

- Locate visitor parking directly adjacent to main entry and administration.
- Provide adequate space adjacent to the building for emergency vehicles.
- Establish separate vehicular circulation routes to service and delivery areas, visitors entry, bus drop-off, student parking, and staff parking.
- Prohibit through traffic on school campuses.
- Provide a secure caged area for off-hour deliveries.

Natural Surveillance

- Locate parking areas in close proximity to school building or activity areas to facilitate natural surveillance.
- Provide windows in classrooms and administration areas that overlook parking areas.
- Provide adequate lighting in drop-off zones and parking areas.
- Utilize zoned parking in limited controlled areas when appropriate.
- Locate bus loading area so that it is visible to administration or adjacent to areas of surveillance.
- Locate access to public transportation in areas that promote natural surveillance.

Territorial Integrity

- Differentiate and identify parking spaces for students, faculty, staff, and visitors.
- Provide designated primary routes and parking lots for after-hours use when applicable.
- Clearly mark transition(s) from public street onto school entry routes and into parking areas.

Separate Pedestrian and Vehicle Traffic

PROVIDE PHYSICAL BUFFER BETWEEN PEDESTRIAN AND VEHICULAR ROUTES

WIDE SIDEWALKS DESIGNED FOR PEAK PERIODS OF USE

CONTROL GATES USED TO DESIGNATE ZONED PARKING IN LARGE HIGH SCHOOL PARKING LOTS

Provide external emergency call points and adequate lighting

External Barriers

UTILIZE SEE THROUGH FENCING IN
RECREATIONAL AREAS

GATE WITH LATCH

WALLS ON THREE
SIDES

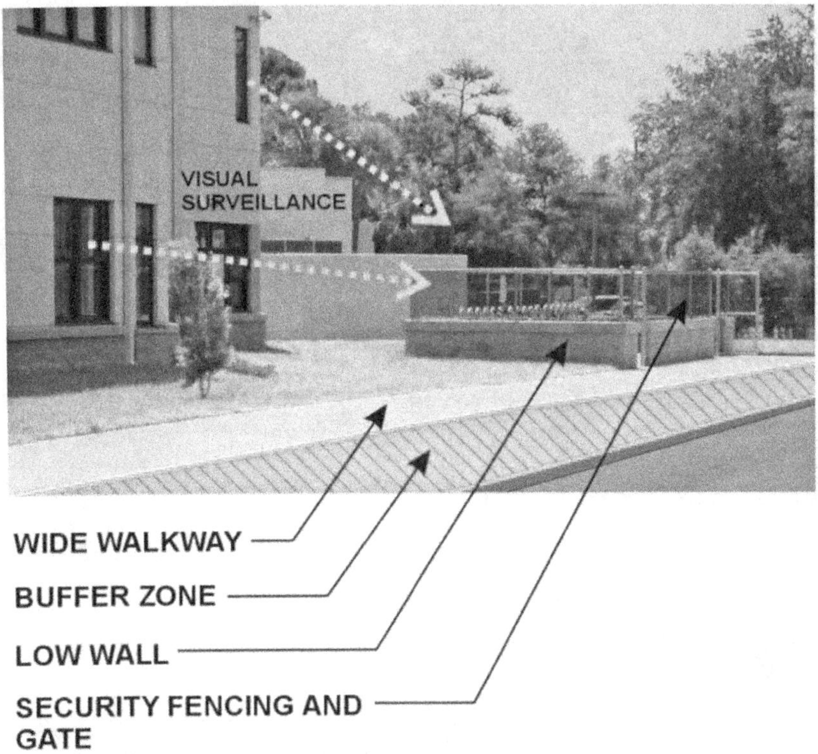
VISUAL
SURVEILLANCE
WIDE WALKWAY
BUFFER ZONE
LOW WALL
SECURITY FENCING AND
GATE

SAFETY ALERT
All visitors and parents
MUST
sign in at the office between
8:00am and 4:00pm.
Thank You
USE STRATEGICALLY PLACED
SIGNAGE WITH BOLD GRAPHICS

Signage

DESIGN SIGNAGE TO
PREVENT CONCEALMENT

ENCLOSE RETENTION POND
WITH FENCING AS
REQUIRED BY CODE

STORMWATER SYSTEM COMPONENTS
UTILIZE GRATING OR METAL REBAR TO PREVENT ACCESS

VISUAL SURVEILLANCE
COORDINATE LOCATION OF SITE UTILITIES TO PROMOTE NATURAL SURVEILLANCE
FENCE SITE UTILITIES TO LIMIT ACCESS

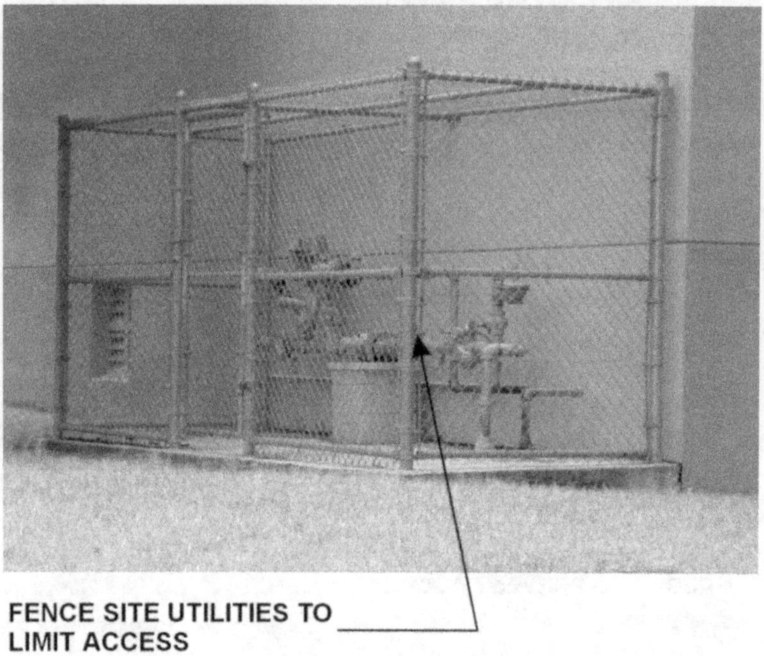
FENCE SITE UTILITIES TO
LIMIT ACCESS

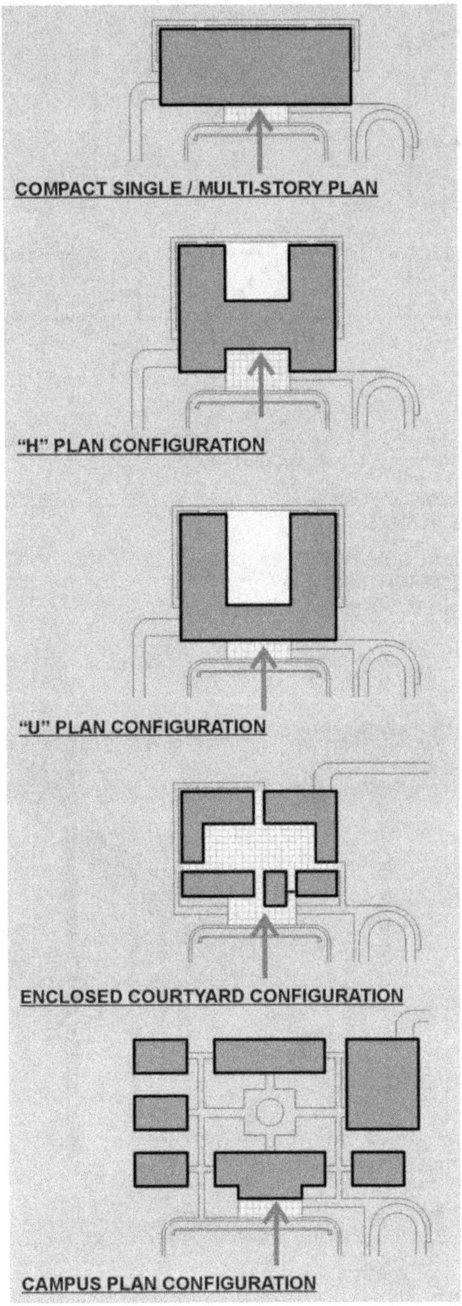

Control Access

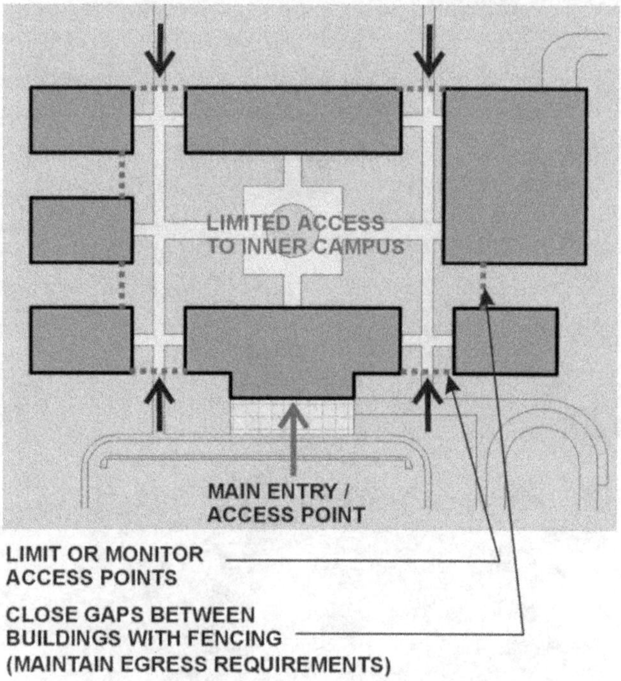
LIMITED ACCESS
TO INNER CAMPUS
MAIN ENTRY /
ACCESS POINT
LIMIT OR MONITOR
ACCESS POINTS
CLOSE GAPS BETWEEN
BUILDINGS WITH FENCING
(MAINTAIN EGRESS REQUIREMENTS)

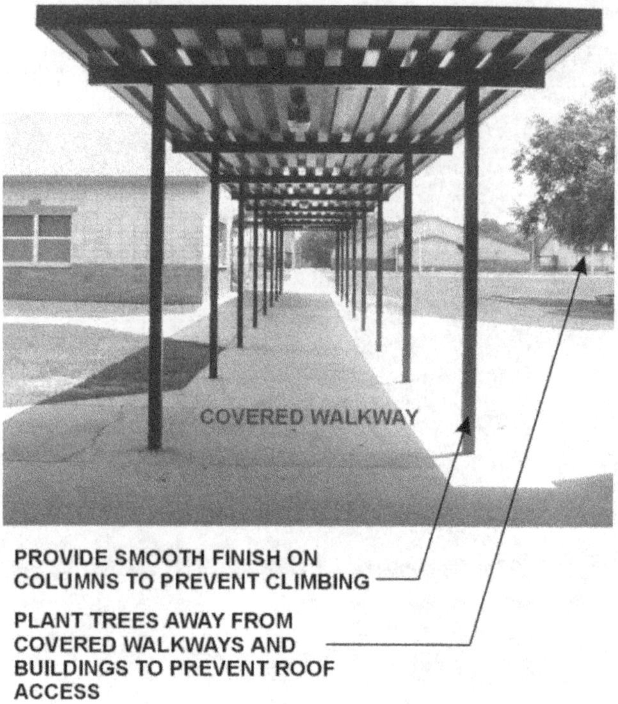
COVERED WALKWAY
PROVIDE SMOOTH FINISH ON
COLUMNS TO PREVENT CLIMBING
PLANT TREES AWAY FROM
COVERED WALKWAYS AND
BUILDINGS TO PREVENT ROOF
ACCESS

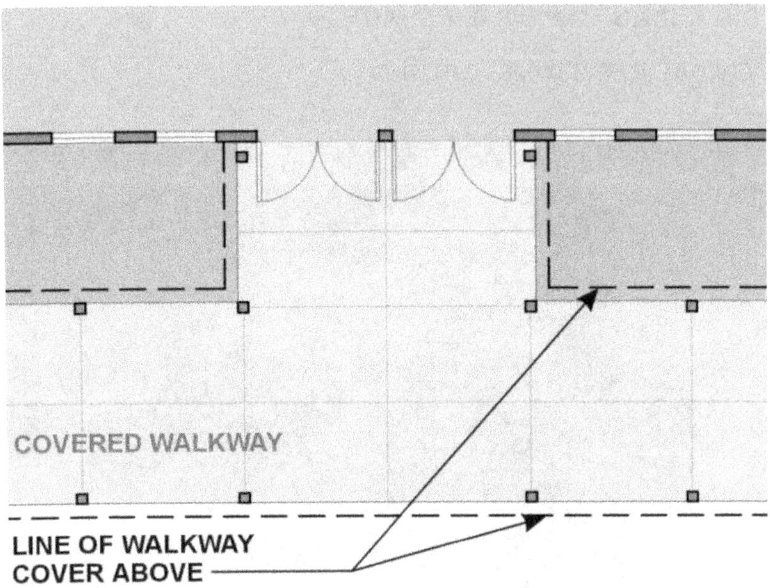
COVERED WALKWAY
LINE OF WALKWAY
COVER ABOVE

MAIN POINT OF ENTRY
LOCATE ADMINISTRATION NEAR
MAIN POINT OF ENTRY AND
PROVIDE EXTENSIVE GLAZING TO
ENHANCE VISUAL SURVEILLANCE

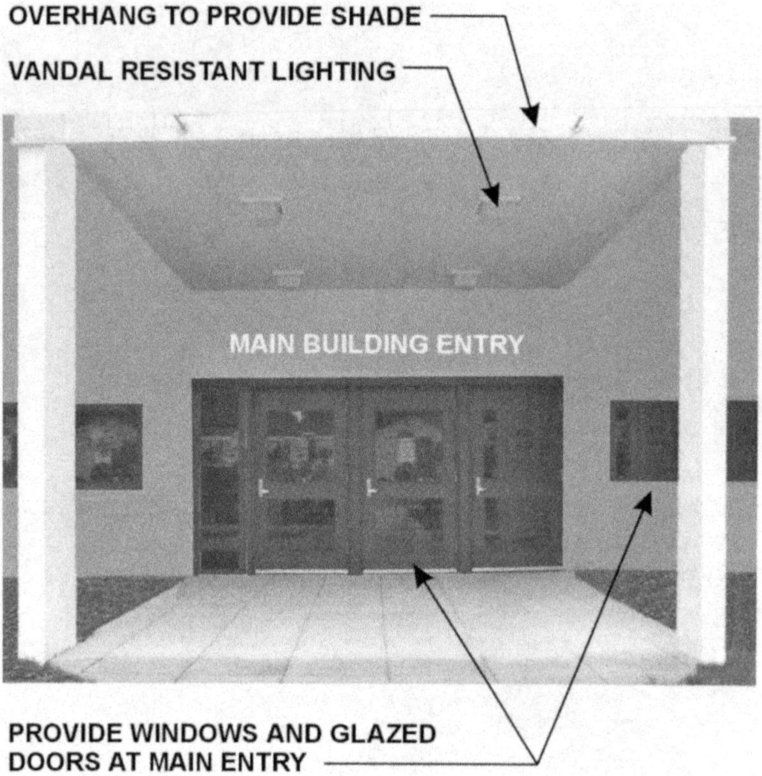
OVERHANG TO PROVIDE SHADE
VANDAL RESISTANT LIGHTING
MAIN BUILDING ENTRY
PROVIDE WINDOWS AND GLAZED DOORS AT MAIN ENTRY

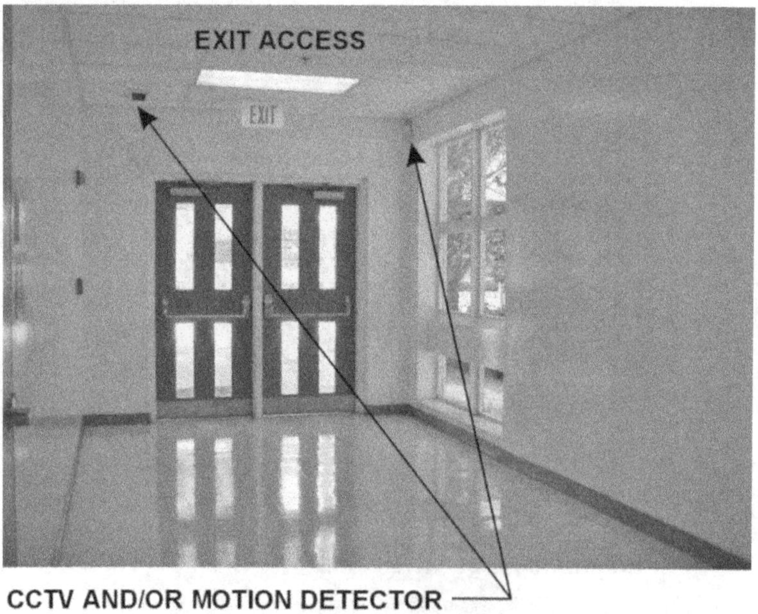
EXIT ACCESS
EXIT
CCTV AND/OR MOTION DETECTOR

SECURITY GATE FOR AFTER HOUR ACCESS CONTROL

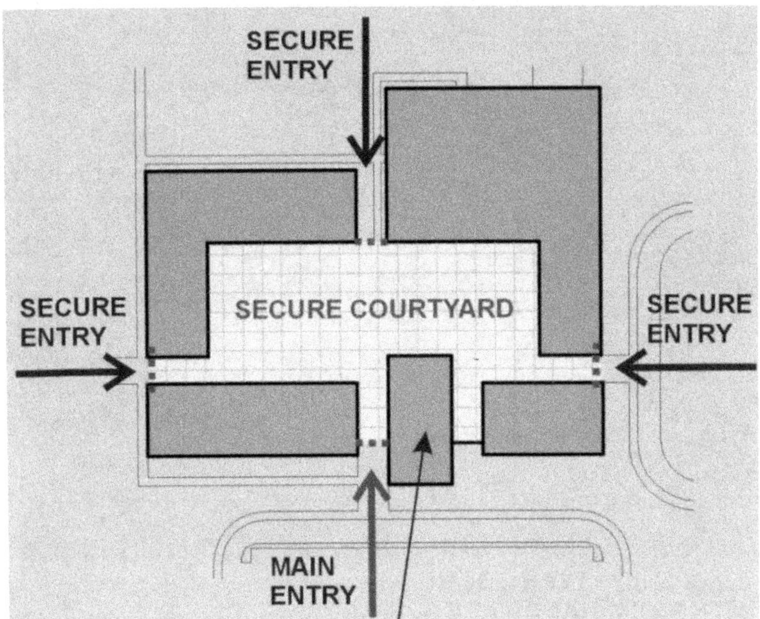

ADMINISTRATION

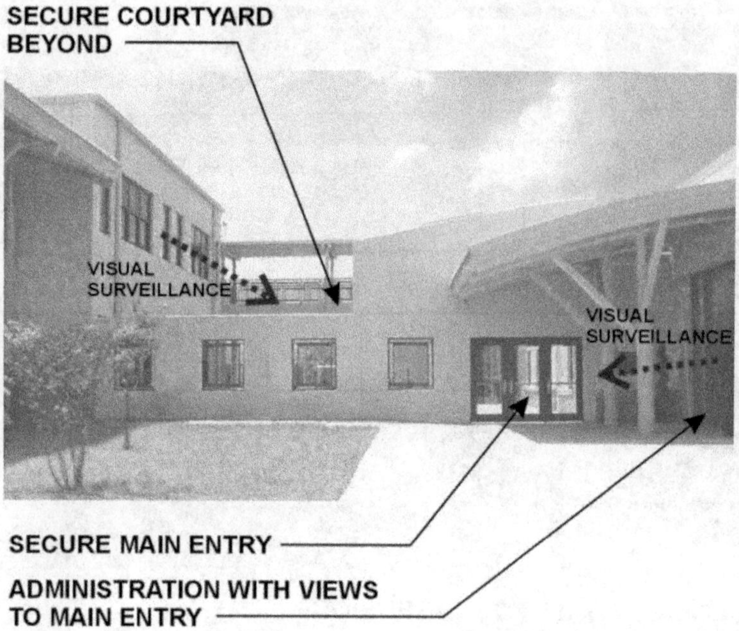
SECURE COURTYARD BEYOND
VISUAL SURVEILLANCE
VISUAL SURVEILLANCE
SECURE MAIN ENTRY
ADMINISTRATION WITH VIEWS TO MAIN ENTRY

EXIT
SECURE COURTYARD SIDE
"EXIT ONLY" TYPE DOORS WITH PANIC HARDWARE AND CLOSURE
ROLL-UP TYPE SECURITY GRILLE

SHADED
COURTYARD
AREA
LANDSCAPING
SEATING
GROUND SURFACE
TREATMENT

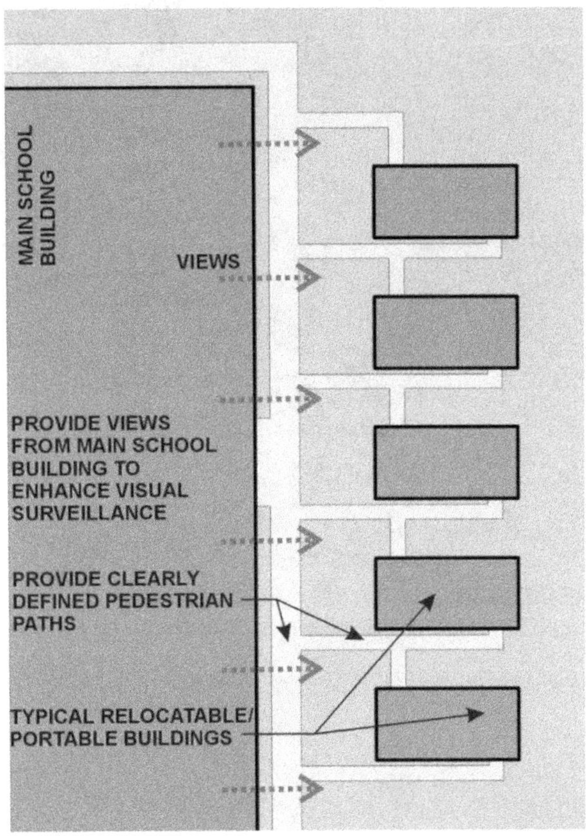
MAIN SCHOOL
BUILDING
VIEWS
PROVIDE VIEWS
FROM MAIN SCHOOL
BUILDING TO
ENHANCE VISUAL
SURVEILLANCE
PROVIDE CLEARLY
DEFINED PEDESTRIAN
PATHS
TYPICAL RELOCATABLE/
PORTABLE BUILDINGS

SCREEN SPACES UNDER
RELOCATABLE/ PORTABLE
BUILDINGS TO PREVENT ACCESS

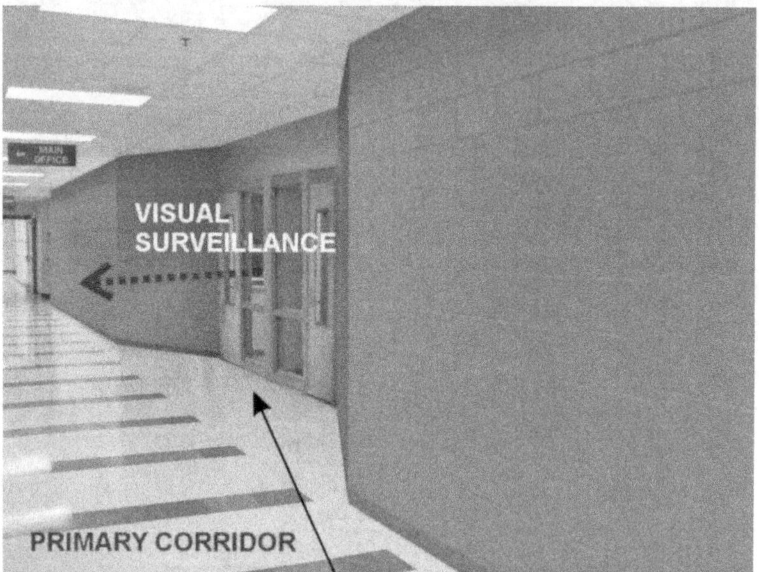

WIDE RECESS WITH
CHAMFERED CORNERS
ENHANCE VISUAL
SURVEILLANCE

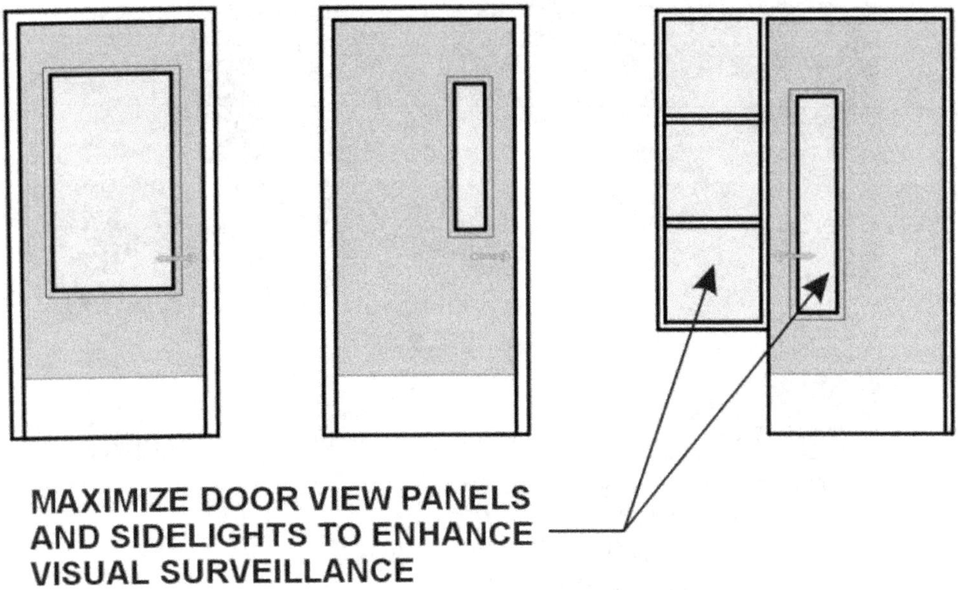
MAXIMIZE DOOR VIEW PANELS
AND SIDELIGHTS TO ENHANCE
VISUAL SURVEILLANCE

EXIT
UTILIZE EXTENSIVE GLAZING AT
MAIN ENTRY TO ENHANCE VISUAL
SURVEILLANCE TO PARKING
LOTS AND PEDESTRIAN ROUTES

STOREFRONT TYPE GLAZING ENHANCES NATURAL SURVEILLANCE FOR COURTYARDS AND OPEN SPACES ON CAMPUS PLAN DESIGNS

HURRICANE SHUTTERS DETER AFTER HOUR ACCESS

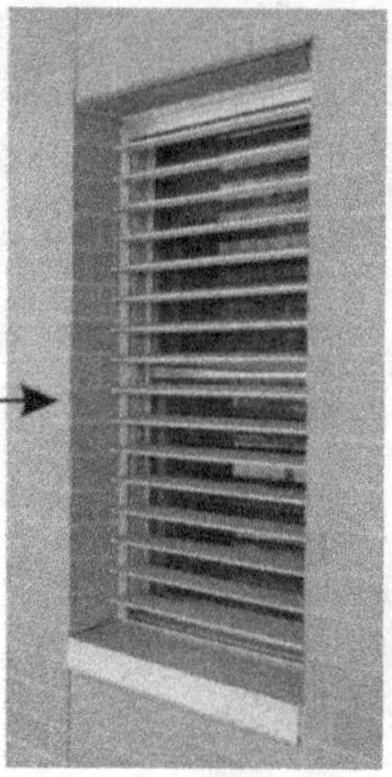

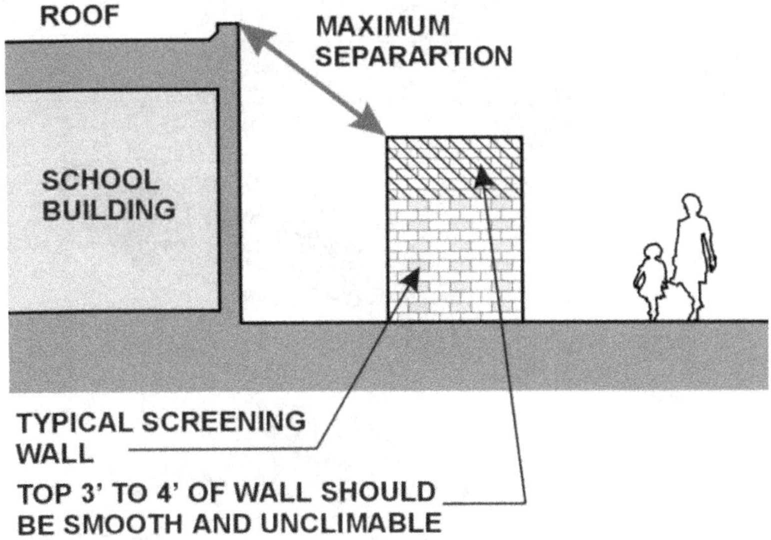
ROOF
MAXIMUM
SEPARARTION
SCHOOL
BUILDING
TYPICAL SCREENING
WALL
TOP 3' TO 4' OF WALL SHOULD
BE SMOOTH AND UNCLIMABLE

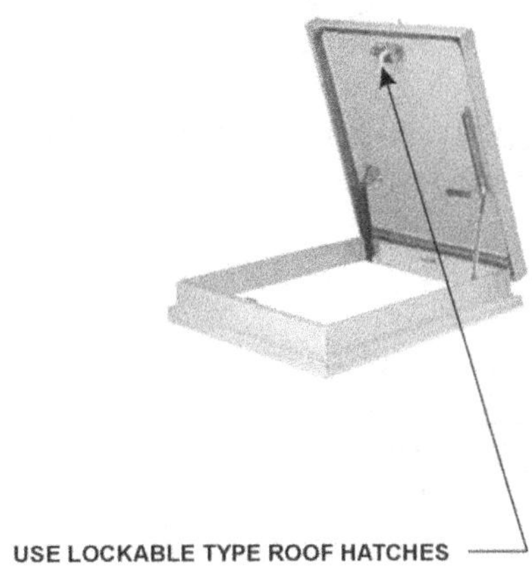
USE LOCKABLE TYPE ROOF HATCHES

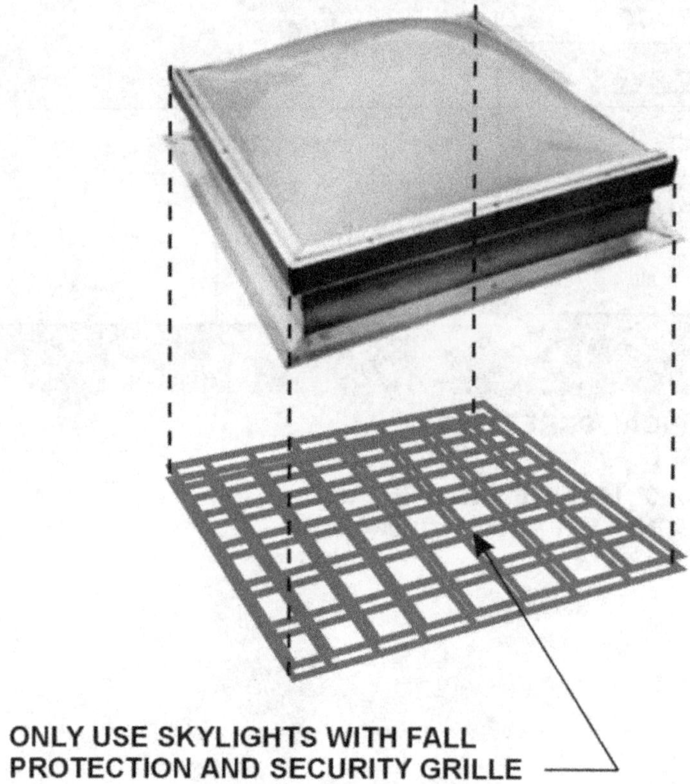
ONLY USE SKYLIGHTS WITH FALL
PROTECTION AND SECURITY GRILLE

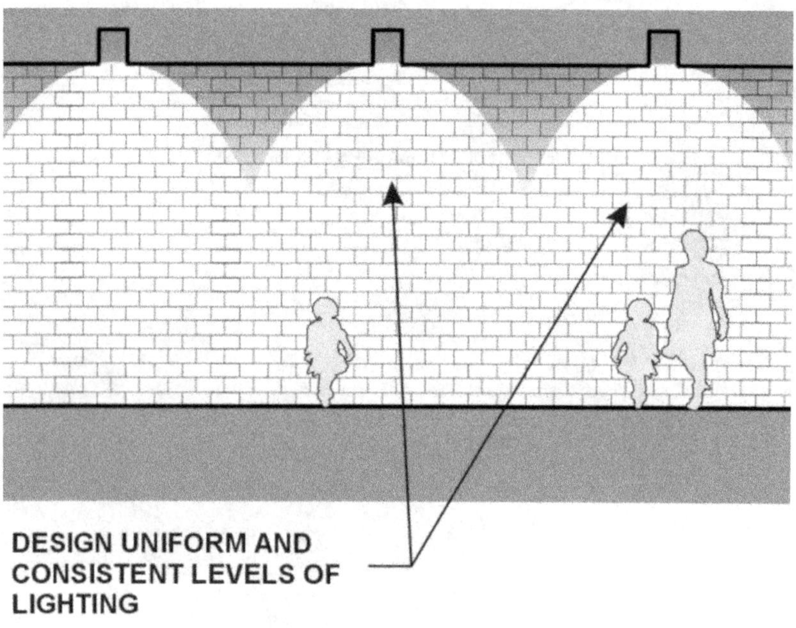
DESIGN UNIFORM AND
CONSISTENT LEVELS OF
LIGHTING

BUILDING
ENTRY
DESIGN UNIFORM CONSISTENT
LEVELS OF LIGHTING
ENTRY CANOPY

COVERED
WALKWAY
UTILIZE VANDAL RESISTANT TYPE
LIGHT FIXTURES

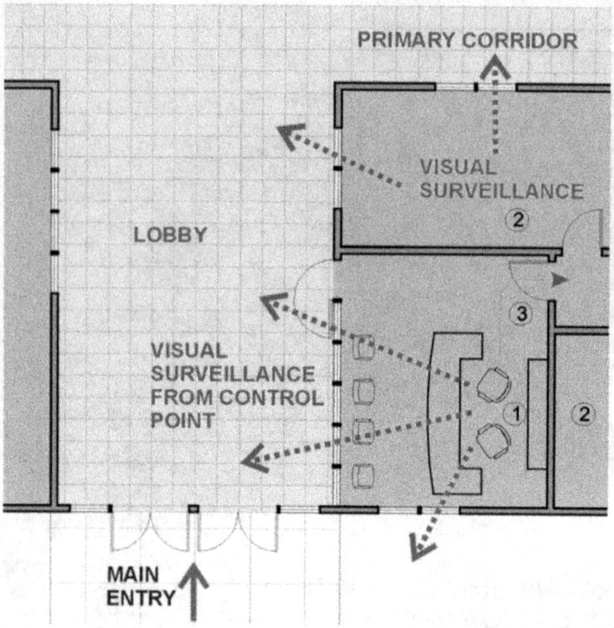

Control Traffic Flow

1) Control point 2) Admin. Area 3) Escape Route

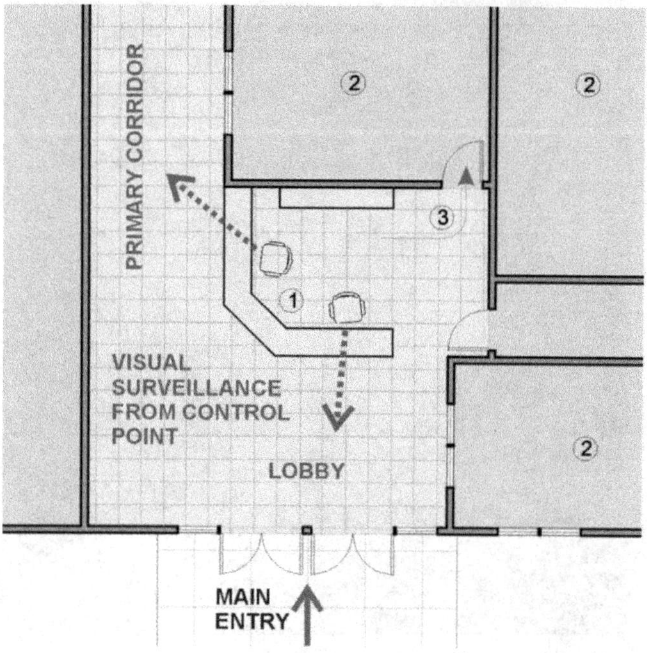

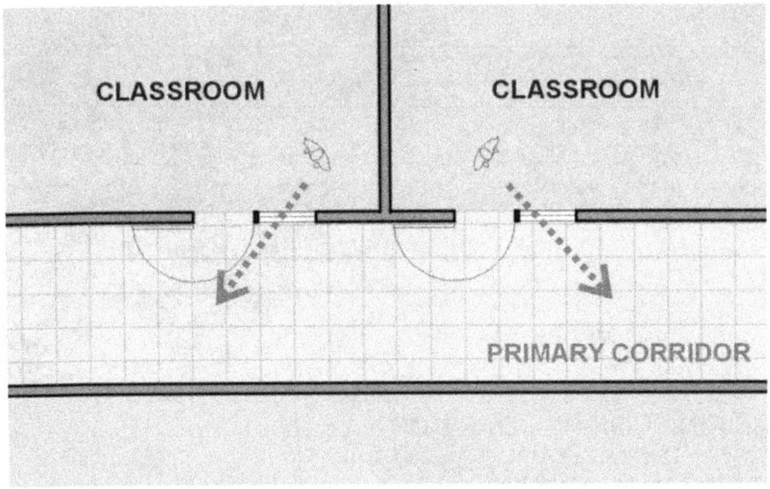
CLASSROOM
CLASSROOM
PRIMARY CORRIDOR

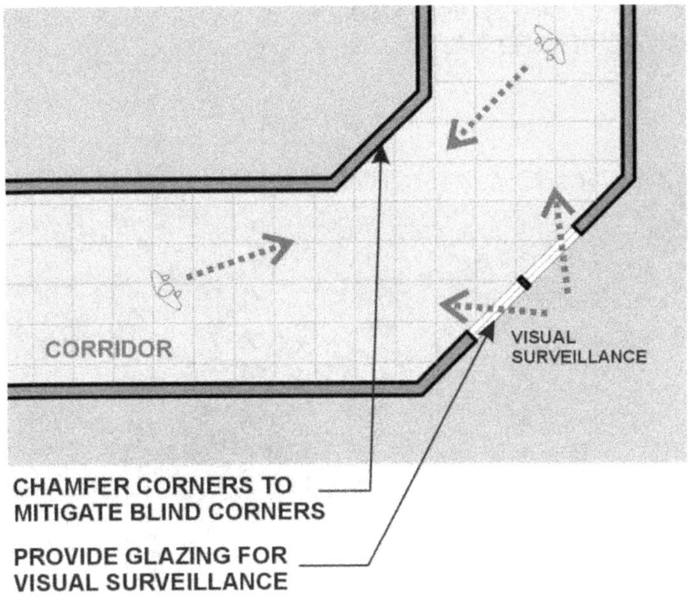
CORRIDOR
VISUAL
SURVEILLANCE
CHAMFER CORNERS TO
MITIGATE BLIND CORNERS
PROVIDE GLAZING FOR
VISUAL SURVEILLANCE

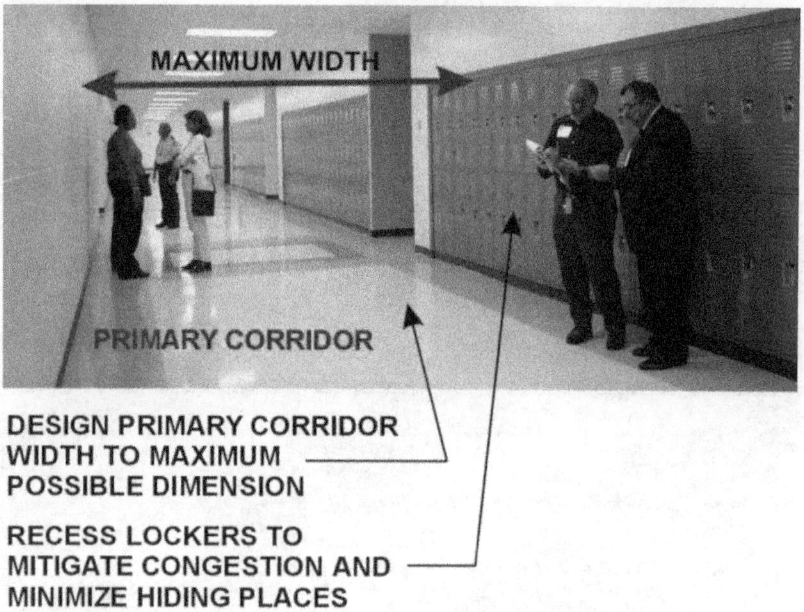
MAXIMUM WIDTH
PRIMARY CORRIDOR
DESIGN PRIMARY CORRIDOR WIDTH TO MAXIMUM POSSIBLE DIMENSION
RECESS LOCKERS TO MITIGATE CONGESTION AND MINIMIZE HIDING PLACES

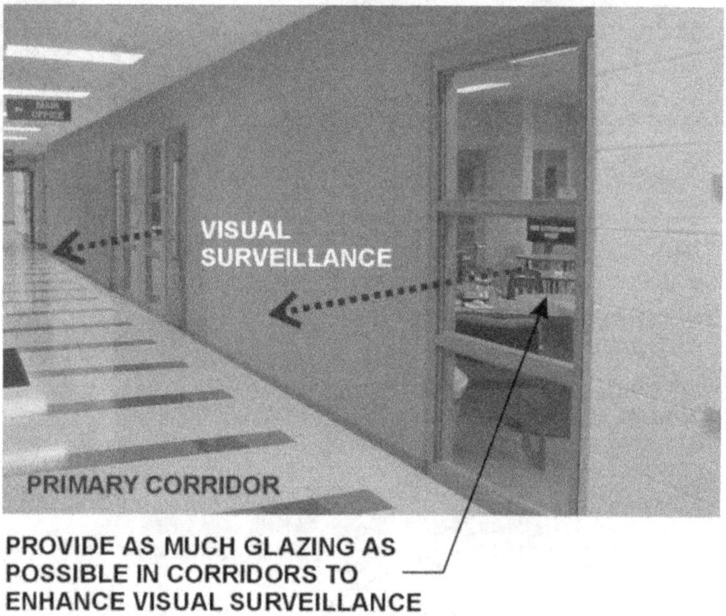
VISUAL SURVEILLANCE
PRIMARY CORRIDOR
PROVIDE AS MUCH GLAZING AS POSSIBLE IN CORRIDORS TO ENHANCE VISUAL SURVEILLANCE

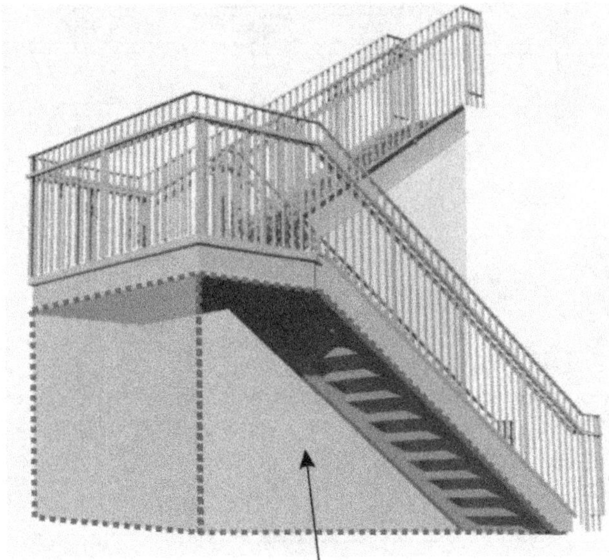

ENCLOSE AREA UNDER STAIRS

OPEN TYPE HANDRAIL AND GUARDRAIL SYSTEMS ALLOW FOR SURVEILLANCE

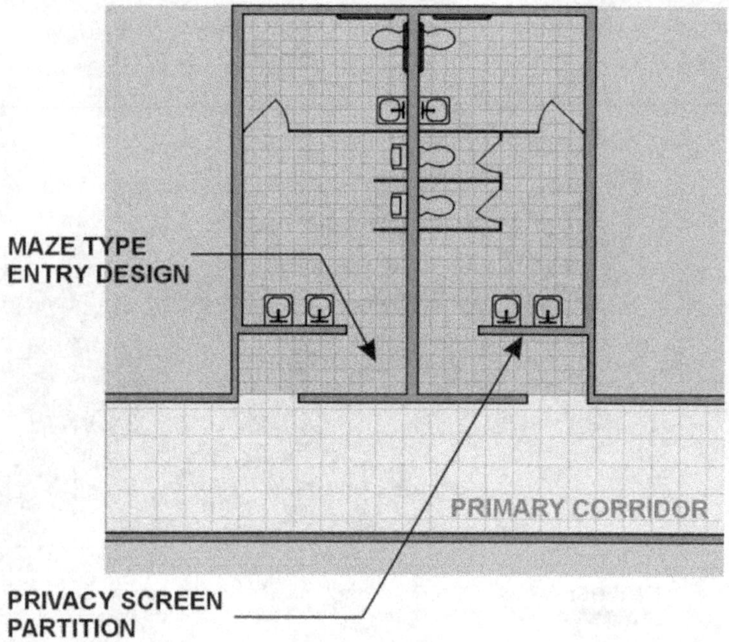

PROVIDE FIRE RATED WALL AND CEILING ASSEMBLIES AS REQUIRED IN GROUP TOILET ROOMS DESIGNED WITHOUT ENTRY DOORS

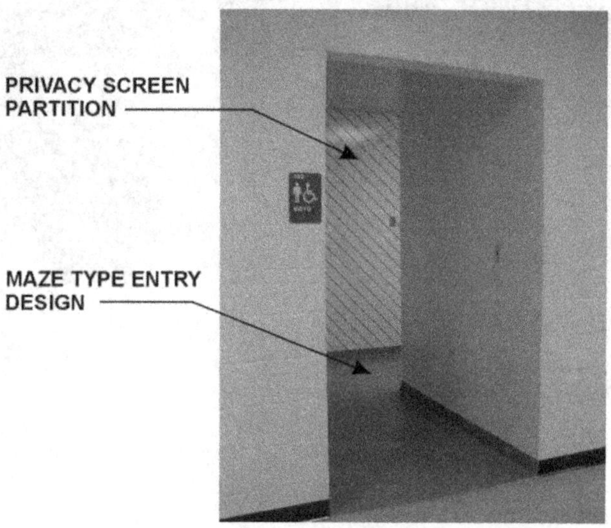

PROVIDE WINDOWS BETWEEN TEACHER'S PLANNING ROOM AND CLASSROOM TO ENHANCE NATURAL SURVEILLANCE

PROVIDE EXTERIOR WINDOWS IN CLASSROOMS TO ENHANCE VISUAL SURVEILLANCE

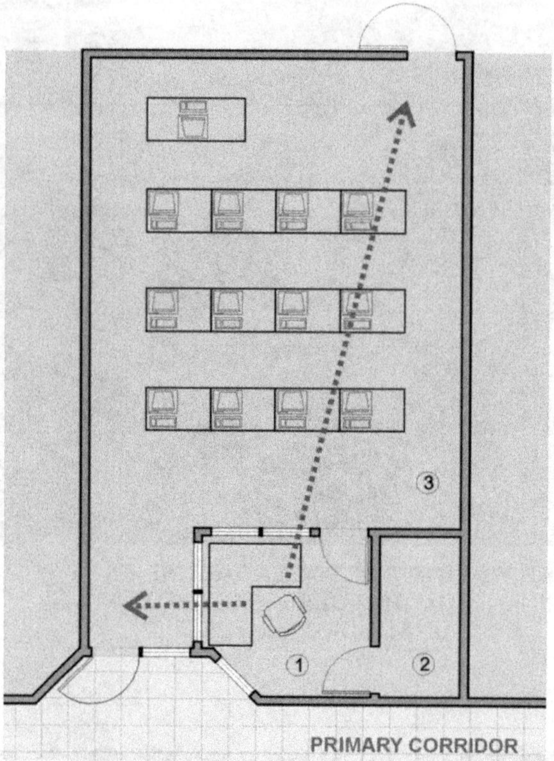

(1) Faculty/Staff (2) Secure Closet (3) Student Work

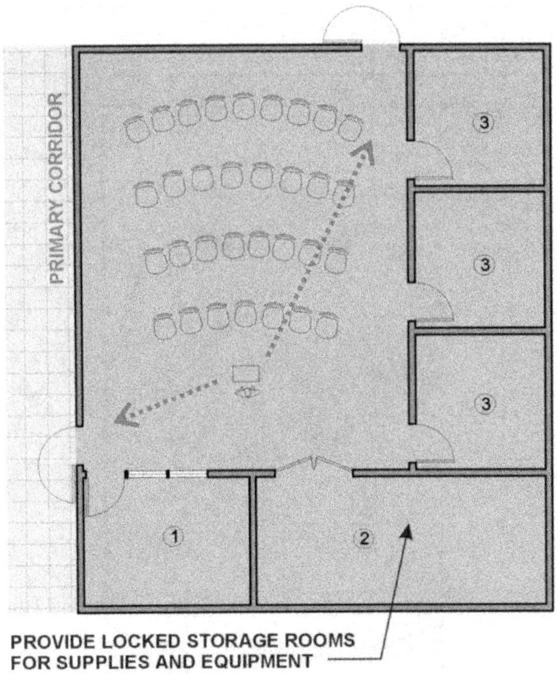

Music Rooms and Similar: (1) Planning Room (2) Storage Room (3) Practice Room

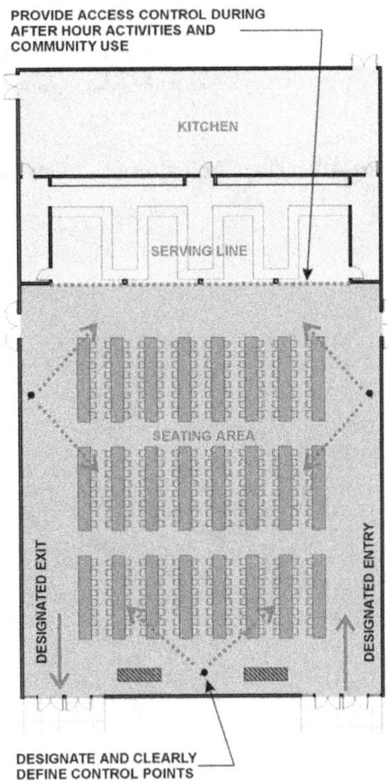

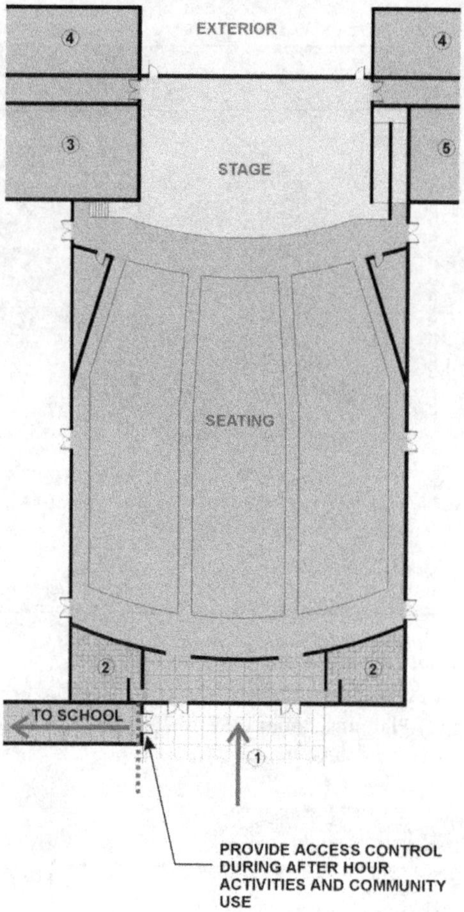

(1) Provide Separate Entrance for After Hour Activities (2) Toilet Facilities (3) Storage (4) Dressing Room (5) Workshop

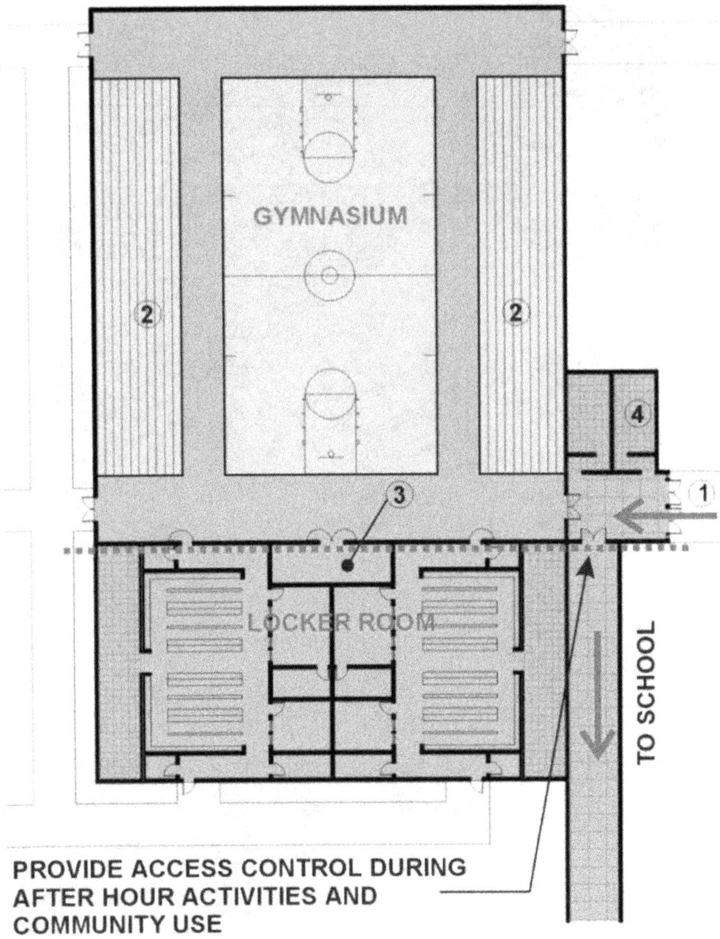

(1) Secure Separate Entrance for After Hour Activities (2) Bleachers (3) Gymnasium Storage (4) Toilet Facilities

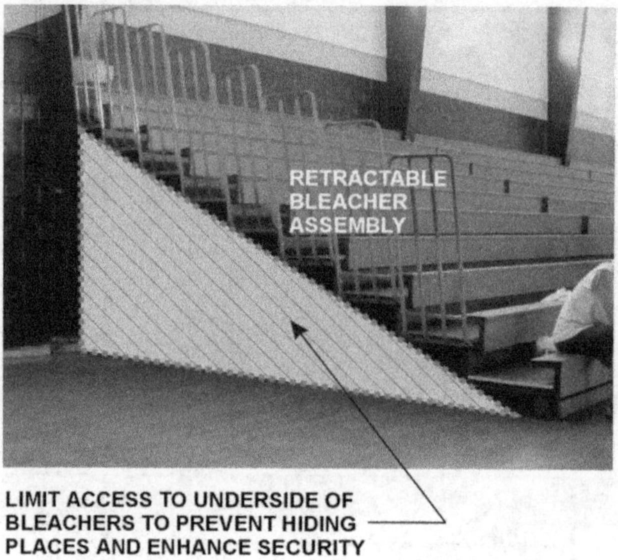

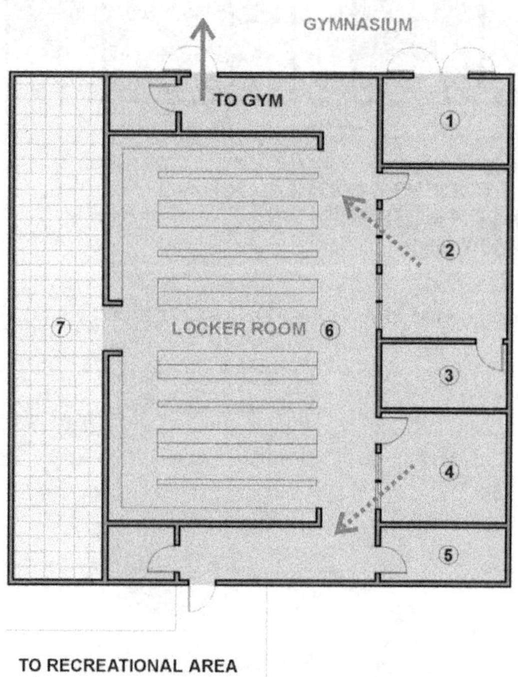

Locker Room Schematic
(1) Gymnasium Storage (2) Coach's Office (3) Coach's Toilet Room (4) Trainer's Office (5) Towel Storage (6) Changing Area and Lockers (7) Toilet and Shower Facilities

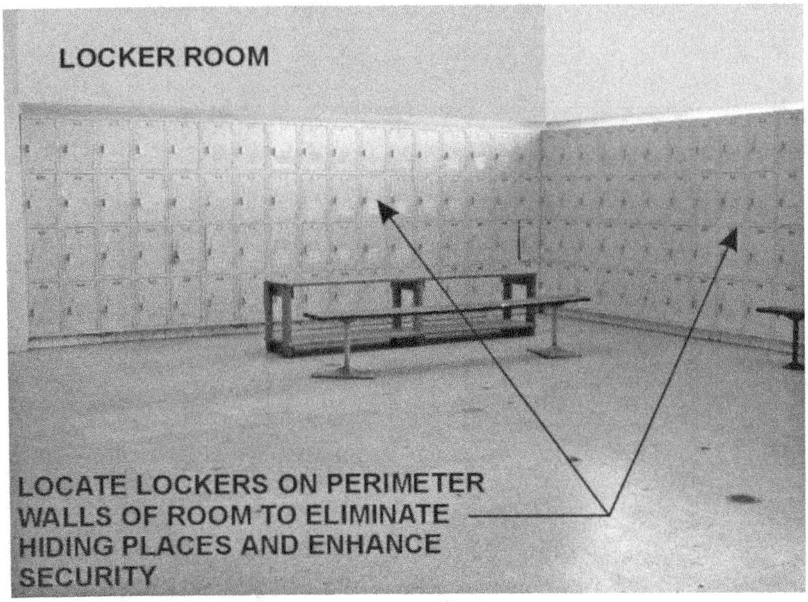

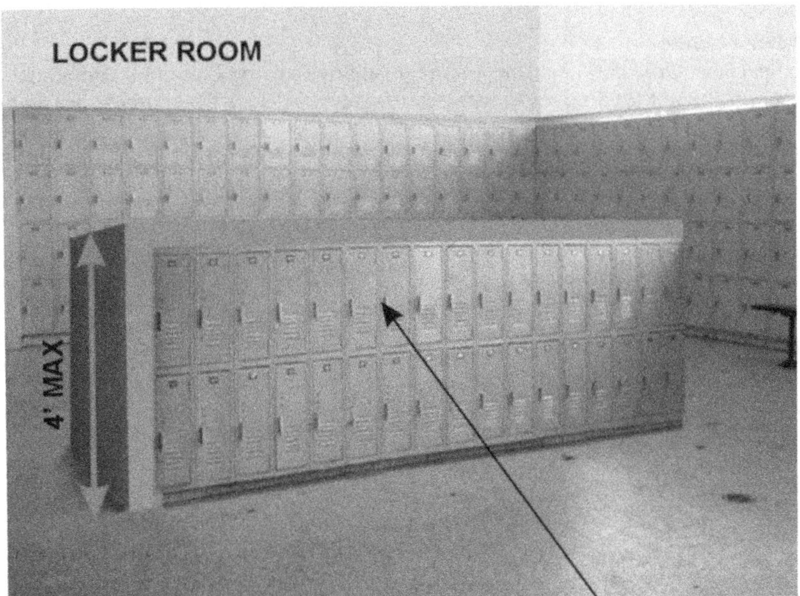

LIMIT LOCKERS LOCATED IN THE CENTER OF ROOM TO FOUR FEET IN HEIGHT TO ENHANCE SURVEILLANCE

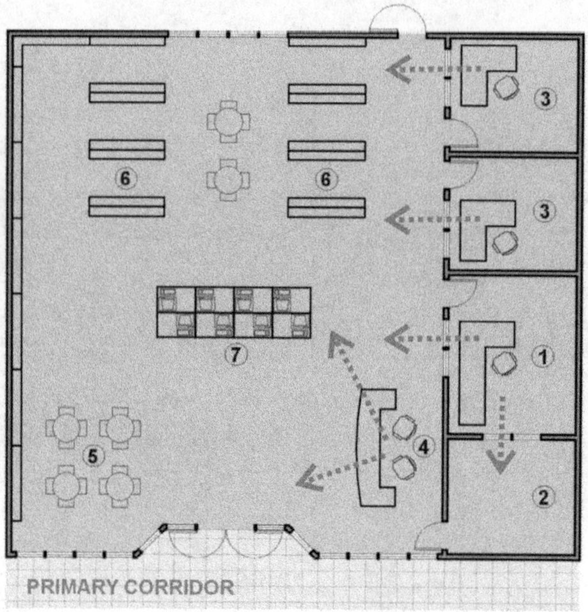

Library/Media Center
(1) Media Specialist Office (2) Lockable Room for Audiovisual and Computer Equipment (3) Work Room (4) Control Point (5) Student Work Area (6) Low Stacks (7) Computer Work Stations

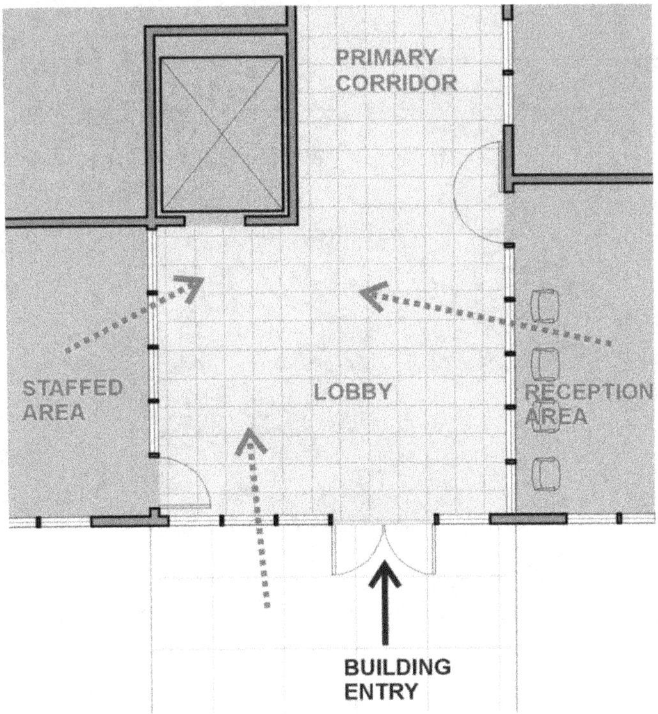
PRIMARY
CORRIDOR
STAFFED
AREA
LOBBY
RECEPTION
AREA
BUILDING
ENTRY

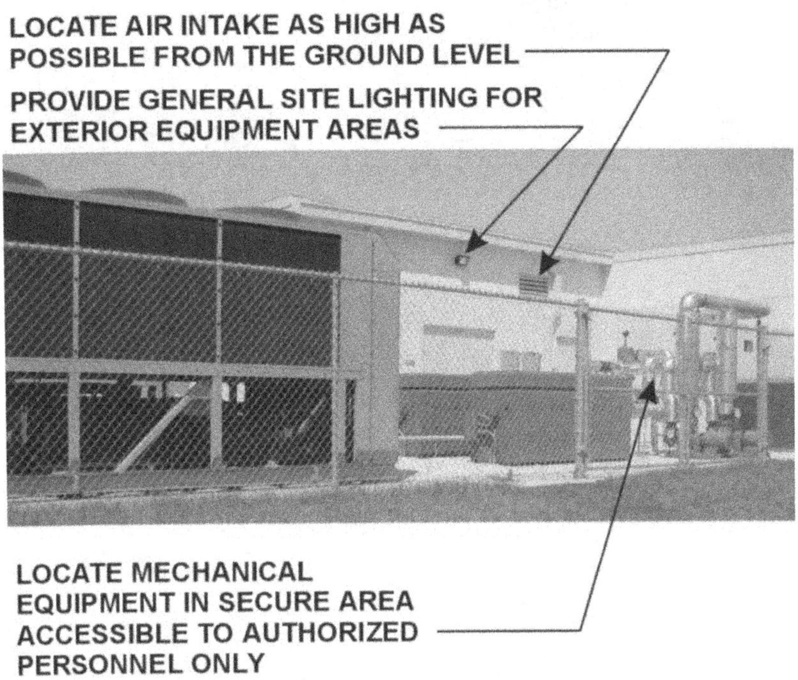
LOCATE AIR INTAKE AS HIGH AS
POSSIBLE FROM THE GROUND LEVEL
PROVIDE GENERAL SITE LIGHTING FOR
EXTERIOR EQUIPMENT AREAS
LOCATE MECHANICAL
EQUIPMENT IN SECURE AREA
ACCESSIBLE TO AUTHORIZED
PERSONNEL ONLY

PROVIDE A SECURE
ENCLOSURE AROUND
CONDENSING UNITS

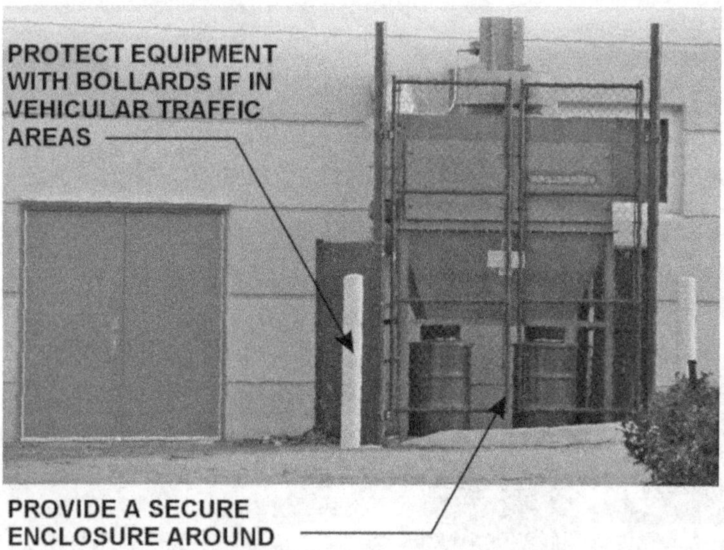
PROTECT EQUIPMENT
WITH BOLLARDS IF IN
VEHICULAR TRAFFIC
AREAS
PROVIDE A SECURE
ENCLOSURE AROUND
MECHANICAL EQUIPMENT

INTERIOR LOCATION

EXTERIOR LOCATION
PROVIDE ROLL-UP TYPE SECURITY GRILLE TO LIMIT ACCESS AFTER SCHOOL HOURS

LOCATE PAY TELEPHONE IN DIRECT LINE OF SIGHT OF FULL-TIME STAFFED WORK STATION

RECESS VENDING MACHINES IN DESIGNATED ALCOVES IN WELL MONITORED AREAS

PROVIDE SECURITY GRILLE TO CONTROL ACCESS TO VENDING MACHINES

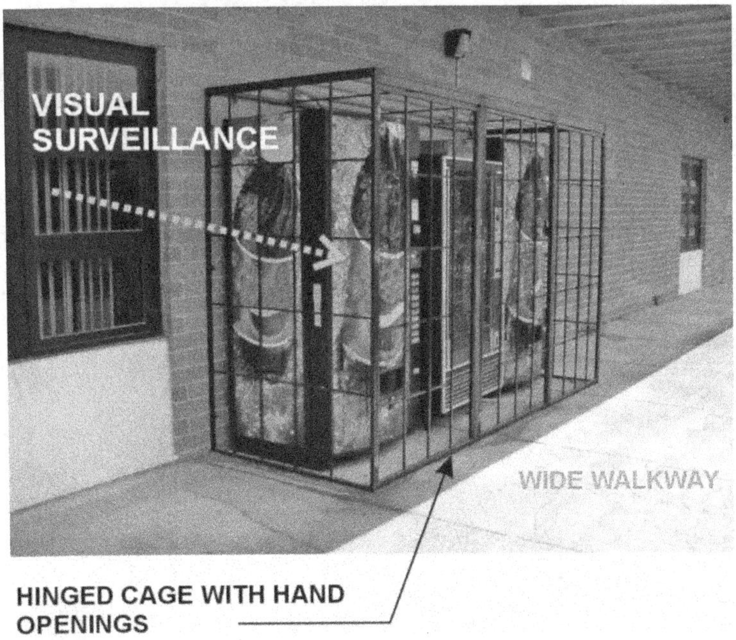
VISUAL
SURVEILLANCE
WIDE WALKWAY
HINGED CAGE WITH HAND
OPENINGS

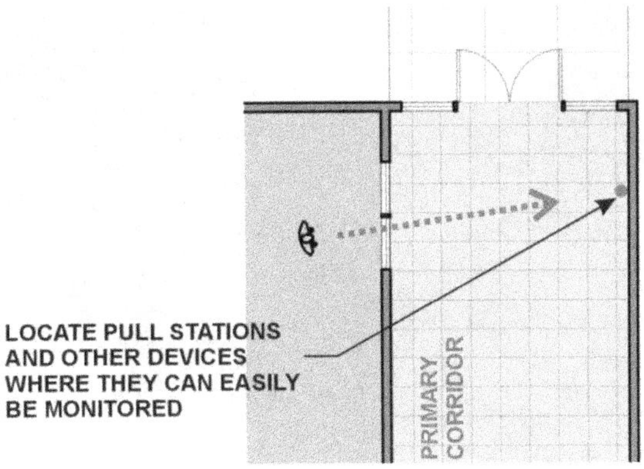
LOCATE PULL STATIONS
AND OTHER DEVICES
WHERE THEY CAN EASILY
BE MONITORED
PRIMARY
CORRIDOR

(1)

(2)

CCTV SHOULD BE INSTALLED IN POTENTIAL PROBLEM AREAS IN NEED OF INCREASED CONTROL

CCTV STRATEGICALLY POSITIONED TO MONITOR VISITOR PARKING AND BUILDING PERIMETER

ELECTRONIC SURVEILLANCE SYSTEM STRATEGICALLY PLACED TO MONITOR BLIND CORNER AT PERIMETER OF SCHOOL BUILDING

Checklist

- Connect with community emergency responders to identify local hazards.
- Review the last safety audit to examine school buildings and grounds.
- Determine who is responsible for overseeing violence prevention strategies in your school.
- Encourage staff to provide input and feedback into the crisis planning process.
- Review incident data.
- Determine the major problems in your school with regard to student crime and violence.
- Assess how the school addresses these problems.
- Conduct an assessment to determine how these problems, as well as others, may impact your vulnerability to certain crises.

DESIGNING FOR EXPLOSIVE AND CHEMICAL, BIOLOGICAL, AND RADIOLOGICAL ATTACKS

Protecting against Explosive, Chemical, Biological, and Radiological Attacks

Designing security into a building requires a complex series of trade-offs. Security concerns need to be balanced with many other design constraints, such as accessibility, initial and life-cycle costs, natural hazard mitigation, fire protection, energy efficiency, and aesthetics.

Because the probability of attack is very small, security measures should not interfere with daily operations of the building. On the other hand, because the effects of attack can be catastrophic, it is prudent to incorporate measures that may save lives and minimize business interruption in the unlikely event of an attack. The measures should be as unobtrusive as possible to provide an inviting, efficient environment that does not attract the undue attention of potential attackers. Security design needs to be part of an overall multi-hazard approach to ensure that it does not worsen the behavior of the building in the event of a fire, earthquake, or hurricane, which are far more prevalent hazards than terrorist attacks.

Because of the severity of the types of hazards discussed, the goals of security-oriented design are by necessity modest. With regard to explosive attacks, the focus is on a damage-limiting or damage-mitigating approach rather than a blast-resistant approach. The goal is to incorporate some reasonable measures that will enhance the safety of the people within the building and facilitate rescue efforts in the unlikely event of attack.

Protection against terrorist attack is not an all-or-nothing proposition. Incremental measures taken early in design may be more fully developed at a later date. With a little forethought regarding, for instance, the space requirements needed to accommodate additional measures, the protection level can be enhanced as the need arises or the budget permits after construction is complete.

The Federal Emergency Management Agency (FEMA) advocates a holistic, multidisciplinary approach to security design by considering the various building systems, including site, architecture, structure, and mechanical and electrical systems and providing general recommendations for the design professional with little or no background in this area.

If you are designing a new building you can strategically site it as far from the perimeter as you can and secure the perimeter against vehicular intrusion using landscaping or barrier methods. Incorporating these security elements into an existing facility requires a different design approach.

Best practices recommended by FEMA for new buildings are listed below. The security concepts involved, however, and the remedies are equally applicable to existing buildings and facilities.

- Place building as far from any secured perimeter as practical.
- Secure the perimeter against vehicular intrusion using landscaping or barrier methods.
- Use lightweight nonstructural elements on the building exterior and interior.
- Place unsecured areas exterior to the main structure or in the exterior bay.
- Incorporate measures to resist progressive collapse.
- Design exterior window systems and cladding so that the framing, connections, and supporting structure have a lateral-load resistance that is equal to or higher than the transparency or panel.
- Place air intakes as far above the ground level as practical.
- Physically isolate vulnerable areas, such as the entries and delivery areas, from the rest of the structure by using floor-to-floor walls in these areas.
- Use redundant, separated mechanical/electrical control systems.

OVERVIEW OF POSSIBLE THREATS

Explosive Threats:

- Vehicle weapon
- Hand-delivered weapon

Airborne Chemical, Biological, and Radiological Threats:

- Large-scale, external, airborne release
- External release targeting building
- Internal release

Although it is possible that the dominant threat mode may change in the future, bombings have historically been a favorite tactic of terrorists. Ingredients for homemade bombs are easily obtained on the open market, as are the techniques for making bombs. Bombings are easy and quick to execute. Finally, the dramatic component of explosions in terms of the sheer destruction they cause creates a media sensation that is highly effective in transmitting the terrorists' message to the public.

EXPLOSIVE ATTACKS

From the standpoint of structural design, the vehicle bomb is the most important consideration. Vehicle bombs are able to deliver a sufficiently large quantity of explosives to cause potentially devastating structural damage. Security design intended to limit or mitigate damage from a vehicle bomb assumes that the bomb is detonated at a so-called critical location. For a vehicle bomb, the critical location is taken to be the closest point that a vehicle can approach, assuming all security measures are in place. This may be a parking area directly beneath the occupied building, the loading dock, the curb directly outside the facility, or a vehicle-access control gate where inspection takes place, depending on the level of protection incorporated into the design.

Another explosive attack threat is the small bomb that is hand delivered. Small weapons can cause the greatest damage when brought into vulnerable, unsecured areas of the building interior, such as the building lobby, mailroom, or retail spaces. Recent events around the world make it clear that there is an increased likelihood that bombs will be delivered by persons who are willing to sacrifice their own lives. Hand-carried explosives are typically on the order of 5–10 pounds of TNT equivalent. However, larger charge weights, in the 50–100 pounds TNT equivalent range, can be readily carried in rolling cases. Mail bombs are typically less than 10 pounds of TNT equivalent.

In general, the largest credible explosive size is a function of the security measures in place. Each line of security may be thought of as a sieve, reducing the size of the weapon that may gain access. Therefore the largest weapons are considered in totally unsecured public space (such as in a vehicle on the nearest public street), and the smallest weapons are considered in the most secured areas of the building (such as in a briefcase smuggled past the screening station).

Two parameters define the design threat: the weapon size, measured in equivalent pounds of TNT, and the standoff. The standoff is the distance measured from the center of gravity of the charge to the component of interest.

The owner in collaboration with security and protective design consultants (engineers who specialize in the design of structures to mitigate the effects of explosions) usually select the design weapon size. Although there are few unclassified sources giving the sizes of weapons that have been used in previous attacks throughout the world, security consultants have valuable information that may be used to evaluate the range of charge weights that might be reasonably considered for the intended occupancy. Security consultants draw upon the experience of other countries, such as Great Britain and Israel, where terrorist attacks have been more prevalent, as well as data gathered by US sources.

To put the weapon size into perspective, it should be noted that thousands of deliberate explosions occur every year within the United States, but the vast majority of them have weapon yields less than 5 pounds. The number of large-scale vehicle weapon attacks that have used hundreds of pounds of TNT during the past twenty years is very small by comparison.

The design vehicle weapon size will usually be much smaller than the largest credible threat. The design weapon size is typically measured in hundreds of pounds rather than thousands of pounds of TNT equivalent.

The decision is usually based on a trade-off between the largest credible attack directed against the building and the design constraints of the project. Further, it is common for the design pressures and impulses to be less than the actual peak pressures and impulses acting on the building. This is the approach that the federal government has taken in its design criteria for federally owned domestic office buildings.

There are several reasons for this choice:

1. The likely target is often not the building under design, but a high-risk building that is nearby. Historically, more building damage has been due to collateral effects than direct attack.
2. It is difficult to quantify the risk of manmade hazards. However, qualitatively it may be stated that the chance of a large-scale terrorist attack occurring is extremely low. A smaller explosive attack is far more likely.
3. Providing a level of protection that is consistent with standards adopted for federal office buildings enhances opportunities for leasing to government agencies in addition to providing a clear statement regarding the building's safety to other potential tenants.
4. The added robustness inherent in designing for a vehicle bomb of moderate size will improve the performance of the building under all explosion scenarios.

DESCRIPTION OF EXPLOSION FORCES

An explosion is an extremely rapid release of energy in the form of light, heat, sound, and a shockwave. The shockwave consists of highly compressed air that wave-reflects off the ground surface to produce a hemispherical propagation of the wave that travels outward from the source at supersonic velocities. As the shockwave expands, the incident or over-pressures decrease. When it encounters a surface that is in line of sight of the explosion, the wave is reflected, resulting in a tremendous amplification of pressure. Unlike acoustical waves, which reflect with an amplification factor of 2, shockwaves can reflect with an amplification factor of up to 13, due to the supersonic velocity of the shockwave at impact. The magnitude of the reflection factor is a function of the proximity of the explosion and the angle of incidence of the shockwave on the surface.

The pressures decay rapidly with time (exponentially), measured typically in thousandths of a second (milliseconds). Diffraction effects, caused by building features, such as reentrant corners and overhangs of the building, may act to confine the air blast, prolonging its duration.

Late in the explosive event, the shockwave becomes negative, followed by a partial vacuum, which creates suction behind the shockwave. Immediately following the vacuum, air rushes in, creating a powerful wind or drag pressure on all surfaces of the building. This wind picks up and carries flying debris in the vicinity of the detonation.

In an external explosion, a portion of the energy is also imparted to the ground, creating a crater and generating a ground shockwave analogous to a high-intensity, short-duration earthquake. The peak pressure is a function of the weapon size or yield and the cube of the distance.

The extent and severity of damage and injuries in an explosive event cannot be predicted with perfect certainty. For instance, two adjacent columns of a building may be roughly the same distance from the explosion, but only one fails because a fragment strikes it in a particular way that initiates collapse. The other, by chance, is not struck and remains in place. Similarly, glass failures may occur outside of the predicted areas due to air-blast diffraction effects caused by the arrangement of buildings and their heights in the vicinity of the explosion. The details of the physical setting surrounding a particular occupant may greatly influence the level of injury incurred. The position of the person, seated or standing, facing toward or away from the event as it happens, may result in injuries ranging from minor to severe. Despite these uncertainties, it is possible to calculate the expected extent of damage and injuries to be expected in an explosive event, based on the size of the explosion, distance from the event, and assumptions about the construction of the building. Additionally, there is strong evidence to support a relationship between injury patterns and structural damage patterns.

Damage due to the air-blast shockwave may be divided into direct air-blast effects and progressive collapse. Direct air-blast effects are damage caused by the high-intensity pressures of the air blast close to the explosion. These may induce localized failure of exterior walls, windows, roof systems, floor systems, and columns.

Progressive collapse refers to the spread of an initial local failure from element to element, eventually resulting in a disproportionate extent of collapse relative to the zone of initial damage. Localized damage due to direct air-blast effects may or may not progress, depending on the design and construction of the building. To produce a progressive collapse, the weapon must be in close proximity to a critical load-bearing element. Progressive collapse can propagate vertically upward or downward from the source of the explosion, and it can propagate laterally from bay to bay as well.

The pressures that an explosion exerts on building surfaces may be several orders of magnitude greater than the loads for which the building is designed. The shockwave also acts in directions that the building may not have been designed for, such as upward pressure on the floor system.

In terms of sequence of response, the air blast first impinges the exterior envelope of the building. The pressure wave pushes on the exterior walls and can cause wall failure and window breakage. As the shockwave continues to expand, it enters the structure, pushing both upward on the ceilings and downward on the floors.

Floor failure is common in large-scale, vehicle-delivered explosive attacks because floor slabs typically have a large surface area for the pressure to act on and a comparably small thickness. Floor failure is particularly common for close-in and internal explosions. The loss of a floor system increases the unbraced height of the supporting columns, which may lead to structural instability.

For hand-carried weapons that are brought into the building and placed on the floor away from a primary vertical load-bearing element, the response will be more localized, with damage and injuries extending a bay or two in each direction. Although the weapon is smaller, the air-blast effects are amplified due to multiple reflections from interior surfaces. Typical damage types that may be expected include the following:

- Localized failure of the floor system immediately below the weapon;
- Damage and possible localized failure for the floor system above the weapon;
- Damage and possible localized failure of nearby concrete and masonry walls;
- Failure of nonstructural elements, such as partition walls, false ceilings, ductwork, and window treatments; and
- Flying debris generated by furniture, computer equipment, and other contents.

More extensive damage, possibly leading to progressive collapse, may occur if the weapon is strategically placed directly against a primary load-bearing element, such as a column.

In comparison to other hazards, such as earthquake or wind, an explosive attack has several distinguishing features:

- The intensity of the localized pressures acting on building components can be several orders of magnitude greater than other hazards. It is not uncommon for the peak pressure on the building from a vehicle weapon parked along the curb to be in excess of 100 psi. Major damage and failure of building components is expected, even for relatively small weapons, in close proximity to the building.
- Explosive pressures decay extremely rapidly with distance from the source. Pressures acting on the building, particularly on the side facing the explosion, may vary significantly, causing a wide range of damage types. As a result, air blast tends to cause more localized damage than other hazards that have a more global effect.
- The duration of the event is very short, measured in thousandths of a second (milliseconds). In terms of timing, the building is engulfed by the shockwave and direct air-blast damage occurs within tens to hundreds of milliseconds from the time of detonation due to the supersonic velocity of the shockwave and the nearly instantaneous response of the structural elements. By comparison, earthquake events last for seconds and wind loads may act on the building for minutes or longer.

Correlation between Damage and Injuries

Three types of building damage can lead to injuries and possible fatalities. The most severe building response is collapse. In past incidents, collapse has caused the most extensive fatalities. For the Oklahoma City bombing in 1995, nearly 90 percent of the building occupants who lost their lives were in the collapsed portion of the Alfred P. Murrah Federal Office Building. Many of the survivors in the collapsed region were on the lower floors and had been trapped in void spaces under concrete slabs.

Although the targeted building is at greatest risk of collapse, other nearby buildings may also collapse. For instance, in the Oklahoma City bombing, a total of nine buildings collapsed. Most of these were unreinforced masonry structures that fortunately were largely unoccupied at the time of the attack. In the bombing of the US embassy in Nairobi, Kenya, in 1998, the collapse of the Uffundi building, a concrete building adjacent to the embassy, caused hundreds of fatalities.

For buildings that remain standing, the next most severe type of injury producing damage is flying debris generated by exterior cladding. Depending on the severity of the incident, fatalities may occur as a result of flying structural debris. Some examples of exterior wall failure causing injuries are listed below.

- In the Oklahoma City bombing, several persons lost their lives after being struck by structural debris generated by infill walls of a concrete-frame building in the Water Resources Building across the street from the Murrah Building.
- In the Khobar Towers bombing in 1996, high-velocity projectiles created by the failed exterior cladding on the wall that faced the weapon impacted most of the

> nineteen US servicemen who lost their lives. The building was an all-precast, reinforced concrete structure with robust connections between the slabs and walls. The numerous lines of vertical support, along with the ample lateral stability provided by the "egg crate" configuration of the structural system, prevented collapse.

Even if the building remains standing and no structural damage occurs, extensive injuries can occur due to nonstructural damage. Typically, for large-scale incidents, these types of injuries occur to persons who are in buildings that are within several blocks of the incident. Although these injuries are often not life threatening, many people can be affected, which impacts on the ability of local medical resources to adequately respond.

An example of nonstructural damage causing injuries is the extensive glass lacerations that occurred in the Oklahoma City bombing within the Regency Towers apartment building, which was approximately 500 feet from the Murrah Building. In this incident, glass laceration injuries extended as far as ten blocks from the bombing. Another example is the bombing of the US embassy in Nairobi, Kenya. The explosion occurred near one of the major intersections of the city, which was heavily populated at the time of the bombing, causing extensive glass lacerations to passersby. The ambassador, who was attending a meeting at an office building across from the embassy, sustained an eye injury as a result of extensive window failure in the building.

It is impractical to design a civilian structure to remain undamaged from a large explosion. The protective objectives are, therefore, related to the type of building and its function. For an office, retail, residential, or light industrial building, where the primary assets are the occupants, the objective is to minimize loss of life. Because of the severity of large-scale explosion incidents, the goals are, by necessity, modest. Moreover, it is recognized that the building will be unusable after the event. This approach is considered a damage-limiting or damage-mitigating approach to design.

To save lives, the primary goals of the design professional are to reduce building damage and to prevent progressive collapse of the building, at least until it can be fully evacuated. A secondary goal is to maintain emergency functions until evacuation is complete.

The design professional is able to reduce building damage by incorporating access controls that allow building security to keep large threats away from the building and to limit charge weights that can be brought into the building. Preventing the building from collapsing is the most important objective.

Historically, the majority of fatalities that occur in terrorist attacks directed against buildings are due to building collapse. Collapse prevention begins with awareness by architects and engineers that structural integrity against collapse is important enough to be routinely considered in design. Features to improve general structural resistance to collapse can be incorporated into common buildings at affordable cost. At a higher level, designing the building to prevent progressive collapse can be accomplished by the alternate-path method (i.e., design for the building to remain standing following the removal of specific elements) or by direct design of components for air-blast loading.

Furthermore, facilitating evacuation, rescue, and recovery efforts through effective placement, structural design, and redundancy of emergency exits and critical mechanical/electrical systems may optimize building design. Through effective structural design, the overall damage levels may be reduced to make it easier for occupants to get out and emergency responders to safely enter.

Beyond preventing collapse and facilitating evacuation/rescue, the objective is to reduce flying debris generated by failed exterior walls, windows, and other components to reduce the severity of injuries and the risk of fatalities. This may be accomplished through selection of appropriate materials and use of capacity-design methods to proportion elements and connections. A well-designed system will provide predictable damage modes, selected to minimize injuries.

Finally, good antiterrorist design is a multidisciplinary effort involving the architect, structural engineer, security professional, and the other design team members. It is also critical for security design to be incorporated as early as possible in the design process to ensure a cost-effective, attractive solution.

CHEMICAL, BIOLOGICAL, AND RADIOLOGICAL PROTECTION

This section discusses three types of airborne hazards.

1. A large exterior release originating some distance away from the building (includes delivery by aircraft).
2. A small localized exterior release at an air intake or other opening in the exterior envelope of the building.
3. A small interior release in a publicly accessible area, a major egress route, or other vulnerable area (such as the lobby, mailroom, or delivery receiving).

Like explosive threats, chemical, biological, and radiological (CBR) threats may be delivered externally or internally to the building. External ground-based threats may be released at a standoff distance from the building or may be delivered directly through an air intake or other opening. Interior threats may be delivered to accessible areas, such as the lobby, mailroom, or loading dock, or they may be released into a secured area, such as a primary egress route.

There may not be an official or obvious warning prior to a CBR event. While you should always follow any official warnings, the best defense is to be alert to signs of a release occurring near you. The air may be contaminated if you see a suspicious cloud or smoke near ground level, hear an air blast, smell strange odors, see birds or other small animals dying, or hear of more than one person complaining of eye, throat, or skin irritation or convulsing.

Chemicals will typically cause problems within seconds or minutes after exposure, but they can sometimes have delayed effects that will not appear for hours or days. Symptoms may include blurred or dimmed vision; eye, throat, or skin irritation; difficulty breathing; excess saliva or nausea.

Biological and some radioactive contaminants typically will take days to weeks before symptoms appear, so listen for official information regarding symptoms. With radioactive "dirty" bombs, the initial risk is from the explosion. Local responders may advise you to either shelter-in-place or evacuate. After the initial debris falls to the ground, leaving the area and washing will minimize your risk from the radiation.

Buildings provide a limited level of inherent protection against CBR threats. To some extent, the protection level is a function of how airtight the building is, but to a greater extent it is a function of the HVAC system's design and operating parameters.

The objectives of protective building design, as they relate to the CBR threat, are first to make it difficult for the terrorist to successfully execute a CBR attack and second, to minimize the impact of an attack (on life, health, property damage, loss of commerce) if it does occur.

In order to reduce the likelihood of an attack, use security and design features that limit the terrorist's ability to approach the building and successfully release the CBR contaminant. Some examples are listed as follows:

- Use security standoff, accessibility, and screening procedures similar to those identified in the explosive threat mitigation section.
- Recognize areas around HVAC equipment and other mechanical systems to be vulnerable areas requiring special security considerations.
- Locate outdoor-air intakes high above ground level and at inaccessible locations.
- Prevent unauthorized access to all mechanical areas and equipment.
- Avoid the use of ground-level mechanical rooms accessible from outside the building. Where such room placement is unavoidable, doors and air vents leading to these rooms should be treated as vulnerable locations and appropriately secured.
- Treat operable ground-level windows as a vulnerability and either avoid their use or provide appropriate security precautions to minimize the vulnerability.
- Inside the building, minimize public access to HVAC return-air systems.

Air Intakes

Air intakes may be made less accessible by placing them as high as possible on the building exterior, with louvers flush with the exterior. All opportunities to reach air intakes through climbing should be eliminated. Ideally, there should be a smooth, vertical surface from the ground level to the intake louvers, without such features as high shrubbery, low roofs, canopies, or sunshades, as these features can enable climbing and concealment. To prevent opportunities for a weapon to be lobbed into the intake, the intake louver should be ideally flush with the wall. Otherwise, a surface sloped at least 45 degrees away from the building and further protected through the use of metal mesh (a.k.a. bird screen) should be used. Finally, closed circuit television (CCTV) surveillance and enhanced security is recommended at intakes.

In addition to providing protection against an airborne hazard delivered directly into the building, placing air intakes high above ground provides protection against ground-based standoff threats because the concentration of the airborne hazard diminishes somewhat with height. Because air-blast pressure decays with height, elevated air intakes also provide modest protection against explosion threats. Furthermore, many recognized sources of indoor-air contaminants (such as vehicle exhaust, standing water, lawn chemicals, trash, and rodents) tend to be located near ground level. Thus, elevated air intakes are a recommended practice in general for providing healthy indoor-air quality.

In the event that a particular air intake does not service an occupied area, it may not be necessary to elevate it above ground level. However, if the unoccupied area is within an otherwise occupied building, the intake should either be elevated or significant precautions (tightly sealed construction between unoccupied/occupied areas, unoccupied area maintained at negative pressure relative to occupied area) should be put in place to ensure that contaminants are unable to penetrate into the occupied area of the building.

Mechanical Areas

Another simple measure is to tightly restrict access to building mechanical areas (such as mechanical rooms, roofs, and elevator equipment access). These areas provide access to equipment and systems (such as HVAC, elevator, building exhaust, and communication and control) that could be used or manipulated to assist in a CBR attack. Including these areas in those monitored by electronic security and by eliminating elevator stops at the levels that house this equipment may provide additional protection. For rooftop mechanical equipment, ways of restricting (or at least monitoring) access to the roof that do not violate fire codes should be pursued.

Return-Air Systems

Similar to the outdoor-air intake, HVAC return-air systems inside the building can be vulnerable to CBR attack. Buildings requiring public access have an increased vulnerability to such an attack. Design approaches that reduce this vulnerability include the use of ducted HVAC returns within public access areas and the careful placement of return-air louvers in secure locations not easily accessed by public occupants. The second objective is to design to minimize the impact of an attack.

For many buildings, especially those requiring public access, the ability to prevent a determined terrorist from initiating a CBR release will be a significant challenge. Compared to buildings in which campus security and internal access can be strictly controlled, public access buildings may require a greater emphasis on mitigation. However, even private access facilities can fall victim to an internal CBR release, whether through a security lapse or perhaps a delivered product (mail, package, equipment, or food). Examples of design methods to minimize the impact of a CBR attack are listed as follows:

- Public access routes to the building should be designed to channel pedestrians through points of noticeable security presence.
- The structural and HVAC design should isolate the most vulnerable public areas (entrance lobbies, mailrooms, load/delivery docks) both physically and in terms of potential contaminant migration.
- The HVAC and auxiliary air systems should carefully use positive and negative pressure relationships to influence contaminant migration routes.

Lobbies, Loading Docks, and Mail Sorting Areas

Vulnerable internal areas where airborne hazards may be brought into the building should be strategically located. These include lobbies, loading docks, and mail sorting areas. Where possible, place these functions outside of the footprint of the main building. When incorporated into the main building, these areas should be physically separated from other areas by floor-to-roof walls. Additionally, these areas should be maintained under negative pressure relative to the rest of the building, but at positive-to-neutral pressure relative to the outdoors. To assist in maintaining the desired pressure relationship, necessary openings (doors, windows, etc.) between secure and vulnerable areas should be equipped with sealing windows and doors, and wall openings due to ductwork, utilities, and other penetrations should be sealed. Where entries into vulnerable areas are frequent, the use of airlocks or vestibules may be necessary to maintain the desired pressure differentials.

Ductwork that travels through vulnerable areas should be sealed. Ideally, these areas should have separate air-handling units to isolate the hazard. Alternatively, the conditioned air supply to these areas may come from a central unit, as long as exhaust/return air from these areas is not allowed to mix into other portions of the building. In addition, emergency exhaust fans that can be activated upon internal CBR release within the vulnerable area will help to purge the hazard from the building and minimize its migration into other areas. Care must be taken that the discharge point for the exhaust system is not colocated with expected egress routes. Consideration should also be given to filtering this exhaust with High Efficiency Particulate Air (HEPA) filtration. For entrance lobbies that contain a security screening location, it is recommended that an airlock or vestibule be provided between the secured and unsecured areas.

Zoning of HVAC Systems

Large buildings usually have multiple HVAC zones, each zone with its own air-handling unit and duct system. In practice, these zones are not completely separated if they are on the same floor. Air circulates among zones through plenum returns, hallways, atria, and doorways that are normally left open. Depending upon the HVAC design and operation, airflow between zones on different floors can also occur through the intentional use of shared air-return/supply systems and through air migrations via stairs and elevator shafts.

Isolating the separate HVAC zones minimizes the potential spread of an airborne hazard within a building, reducing the number of people potentially exposed if there is an internal release. Zone separation also provides limited benefit against an external release, as it increases internal resistance to air movement produced by wind forces and chimney effect, thus reducing the rate of infiltration. In essence, isolating zones divides the building into separate environments, limiting the effects of a single release to an isolated portion of the building. Isolation of zones requires full-height walls between each zone and the adjacent zones and hallway doors.

Another recommendation is to isolate the return system (i.e., no shared returns). Strategically locate return-air grilles in easily observable locations and preferably in areas with reduced public access. Both centralized and decentralized shutdown capabilities are advantageous.

To quickly shut down all HVAC systems at once in the event of an external threat, a single-switch control is recommended for all air exchange fans (including bathroom, kitchen, and other exhaust sources). In the event of a localized internal release, redundant decentralized shutdown capability is also recommended. Controls should be placed in a location easily accessed by the facility manager, security personnel, or emergency response personnel. Duplicative and separated control systems will add an increased degree of protection. Further protection may be achieved by placing low-leakage automatic dampers on air intakes and exhaust fans that do not already have back-draft dampers.

Positive Pressurization

Traditional good engineering practice for HVAC design strives to achieve a slight overpressure of 5–12 Pa (0.02 inch < N > 0.05-inch w.g.) within the building environment, relative to the outdoors. This design practice is intended to reduce uncontrolled infiltration into the building. When combined with effective filtration, this practice will also provide enhanced protection against external releases of CBR aerosols.

Using off-the-shelf technology (such as HEPA), manually triggered augmentation systems can be put into place to overpressure critical zones to intentionally impact routes of contaminant migration and/or to provide safe havens for sheltering-in-place. For egress routes, positive pressurization is also recommended, unless of course, the CBR source is placed within the egress route. Design parameters for such systems will depend upon many factors specific to the building and critical zone in question. Care must be taken that efforts to obtain a desired pressure relationship within one zone will not put occupants in another zone at increased risk. Lastly, the supply air used to pressurize the critical space must be appropriately filtered (see Filtration Systems below) or originate from a non-contaminated source in order to be beneficial.

Air Tightness

To limit the infiltration of contaminants from outside the building into the building envelope, building construction should be made as airtight as possible. Tight construction practices (weatherization techniques, tightly sealing windows, doors, wall construction, continuous vapor barriers, sealing interface between wall and window/door frames) will also help to maintain the desired pressure relationships between HVAC zones. To ensure that the construction of the building has been performed correctly, building commissioning is recommended throughout the construction process and prior to taking ownership to observe construction practices and to identify potential airflow trouble spots (cracks, seams, joints, and pores in the building envelope and along the lines separating unsecured from secured space) before they are covered with finish materials.

Filtration Systems

To offer effective protection, filtration systems should be specific to the particular contaminant's physical state and size. Chemical vapor/gas filtration (a.k.a. air cleaning) is currently a very expensive task (high initial and recurring costs) with a limited number of design professionals experienced in its implementation. Specific expertise should be sought if chemical filtration is desired. Possible application of the air cleaning approach to collective protection zones (with emergency activation) can assist in significantly reducing the cost, though the protection is limited to the reduced size of the zone.

Most traditional HVAC filtration systems focus on aerosol-type contaminants. The CBR threats in this category include radioactive "dirty" bombs, bio-aerosols, and some chemical threats. Riot-control agents and low-volatility nerve agents, for example, are generally distributed in aerosol form; however, a vapor component of these chemical agents could pass through a filtration system. HEPA filtration is currently considered adequate by most professionals to achieve sufficient protection from CBR particulates and aerosols. However, HEPA filtration systems generally have a higher acquisition cost than traditional HVAC filters, and they cause larger pressure drops within the HVAC system, resulting in increased energy requirements to maintain the same design airflow rate.

Due to recent improvements in filter media development, significant improvements in aerosol filtration can be achieved at relatively minimal increases in initial and operating costs. Also important is that incremental increases in filtration efficiency will generally provide incremental increases in protection from the aerosol contaminant. In 1999, the ASHRAE released Standard 52.2-1999. This standard provides a system for rating filters that quantifies filtration efficiency in different particle-size ranges to provide a composite efficiency value named the Minimum Efficiency Reporting Value (MERV).

MERV ratings range between one and twenty, with a higher MERV indicating a more efficient filter. Using the MERV rating table, the desired filter efficiency may be selected according to the size of the contaminant under consideration. For example, a filter with a MERV of 13 or more will provide a 90 percent or greater reduction of most CBR aerosols (generally considered to be at least 1–3 μm in size or larger) within the filtered air stream with much lower acquisition and maintenance costs than HEPA filtration.

Efficiency of filtration systems is not the only concern. Air can become filtered only if it actually passes through the filter. Thus, filter-rack design, gasketing, and good-quality filter sources should all play a role in minimizing bypass around the filter. The use of return-air filtration systems and the strategic location of supply and return systems should also be carefully employed to maximize effective ventilation and filtration rates.

Detection Systems

Beyond the measures discussed above, there is the option of using detection systems as part of the protective design package. In general, affordable, timely, and practical detection systems specific to all CBR agents are not yet available. However, for aerosol contaminants, nonspecific detection equipment can be employed to activate response actions should a sudden spike in aerosol concentration of a specific size range be detected. If the spike were detected in an outdoor intake, for example, this could trigger possible response options, such as damper closure, system shutdown, bypass to alternate air intake, or rerouting the air through a special bank of filters. Such protective actions could occur until trained personnel could perform an investigation (check with adjacent alarms and review security tape covering outdoor-air intake). Unless foul play was discovered, the entire process could be completed within 10 minutes or less and without alarming occupants.

The initial cost of such a system is relatively modest (depending upon the number of detectors and response options incorporated into the design), but the maintenance requirements are relatively high. Similar monitoring systems could be employed to trigger appropriate responses in high-threat areas, such as mailrooms, shipping/receiving areas, or entrance lobbies. The approach could also be expanded to incorporate some of the newer chemical detection technologies, though the low-threshold requirements may generate a substantial number of false positives. As technology progresses, detector availability and specificity should continue to expand into the general marketplace.

Emergency Response Using Fire/HVAC Control Center

Certain operations that are managed at the Fire Control Center can play a protective role in the response to a CBR incident. Examples of such operations and how they could be used are given as follows.

Purge fans. These can be used to purge an interior CBR release or to reduce indoor contaminant concentrations following building exposure to an external CBR source. (Note: In practice, some jurisdictions may recommend purging for chemical and radiological contaminants but not for biological contaminants, which may be communicable and/or medically treatable.)

Communication systems. Building communication systems that allow specific instructions to be addressed to occupants in specific zones of the building can play a significant role in directing occupant response to either an internal or external release.

Pressurization fans. These provide two functions. First, the ability to override and deactivate specific positive-pressure zones may be beneficial in the event that a known CBR source is placed into such an area. Second, areas designated for positive pressurization (generally for smoke protection) may also become beneficial havens for protection from internal and external CBR releases, if they are supplied by appropriately filtered air.

HVAC controls. The ability to simultaneously and individually manipulate operation of all HVAC and exhaust equipment from a single location may be very useful during a CBR event. Individuals empowered to operate such controls must be trained in their use. The provision of simple floor-by-floor schematics showing equipment locations and the locations of supply and return louvers will aid the utility of this control option.

Elevator controls. Depending upon their design and operation, the ability to recall elevators to the ground floor may assist in reducing contaminant migration during a CBR event.

Evolving Technologies

Many of the challenges relating to CBR terrorism prevention will be facilitated with the introduction of new technologies developed to address this emerging threat. As vendors and products come to market, it is important that the designer evaluate performance claims with a close level of scrutiny. Vendors should be willing to guarantee performance specs in writing and provide proof of testing (and show certified results) by an independent, reputable lab, and the testing conditions (such as flow rate, residence time, and incoming concentrations) should be consistent with what would be experienced within the owner's building. For CBR developments, proof of federal government testing and acceptance may be available.

Chemical, Biological, and Radiological Protective Measures

- Place air intakes servicing occupied areas as high as practically possible (minimum 12 feet above ground). The GSA may require locating at fourth floor or above, when applicable.
- Restrict access to critical equipment.
- Isolate separate HVAC zones and return-air systems.
- Isolate HVAC supply and return-air systems in unsecured areas.
- Physically isolate unsecured areas from secured areas.
- Use positive pressurization of primary egress routes, safe havens, and/or other critical areas.
- Commission building throughout construction and prior to taking ownership.
- Provide redundant, easily accessible shutdown capabilities.
- For higher levels of protection, consider using contaminant-specific filtration and detection systems.
- Incorporate fast-acting, low-leaking dampers.
- Filter both return air and outdoor air for publicly accessible buildings.
- Select filter efficiencies based upon contaminant size. Use reputable filter media installed into tight-fitting, gasketed, and secure filter racks.
- For higher-threat areas (mailroom, receiving, reception/screening lobby):
  - Preferably locate these areas outside the main building footprint.
  - Provide separate HVAC, with isolated returns capable of 100 percent exhaust.
  - Operate these areas at negative pressure relative to the secure portion of the building.
- Use airtight construction, vestibules, and air locks if there is a high traffic flow.
- Consider installation of an emergency exhaust fan to be activated upon suspected internal CBR release.
- Lock, secure, access-log, and control mechanical rooms.
- In public access areas, use air diffusers and return-air grilles that are secure or under security observation.
- Zone the building communication system so that it is capable of delivering explicit instructions and has backup power.
- Create safe zones using enhanced filtration, tight construction, emergency power, dedicated communication systems, and appropriate supplies (food, water, first aid, and personal protective equipment)

INTRUSION DETECTION AND SECURITY-ACCESS SYSTEM

Main cabinet

A. Steel construction #14 gauge, 36″ × 36″ × 6″ cabinet with hinged, lockable cover, 1/2″ thick plywood backboard, painted light gray.
   1. Also allow a 4′ × 4′ empty space above or to either side of the main cabinet at a centerline of 7′ for additional security equipment.

B. Main cabinet shall be surface mounted.

C. The main security cabinet shall be located in the video control room; however, section 2.3 A. has priority.

D. The location of the main, head-end, security terminal cabinet shall not exceed 95′ from the antenna weather head.

Building and floor distribution cabinets

E. Steel construction #14 gauge, 12″ × 18″ × 6″ cabinet with hinged lockable cover, 1/2″ thick plywood backboard, painted light gray.

F. All distribution cabinets shall be surface mounted.

Conduit and boxes

G. Criteria for security antenna stub-out location:
   1. This location will be based on the radio frequency assigned to the site and the direct line of sight from the antenna to the repeater location and will be selected at time of review.
   2. The security system radio has a direct line of sight antenna to the repeater located at latitude (26, 35, 20 N) and longitude (080, 12, 43 W) US 441 and Lantana Road.

H. Provide and install a 3″ raceway to outside of building at the antenna mounting structures designated location and terminated with a weather head.
   1. The penetrating portion of conduit shall be a contiguous 10′ piece of rigid, with 6′ firmly supported at two joist points below the roof penetration and 4′ above the roof.
   2. The raceway shall continue and terminate against the wall below the ceiling 90″ AFF over the main head-end terminal cabinet capped with an EMT set screw connector and fiber bushing.

I. Provide and install an antenna mounting structure located next to the antenna stub-out (refer to B above).
   1. Design the antenna mounting structure to support an antenna array consisting of six (6) 6-element Yagi 9dBd gain antennas model #460-6.
   2. Structures shall meet ASCE-78 requirements.
   3. Top of structures shall be no more than 49′ or 15 meters above grade.

J. Provide and install a building distribution cabinet for each building.
   1. Provide and install a separate 2″ raceway from the main head-end cabinet for each building distribution cabinet.

2. For buildings with more than one level, provide a floor distribution cabinet for each additional floor level.
3. Feeding each of these cabinets with a separate 2″ conduit, from the building distribution cabinet.
4. If any floor level area is greater than 20,000 sq. ft., additional floor distribution cabinets shall be provided; with no one cabinet feeding more than 20,000 sq. ft.
5. Raceway shall not exceed 400′ without a pull box.

K. Provide and install 3/4″ conduits from the distribution cabinets and distribute to feed all junction and mounting boxes in the distribution area.
1. Each 3/4″ conduit shall feed no more than sixteen wall sensor-mounting box locations.
2. Each of these conduit runs shall leave and return to the cabinet (LOOPED).

L. Provide and install a separate 3/4″ conduit from the main security terminal cabinet to the fire alarm main terminal cabinet.

M. Provide and install a separate 3/4″ conduit from the nearest security distribution cabinet to all walk-in refrigeration units, and terminate this conduit to the unit's temperature sensor relay enclosure.

N. Provide and install a separate 3/4″ conduit from the nearest security distribution cabinet to the emergency electric generator control panel.

O. Provide and install recessed mounted 4″ × 4″ × 2–1/8″ boxes with flush single gang ring and single gang cover; mounted with the opening vertical, located 12″ to the side of the intercom speaker or clock; in rooms where clocks or speakers are designated.
1. If a projection screen will block the security device in this location, move the security junction box to clear the blockage to either side of the projection screen.

P. Provide and install recessed mounted 4″ × 4″ × 2–1/8″ sensor-mounting boxes with single gang ring and single gang cover; mounted with the opening vertical, 90″ above the floor and 25′ from any exterior doors or exterior glass (susceptible to entry) at the following locations:
1. All interior corridors;
2. Rooms that do not have a clock or speaker but have exterior doors or glass; and
3. Electrical rooms, mechanical rooms, storage rooms, and the like, that may allow access through them to the interior of the building.

Q. Provide and install recessed mounted 4″ × 4″ × 2–1/8″ sensor-mounting boxes with single gang ring and single gang cover; mounted with the opening vertical, 90″ above the floor and 15′ from the entry/exit doors in the following areas:
1. Cafeteria manager's office;
2. Media CCTV studio;
3. Media CCTV storage room;
4. PE storage room;
5. PE office;
6. Main telephone equipment room;
7. Clinic;
8. All corridors, approximately 65′ apart; and

9. All classrooms, workrooms, administrative areas, adjacent to all the video-surveillance j-boxes, and other areas as defined in the plan specific review process.

R. Provide and install an additional 4–11/16″ × 4–11/16″ × 2–1/8″ box above suspended ceiling directly above each wall sensor-mounting box.
   1. Locate box no higher than 36″ above the ceiling, in the same room as the wall-mounted boxes.

S. Provide and install two recessed mounted Knox-Boxes with UL listed tamper switches, 1/4″ plate steel housing, 1/8″ thick door with interior gasket seal. Both box and lock shall have 2″ steel dust cover with tamper seal mounting capability. Finish color shall be aluminum for police and black for fire. The Knox Box shall be mounted at 48″ above the finished floor.
   1. The following Knox Box is recommend: Manufactured by The Knox Company, Model 3221 for Fire Department and 3225 for Police Department with the recessed mounting kit (3240). All recessed mounting kit shell housings, including the cover plate and screw heads, are flush with the finish wall. The housing must be plumbed to ensure vertical alignment of the vault.
   2. Provide and install a 3/4″ conduit from the back of the Knox Box to a 4–11/16″ × 4–11/16″ × 2–1/8″ box with cover, located above the nearest removable ceiling tile. Continue the conduit run to the nearest security distribution terminal cabinet.
   3. The school district's project manager shall coordinate with the Department of School Police and the contractor to transmit the Knox authorization forms, which will be provided by the Department of School Police to assure it is keyed to the Master. The Knox authorization forms for the fire department Knox Box must be obtained from the fire department having jurisdiction.

T. Provide and install a separate 3/4″ conduit from the main security cabinet to the main telephone room terminal board.

U. Provide and install a separate 3/4″ conduit from the back of the fire department's Knox Box to a 4–11/16″ × 4–11/16″ × 2–1/8″ box with cover, located above the nearest removable ceiling tile. Continue the conduit run to the nearest security distribution terminal cabinet.

V. Provide and install a 3/4″ conduit between all security distribution cabinets and all card access distribution cabinets at all terminal cabinet locations.

W. Provide and install six additional recessed mounted 4″ × 4″ × 2–1/8″ sensor-mounting boxes with flush single gang ring and single gang cover; with the opening mounting vertical with 3/4″ conduit and run to nearest security junction box. (Estimate 100′ of conduit for each run.) Wall-mount locations to be directed in field during construction prior to final above-ceiling inspection.

Power feed

X. Provide and install a dedicated 120-volt normal power duplex outlet located adjacent to the upper portion of the main security system's cabinet.

Y. Provide and install a dedicated 120-volt duplex outlet fed from the optional branch of the generator power source, adjacent to the normal power receptacle.

Execution

Installation

A. Install system in accordance with NECA "Standard of Installation" and section 16111.
B. Obtain a detail book from the School Police Department Security Section for system specifics.
C. Permanently label all conduits as to plan room number destination, at all terminal cabinets.
D. Paint all security system junction-box covers black.
E. Install 1/2″ (black round indicators) of paper construction on ceiling tile grid work at all locations where security system boxes are located above the drop ceiling.
F. Permanently label all the security system terminal cabinets "security system."
G. Install 200-lb.-strength pull-string throughout the raceway system.
H. All junction boxes mounted above ceiling shall be mounted with the opening facing down, and shall have a reasonable immediate access pathway provided. Note: removal of a light fixture or other similar ceiling equipment is not considered as a reasonable access pathway.
I. The security system raceway shall be a separate raceway system and shall not interconnect with or be used by any other system without the authorization of the Department of School Police Security System Section or per DMS sections 16722, 16723, 16724, and 16725.
J. All conduit runs shall be as direct as possible in order to save on wiring cost and to reduce poor performance due to cable voltage loss.

2. VIDEO SURVEILLANCE SYSTEM

General

System description

The video-surveillance system shall include the furnishing and installation of main and distribution cabinets, conduit system, and power feeds.

Main cabinet

A. Steel construction #14 gauge, 36″ × 36″ × 6″ cabinet with hinged, lockable cover, 1/2″ thick plywood backboard, painted light gray and shall be surface mounted.

Distribution cabinets

B. Steel construction #14 gauge, 12″ × 18″ × 6″ cabinet with hinged lockable cover, 1/2″ thick plywood backboard, painted light gray and shall be surface mounted.

Conduit and boxes

C. Provide and install a building distribution cabinet for each building.
   1. Provide and install a separate 2″ raceway from the main "head-end" cabinet for each building distribution cabinet.
      a. For buildings with more than one level, provide a floor distribution cabinet for each additional floor level.
      b. Feeding each of these cabinets with a separate 1 1/2″ conduit from the building distribution cabinet.
      c. If any floor level area is greater than 20,000 sq. ft., additional floor distribution cabinets shall be provided; with no one cabinet feeding more than 20,000 sq. ft.
      d. Raceway shall not exceed 400′ without a pull box.

D. Provide and install feed conduits from each distribution cabinet to feed the mounting boxes designated for the area. Size each of the separate feeds as follows:
   1. 3/4″ for a maximum of two mounting boxes.
   2. 1″ for a maximum of four mounting boxes.
   3. 1–1/4″ for a maximum of six mounting boxes
   4. 1–1/2″ for a maximum of eight mounting boxes

E. Provide and install 4–11/16″ × 4–11/16″ × 2–1/8″ mounting boxes with covers at designated interior locations. If the area location has removable ceiling tile, the box shall be located above the tile.
   1. If the location has a structure of fixed ceiling material, then flush-mount the box with a double gang mud ring and cover.

F. Provide and install 4–11/16″ × 4–11/16″ × 2–1/8″ flush-mounted mounting boxes with double gang mud ring and weatherproof covers; mounted with the opening vertical, at all designated exterior locations fed from additional 4–11/16″ × 4–11/16″ × 2–1/8″ boxes at the interior feed penetrations through the exterior walls.

G. Provide and install a 1″ conduit from the main Video Surveillance System (VSS) cabinet to the main "head-end" security cabinet.

H. Provide a 3/4″ conduit from the main VSS cabinet to the main card access system cabinet.

I. Provide and install a 2″ conduit from the main video-surveillance terminal cabinet to a brook's box located at designated entry gates.
   1. All entry gates on school site will have a 2″ conduit run.
   2. Exact location to be determined at the plan review.

J. Provide and install an additional 4–11/16″ × 4–11/16″ × 2–1/8″ box above suspended ceiling directly above each wall video-mounting box. Locate J-box no higher than 36″ above the ceiling grid and in the same room as the wall-mounted boxes.

Power feed

K. Provide and install a dedicated 120-volt duplex outlet fed from the optional branch of the generator power source, adjacent to the normal power receptacle for the main cabinet.

L. Provide and install a dedicated 120-volt duplex outlet fed from the optional branch of the generator power source, adjacent the lower portion of the entry terminal of each building.

Executions

Installation

A. Install system in accordance with NECA "Standard of Installation" and section 16111.
B. Obtain a detail book from the School Police Department Security Section for system specifics.
C. Permanently label all conduits as to plan room number destination, at all terminal cabinets.
D. Paint all video-surveillance system junction-box covers gold.
E. Install 1/2″ (gold round indicators) of paper construction on ceiling tile grid work at all locations where video-surveillance system boxes are located above the drop ceiling.
F. Permanently label all the video-surveillance system terminal cabinets "video-surveillance system."
G. Install 200-lb.-strength pull-string throughout the raceway system.
H. All junction boxes mounted above ceiling shall be mounted with the opening facing down, and shall have a reasonable immediate access pathway provided. Note: removal of a light fixture or other similar ceiling equipment is not considered as a reasonable access pathway.
I. The "Video-Surveillance System Raceway" shall not be used as part of any other raceway.
J. Provide and install six additional 4–11/16″ × 4–11/16″ × 2–1/8″ mounting boxes with flush single gang ring and single gang cover; with the opening mounting vertical with 3/4″ conduit run to the nearest video terminal cabinet or junction box. (Estimate 100′ of conduit for each run.) Mounting locations to be directed in field during construction prior to final above-ceiling inspection.

Location

K. Provide video junction boxes in the following locations:
   1. At all card reader locations;
   2. Cafeteria main dining room;
   3. Bus loop;
   4. Main office reception area;
   5. In all corridors;
   6. Student pick-up/drop-off;
   7. Parking areas;
   8. Gymnasium and gymnasium lobby;
   9. Auditorium and auditorium lobby; and
   10. Other areas as defined in the project specific review process.

3. EMERGENCY RADIO COMMUNICATION SYSTEMS

Conduit and boxes
Power wiring

System description

- Emergency radio communications systems shall include the furnishing and installation of raceway systems and power feeds for the emergency broadcast receiver, radio communications repeater station, hurricane shelter communications, and wireless propagation enhancement systems.

Conduit and boxes

A. All sites: If the Emergency Broadcast Receiver/Repeater is located at the main security terminal cabinet then the following requirement (item A only) is not required. If the Emergency Broadcast Receiver/Repeater is not located at the main security terminal cabinet then provide and install a 1 1/2″ raceway to outside of building through the roof at the designated Emergency Broadcast Receiver/Repeater antenna location and terminated with a weather head. The penetrating portion of conduit shall be a contiguous 10′ piece of rigid, with 6′ firmly supported at two joist points below the roof penetration and 4′ above the roof. Continue this raceway and terminate into a 6″ × 6″ × 4″ box located just above the ceiling at the designated Emergency Broadcast Receiver/Repeater "head-end" room location. Provide and install two 3/4″ conduits run from the 6″ × 6″ × 4″ box and terminate into two flush-mounted 4–11/16″ × 4–11/16″ × 1 1/2″ boxes with single gang mud rings and single gang covers 48″ AFF in their respective locations in the designated "head-end" room.

B. All buildings: Provide and install a 1 1/2″ raceway to outside of building through the roof at the buildings designated Wireless Propagation Enhancement location and terminate with a weather head. The penetrating portion of conduit shall be a contiguous 10′ piece of rigid, with 6′ firmly supported at two joist points below the roof penetration and 4′ above the roof. Continue this raceway and terminate into a 6″ × 6″ × 4″ box located just above the ceiling at the designated Wireless Propagation Enhancement "head-end" room location. For each floor of the respective building, provide and install one 3/4″ conduit run from the 6″ × 6″ × 4″ box and terminate into one flush-mounted 4–11/16″ × 4–11/16″ × 1–1/2″ box with single gang mud rings and single gang covers 48″ AFF in their respective locations in the designated "head-end" room.

C. HURRICANE SHELTERS:

   1. Provide and install an antenna mounting structure located next to the Hurricane antenna stub-out location. Design the antenna mounting structure to support an antenna array of six 6-element Yagi 9dBd gain antennas model #460-6. Structure shall meet ASCE-78 requirements. Top of structure shall be at least 49′ (15 meters) above grade.

2. Provide and install a 2″ raceway to outside of building through the roof at the designated Hurricane Shelter antenna location and terminate with a weather head. Continue this raceway and terminate into a 6″ × 6″ × 4″ box located just above the ceiling at the designated Hurricane Shelter "head-end" room location. Provide and install (2) 3/4″ conduits run from the 6″ × 6″ × 4″ box and terminate into two flush-mounted 4–11/16″ × 4–11/16″ × 1–1/2″ boxes with single gang mud rings and single gang covers. Locate these boxes 80″″ AFF, parallel with 4″ between each in the designated "head-end" room.

Power feed

D. Provide and install a dedicated duplex 120-volt power receptacle fed from the life safety branch of emergency generator power source in each of the designated emergency broadcast receiver, radio communications repeater station, hurricane shelter communications, and wireless propagation enhancement systems "head-end" locations. Locate adjacent to the "head-end" single gang boxes.

Execution

Installation

A. Install systems in accordance with NECA "Standard of Installation" and section 16111.
B. Obtain a detail book from the School Police Department Security Section for system specifics.
C. Install a 200-lb.-strength pull-string throughout the raceway systems.
D. All junction boxes mounted above ceiling shall be mounted with the opening facing down, and shall have a reasonable immediate access pathway provided. Note: removal of a light fixture or other similar ceiling equipment is not considered as a reasonable access pathway.
E. Antenna raceways shall not exceed 100′ from the weather head to the designated "head-end" locations.

Card access system

A. Main cabinet
B. Terminal cabinet
C. Conduit and boxes
D. Power wiring

Main cabinet

A. Steel construction #14 gauge, 36″ × 36″ × 6″ cabinet with hinged, lockable cover, 1/2″ thick plywood backboard, painted light gray and shall be surface mounted.

DISTRIBUTION "BUILDING AND FLOOR" CABINETS

B. Steel construction #14 gauge, 24″ × 24″ × 6″ cabinet with hinged lockable cover, 1/2″ thick plywood backboard, painted light gray.

Conduit and boxes

C. Provide and install the building and floor distribution cabinets for each building according to the following criteria:
   1. There must be one of these cabinets within 300′ of a controlled device.
   2. Each cabinet can feed no more than eight controlled devices and the cabinet must be located on the same floor as the controlled devices.
   3. The main cabinet can serve as the distribution cabinet for its area.
   4. Install a 2″ raceway from the main cabinet to the next building and floor distribution cabinet.
   5. After all distribution cabinets are fed with this raceway, return the raceway to the main cabinet.
   6. Raceway shall not exceed 400′ without a pull box.
   7. For buildings with more than one floor provide a 2″ raceway from the first floor cabinet to feed each floor's cabinet.
   8. If floor two or three has more than one cabinet, then feed cabinet (1) on that floor from the first floor cabinet with a 2″ then proceed to cabinet (2) on that floor with a 2″, and from there to the next floor cabinet.

D. Provide and install 3/4″ conduits from the distribution cabinets and distribute to feed the junction and mounting boxes designated for the area. Each separate 3/4″ feed will supply no more than two Controlled Device/Card Reader Feed locations.

E. Provide and install an elevator interface cabinet, 12″ × 12″ × 4″ with hinged lockable cover and a 1/2″ plywood backboard painted light gray with a ten-lug terminal strip mounted on the board at all designated elevator control panel locations. Feed this interface box from the 3/4″ card access raceway. Provide and install a separate 3/4″ raceway from the box into elevator control panel. Label this box "Card Access/Elevator Interface." The elevator contractor shall extend his control wiring from the elevator control panel to this interface box.

F. At all elevator stops provide 4–11/16″ × 4–11/16″ × 2–1/8″ flush-mounted box with single gang mud ring with cover. Mount the box adjacent to the elevator call button box and use a 3/4″ conduit to connect the two boxes together. Run a 3/4″ conduit from the 4–11/16″ × 4–11/16″ × 2–1/8″ box back to the nearest card access terminal cabinet and terminate.

G. At each controlled door, install the electric hardware's power supply above drop ceiling within 50′ of the hardware. In "separate" raceway, supply 120-volt feed to the power supply. Provide and install a 3/4″ conduit from the supply box to a neutral 6″ × 6″ × 4″ distribution box located next to the supply. Feed this box from the 3/4″ card access system raceway. Also from this neutral box, provide and install an additional 3/4″ raceway along with (8) 18AWG and (2) 12AWG conductors run from the supply box to feed the electrical power transfer device.

From the supply, using that wiring, connect the electric hardware through the electrical power transfer device.

H. Provide and extend raceway to feed 4–11/16″ × 4–11/16″ × 2–1/8″ flush-mounted boxes with single gang mud ring and weatherproof covers; mounted with the opening vertical, at all designated card reader locations. Locate to the strike side of single doors and as designated for double doors, elevator control and gates. Center 4′ above finished floor.

I. Provide and install a 2″ conduit from the main card access terminal cabinet to a brook's box located at designated entry gates. Exact location to be determined at the plan review.

J. Provide and install a 4–11/16″ × 4–11/16″ × 2–1/8″ card access feed junction box with cover, at the interior side of all designated card access door locations. If the area location has removable ceiling tile, the box shall be located above the tile. If the location has a structure of fixed ceiling material, then flush-mount the box with a square to round mud ring and cover.

K. Provide and install a 4–11/16″ × 4–11/16″ × 2–1/8″ flush-mounted box with single gang mud ring and weatherproof cover; mounted with the opening vertical at 60″ AFF as the designated "TeleEntry" mounting box.

L. Provide and install a 4–11/16″ × 4–11/16″ × 2–1/8″ flush-mounted box with single gang mud ring and weatherproof cover; mounted with the opening vertical at 60″ AFF as the designated "Master Key Control Box" mounting box. This mounting box must have 24″ × 24″ free space for mounting the key box.

M. Provide and install a 3/4″ conduit from the main telephone room terminal board to the main card access cabinet.

N. Provide and install six additional recessed mounted 4–11/16″ × 4–11/16″ × 2–1/8″ mounting boxes with flush single gang ring and single gang cover; with the opening mounting vertical with 3/4″ conduit run to nearest card access junction box. (Estimate 100′ of conduit for each run.) Wall-mount locations to be directed in field during construction, prior to final above-ceiling inspection.

Power feeds

O. Provide a double duplex, dedicated 120-volt power receptacle fed from the optional branch of emergency generator power source, adjacent to the lower portion of the main terminal cabinet and each distribution cabinet.

P. Provide and install dedicated 120-volt power feeds from the optional branch of emergency generator source to all Electric Hardware Power supplies.

Installation

A. Install system in accordance with NECA "Standard of Installation" and section 16111.

B. Obtain a detail book from the School Police Department Security Section for system specifics.

C. Permanently label all conduits as to plan room number destination, at all terminal cabinets.
D. Paint all card access system junction-box covers tan.
E. Install 1/2″ (tan round indicators) of paper construction on ceiling tile grid work at all locations where card access system boxes are located above the drop ceiling.
F. Permanently label all the card access system terminal cabinets "card access system."
G. Install 200-lb.-strength pull-string throughout the raceway system.
H. The card access system raceway shall be a separate raceway and shall not interconnect with or be used by any other system without the authorization of the School Police Department Security Section or per LCCCG sections 16722, 16723, 16724, and 16725.
I. All junction boxes mounted above ceiling shall be mounted with the opening facing down, and shall have a reasonable immediate access pathway provided. Note: removal of a light fixture or other similar ceiling equipment is not considered as a reasonable access pathway.
J. All conduit runs shall be as direct as possible in order to save on wiring costs and to reduce poor performance due to cable loss.
K. Refer to LCCCG section 8700 for Card Access Door preparation.

Location

L. Provide a card reader/controlled device at each of the following locations. These are the minimum number of locations, and more locations could be added in the plans review process:
   1. Administrative main entry;
   2. Main entry doors to facility, principal's office;
   3. Bookkeeper's office;
   4. Media center;
   5. A/V storage room;
   6. CCTV studio area;
   7. Cafeteria main entries;
   8. Cafeteria receiving area;
   9. Cafeteria dry food storage area;
   10. Cafeteria manager's office;
   11. Elevator entry on all floors;
   12. Gymnasium entries;
   13. Auditorium entries;
   14. Custodial receiving;
   15. Video-Surveillance Control Room Door;
   16. School Police Office; and
   17. Other areas as defined in the plan specific review process.

FLORIDA DEPARTMENT OF EDUCATION DESIGN AND CONSTRUCTION PROCESS

Executive Summary

Introduction

This is a summary of specific and general findings and recommendations based upon responses to survey instruments, field investigations at schools and community colleges throughout Florida, and a review of the literature on safety and security in schools and community colleges. The "Methodology" and "Survey Results and Related Data Analysis" sections below provide a detailed description of the approaches used to gather the data and the results of that effort. Accompanying some of the findings here are general recommendations; others are spelled out in the *Guidelines* that precede this section and are highlighted in the bullet points associated with the graphics and the text.

Specific Findings and Recommendations

Findings:

Knowledge, Use, and Assessment of CPTED, Safe School Design Principles, and the Existing Guidelines

Survey respondents for this research were, by and large, much more aware of CPTED principles than were respondents to the survey conducted as part of the research preceding publication of the 1993 *Guidelines*. Awareness and use of the principles is credited to increased training through reference organizations and through articles in the trade and professional literature.

Despite higher knowledge and awareness levels of those principles generally, survey respondents and those interviewed in the field were considerably less aware of 1993 *Guidelines*. Design professional respondents were much more likely than other respondents to indicate that either Safe School Design principles or the *Guidelines* have been incorporated into the design and construction of public schools and community colleges.

Most Important Features Relative to Safety and Security Issues

When asked to identify the "most important" Safe School Design features relative to safety and security, the responses from the school district, community college, and design professional groups centered around *surveillance*, *access control*, and *territoriality* issues, in this order.

Does Safe School Design Make Schools Safer?

Although some design professionals are critical of elements of the *Guidelines*, the great majority of all respondents believe that the use of Safe School Design principles and the *Guidelines* makes schools and community colleges safer places. Some design professionals report difficulty fulfilling the design intent of "slippery finishes," "audio/motion detection systems," and providing for the "separation of after-school activities" from the core educational facility.

Recommendation:

Continue education programs focused on CPTED and Safe School Design principles as contained in the *Florida Safe School Design Guidelines* for school district risk/facility managers, community college risk/facility managers, and design professionals. Wherever possible, partner with organizations, such as the Florida Association of School Resources Officers (FASRO), to present educational programs about the *Guidelines* and their impacts on promoting school safety and security.

What Are Serious Crimes on Public School and Community College Campuses?

Finding:

Principals report that *fighting*, *disorderly conduct*, and *vandalism* are the top three crimes on their campuses in terms of numbers of incidents while community college respondents say that *larceny/theft*, *vandalism*, and *breaking and entering* are the three crimes they perceive to be the most serious based on event frequency.

Preferred School Design Type

Finding:

A sizeable majority of respondents overall favor the one-story, centrally organized building group configuration over other plan types. However, only a plurality of principals and community college respondents favored this arrangement over others.

Critical Areas of School Design

Finding:

Relative to critical areas of school design, *corridor surveillance* issues are the most important concerns among the respondent groups generally. These are followed by *perimeter enclosure* issues and by the perceived need to *minimize niches*. For public schools in particular, corridor surveillance is considered to be especially important. Community college risk/facility managers reported that their most critical areas of school design related to security are *exterior lighting*, *alarm systems,* and *interior lighting*, in this order.

What Single Policy or Procedure Change Would You Recommend?

Finding:

All respondents were asked the open-ended question, "What single policy or procedural change would they recommend if funding were available?" Central themes that emerge from this very mixed collection of open-ended responses tend to center around the fundamental problems of *surveillance* and *access control*, which are interwoven issues.

Public Access after Hours

Finding:

With the notable exception of community college respondents who were split almost evenly (50 percent "Yes," 45 percent "No," and 5 percent nonresponsive), significant majorities within the other three respondent groups (ranging from 65 to 69 percent) believe that public access does make their institutions more prone to criminal activity.

Posting Signs for Access Control

Finding:

Most respondents (64 percent) who answered this question report that their school, district, or community college had policies in place governing the posting of access-control signs during school hours; however, a significantly lower number (33 percent) report having policies for signage to deal with *after-hours* access.

Crime by Location

Finding:

Several distinct peaks and valleys are evident. On the *high* side of the spectrum, respondents overall identify *parking lots, off-grounds/adjacent buildings, locker rooms,* and *restrooms* as the top four places for crimes and report *low* criminal activity on *rooftops of covered walkways*, on *building rooftops*, in *lobby/reception areas*, and at *main entrances*. The most crimes reported in parking lots are trespassing and vandalism. The most crimes reported in off-grounds/adjacent buildings are use of alcohol, tobacco, and drugs and fighting. For locker rooms, the most crimes reported are larceny/theft and fighting. The most crimes in restrooms are vandalism and the use of alcohol, tobacco, and drugs.

Crime by Time

Finding:

All respondent groups for this question perceive that significantly more crimes occur *before school, after school, during the evening*, and *on weekends* than during *normal classroom hours* (including breaks between classes). This perception is consistent with earlier results showing that respondent groups (with the exception of community college risk/facility managers) overwhelmingly report the perception that after-hours access to public school campuses makes them more prone to crime.

Other Serious Concerns on Public School and Community College Campuses

Finding:

Principals, school district risk/facility managers, community college risk/facility managers, and school resource officers were asked whether they had serious concerns relative to "*gang-related activities, hate crimes, bomb threats, terrorists, and violence in the workplace*." Of the listed concerns, "bomb threats" registered the highest number of affirmative responses, with slightly over a fourth (26 percent, or 40 individuals) of all those who answered this question saying that these were a serious concern at their schools. The second most serious concern was "violence in the workplace" which registered a positive response from 24.2 percent of the respondents (thirty-four individuals) who answered this question, and the third was *gang violence*, which was viewed by 19.3 percent of the respondents as a serious concern.

What Design Changes Would You Implement in Your School/Community College to Make It Safer?

Finding:

All of the respondent groups put *surveillance* issues clearly at the forefront, followed by *access control* (with the exception of school resource officers who combined *access control and territoriality* issues second), and *territoriality* third. Some respondents combined all of these, as well as *guardianship* and *management* issues, in their comments.

General Findings and Recommendations

Surveillance and CCTV

Finding:

There is significant and growing interest in the use of closed circuit television (CCTV) to accomplish surveillance in both interior and exterior spaces at schools and community colleges.

Recommendation:

Continue emphasis on programs that focus on designs for natural surveillance, which should be the first strategy for surveillance. Nevertheless, a framework for the application and design integration of CCTV for security purposes should be developed for Florida public schools. A model could be the National Institute of Justice report, the Appropriate and Effective Use of Security Technologies in US Schools, NCJ 178265, September 1999.

Small-Scale Design Elements

Finding:

Specific, small-scale elements may be important contributors to school and community college security. Examples include the design and placement of access-control signs and alarming secondary building entrances to warn administrators and school resource officers when doors are propped open.

Recommendation:

Support research into small-scale design changes in schools and community colleges.

Community Colleges and K–12

Finding:

Even though they are subject to the same Safe School Design Requirements found in Florida Building Code requirements, community colleges and public schools present very different security design profiles in that they have very different student populations and, for the most part, different campus plans and organization. As a consequence, they tend to have different crime and crime prevention issues.

Recommendation:

Future research should concentrate on community colleges and public school as separate entities insomuch as security design is concerned.

After-Hours Access

Finding:

Respondents to our surveys, and especially those from public schools, report overwhelmingly that after-hours access to school campuses makes them more prone to crime, even though the co-location of public facilities is bound to increase in the future. School designers report that they have problems fulfilling the design intent of this aspect of Safe School Design. Moreover, most respondents believe that more crime takes place after normal daytime class hours than during that period.

Recommendation:

Increased research and continuing education should be devoted to the after-hours access issue.

School Crime and Location Data

Finding:

There is a lack of Florida data that connects school design, location, and criminal acts.

Recommendation:

School Environment Safety Incident Reporting System (SESIR) data should include references to the specific locations where crimes are committed in schools.

Student Interviews

Finding:

Field interviews and survey research about crime in Florida public schools and community colleges suggest that school children and community college students may be the best sources of information about crime on their campuses.

Recommendation:

Student interviews should be incorporated into research conducted for the updating and revision of future *Florida Safe School Design Guidelines*.

Section V

Appendices and References

APPENDIX 1

CRISIS MANAGEMENT TEAM ACTION CHECKLISTS

Action Checklist: Crisis Manager (Superintendent/Delegated Authority)

Key roles: 1. Represent SS to external world 2. Delegate authority to CMT decision makers/team 3. Communicate/negotiate with stakeholders who have an interest in the event 4. Provide policy or position guidance to CMT
Actions
Seek briefing from CMT leaders/decision makers and discuss initial assessment: • What has happened? • Why has it happened? • What is the potential for escalation? Is incident contained? • How has it happened? • Is anyone injured? • When did it happen? • Where has it happened?
Confirm the need to activate the CMT: • Overreaction better than under-reaction • Play it safe in case of escalation
Discuss management strategy with CMT leaders. Monitor response; be updated by CMT leaders and liaise with other SS facilities and Mayor's Office, as required: • Agree upon follow-up schedule
If CMT activated, then inform Mayor's Office as the situation warrants.
If possible, attend initial briefing with CMT: • Brainstorm—identify issues, assign and track actions. Determine SS responsibility for incident and posture that district should adopt. • Remember role and focus of CMT: — PROTECT WELFARE AND SAFETY OF PEOPLE FIRST—THIS IS PARAMOUNT — District response — Support joint venture partners/government/media strategy/local response team
Review media strategy with External Affairs and advise CMT leaders. Consider the following: • Prime district spokesperson • Timely and appropriate press statements • Media reaction to date • Is incident on TV/news? • Is a press conference likely to be required? — When, where? — Have Q&As been developed?
Ensure initial media statement drafted (and released as soon as possible as appropriate).
Confirm statutory notifications have been made and updates are being provided:
Confirm SS internal communications are effective and timely.

Identify and conduct executive liaison with all relevant stakeholders: • Partners with vested interest • Government entities • Parents • Local community • Political entities
Attend regular CMT updates as appropriate.
Make strategic decisions to minimize impact on SS image, operability, and liability.
Monitor the effectiveness and morale of the CMT.
Ensure that SS decisions are implemented.
If city/county CMT(s) are(is) in place, ensure CMT coordinator liaises with them: • Is it supporting or micromanaging?
Any signs of stress within the CMT? • SS employee assistance program (EAP) • Consider member substitutions or relief
Is composition of CMT appropriate? • Too many/too few? • Does CMT need to remain in place all the time?
Ensure student/employee support programs activated? • What programs are being initiated to take care of students/employees? • Support to family/significant others? • Financial
Is normal business able to continue? • If not, confirm strategy for continuity of operations • What actions are being taken to advise parents/community of potential disruption? • How long before business operations resume and in what capacity?
Sample strategic issues: • What are the district's real liabilities? • Who is responsible for this incident? • What posture should asset/district take? • What are the impacts and effects? • What other strategic issues?

Action Checklist: Crisis Management Team Leader (Executive Director, Business Affairs/Delegated Authority)

Key roles: 1. Evaluate overall execution of CMT 2. Communicate activities to superintendent, and remote locations, as applicable 3. Provide additional input and guidance as requested 4. Keep superintendent informed
Action
Make initial assessment (in consultation with security and manager of Energy, Environmental, Health, and Safety): • What has happened? • Why has it happened? • What is the potential for escalation? Is incident contained? • How has it happened? • Is anyone injured? • When did it happen? • Where has it happened?
Determine the need to activate the CMT. Consider the following: • Full or partial activation required • Other potential emergency support groups
Assess the incident using the Crisis Triggers (section 2.1); initiate the CMT activation if required.
Contact the superintendent and provide an update. Confirm intentions for management of incident: • Monitor only • Monitor and liaise with CMT as required • Mobilize, lead, and liaise with CMT
Contact on-scene SS incident commander and get an update. Ensure response is appropriate and fully resourced. • What does the site need? • Determine call schedule with site
Set and agree on priorities with CMT: • Take care of people • Protect environment • Protect assets • Business continuity as efficiently and safely as possible
Determine support required by CMT: • Additional response staff • Employee assistance programs • Liaise with safety officials • Support staff/administrative staff • Other resource specialists

Attend initial briefing with CMT: • Brainstorm—identify issues, assign and track actions • Determine SS responsibility for incident and posture that district should adopt • Agree on objectives developed by CMT • Remember role and focus of CMT: — Protect welfare and safety of people — District response — Government/media strategy/local response team • No shortcuts of emergency management structure
Agree on media strategy with External Affairs and superintendent. Consider the following: • Prime district spokesperson • Timely and appropriate press statements • Media reaction to date • Is incident on TV/News? • Is a press conference likely to be required? — When, where? — Q&As
Ensure initial media statement drafted (and released as soon as possible, if appropriate).
Ensure a Stakeholder Management Plan is developed and implemented: • Prioritize • Ensure statutory notifications are made and updates provided
Ensure Planning takes into account escalation and worst-case scenarios.
Ensure SS internal communications are effective and timely.
Is Telephone Response Team in place, briefed, and ready to take calls? • Ensure switchboard operators are briefed on how to handle calls
Identify and conduct executive liaison with all affected constituencies: • Partners with vested interest • Government agencies • Parents • Local community • Political entities • Other stakeholders
Ensure that External Affairs keeps Telephone Response Team fully briefed and that the Telephone Response Team has the latest press releases.
Hold regular CMT updates: • Time-out phones off hook
Make strategic decisions to minimize impact on SS image, operability, and liability.
Monitor the effectiveness and morale of the CMT.
Determine the need for alternates and the rotation of team members.
Ensure that arrangements are made to continue normal business operations, if appropriate:

• Confirm need/implementation of business continuity plans as appropriate
If city/county CMT is in place, coordinate and confirm mutual information and support needs and expectations.
Are Information Boards being kept up to date?
Is there an auditable trail of information?
Are personnel matters being sympathetically handled? • Are student/employee/family support programs being implemented, as appropriate?
Are you getting good feedback from External Affairs over what questions are being asked by callers?
Any signs of stress in CMT, or other responding SS units? • Brief Employee Assistance/Counseling Program(s) and ensure teams are aware of the numbers to call
Is composition of CMT appropriate? • Too many/too few? • Does CMT need to remain in place all the time?
Is normal business able to continue? If not, what actions are being undertaken to address?
Ask yourself—How am I holding up?
Continually review all of the above.
Strategic issues for consideration: • Is SS effectively positioning itself for a successful outcome? • Will there be an inability to meet commitments/delivery? • Impact on student body/staffing (long, medium, and short term) • Impact on suppliers or other service providers? • Other direct business interruptions? • Any indirect constraints on business? • Risk/need to shut down similar operations? • Other threat to operations • How long for repair/replacement • Any regulatory impacts? • Additional load on management? Managing the incident/recovery

Action Checklist: Crisis Management Team Coordinator (Manager, Energy, Environmental, Health, and Safety/Delegated Authority)

Key roles: 1. Manage Crisis Center facilities 2. Facilitate meetings 3. Ensure all actions are documented and tracked 4. Provide resources to CMT 5. Keep CMT up to date on objectives and focused on appropriate issues 6. Liaise with other coordinators and teams
Action
Set up Crisis Center facilities (see table 3.2 activation checklist).
Evaluate status of emergency and emergency response system needs. • Understand priorities • Mobilize recorder if appropriate • Brief log keeper and recorder • Ensure faxes, telephones, photocopiers are all in working order, with a plentiful supply of paper, and so on • Set up PCs, log on, and open e-mail account (send test messages to CRT/CMT, as appropriate) • Hang information displays on wall • Produce issues/actions tracking sheet and hang on wall for meetings and/or use PC projectors • Issues/actions/responsibility/target completion • Track status throughout and follow up as required • Ensure action/event log is displayed on main screen and data entered by log keeper • Ensure all office supply requirements are available • Distribute incident checklists and office supplies • Ensure white boards have been set up correctly and that markers are available • Check that clock time is correct and working properly • Ensure CMT has mobilized the Telephone Response Team (TRT)—minimum of two members for all incidents. Review incident and increase numbers of TRT if appropriate. • When TRT is manned, direct switchboard to transfer all incident-related calls to TRT • Arrange for copies of Crisis Management Plan for each team member • Check miscellaneous documentation for relevant supplementary information • Ensure auditable trail of information is maintained — CMT members to use bound notebooks for all notes — Enter name, date, and time when coming on shift • Contact and liaise with Logistics. — Organize staff relief — Food/water/drinks, and so forth — Accommodation — Taxis, and so on • Contact security for any advice and/or for additional support/assistance

Brief switchboard operator and the security staff: • Outline of incident • Press/media inquiries • Use of TRT as filter for CMT (no calls to be transferred directly to the CMT) • Ensure switchboard operator and security know the transfer numbers for the TRT • Ensure security is sufficient and consider restricting staff access to the Crisis Center
Facilitate initial CMT briefings: • Review setup checklist and ensure team is fully briefed • See initial briefing checklist • Ensure the school/facility is represented/aware • Brainstorm issues and help identify actions
Initiate and maintain contact as required with other team coordinator(s): • Ensure CMT remains focused on business response—no micromanaging • Share issues and plans for next period • Ensure call schedule is appropriate • Ensure auditable trail of information is being maintained • Encourage use of e-mail to pass updates
Ensure that the CMT and support staff are regularly briefed and kept up to date: • Ensure initial briefing follows checklist • Ensure personnel members review checklists and understand roles and responsibilities • Issue press statements to Telephone Support Team
If mobilized, initiate and maintain contact with city/county CMT coordinator: • Share issues and plans for next period • Ensure call schedule is appropriate • Determine CMT issues/concerns/themes
Monitor the level of resources required to support the CMT: • Ensure all personnel members involved in the Emergency Management have everything they require • Ensure that the CMT is operating as designed • Coordinate the collection, display, and dissemination of relevant information • Maintain a display of the CMT workload, actions, and priorities • Identify and mobilize additional technical support/expertise required
Ongoing checks: • Individual team efficiency checks: are they working effectively? • Team interaction or communication issues, breakdowns • Is communication logging and recording being maintained? • Are call logging sheets being used by TRT, and are actions being closed out? • Are communication processes working? Potential bottlenecks or overload? • Are teams maintaining their focus? • Does CMT need to be completely manned full time? • Can CMT achieve aims by meeting at regular intervals (determine frequency)?

Coordinating ongoing health checks: • Stress monitoring, support, and management • Provision of a quiet room or area with facilities to unwind • Consider the need for specialist counseling or support agencies • Additional Human Resource specialist support
Resource drawings, plans, charts, and photographs: • Ensure copies of emergency plans are available • Are approved actions being undertaken? • Identify outstanding issues, hurdles, and challenges, and bring to CMT's attention for evaluation
CMT staffing review forward planning: • Establish or estimate anticipated incident duration, hours, and days • Conduct a staffing review with the CMT leader • Coordinate a strategy for mobilizing the staff to the incident site • Request shift rotations for extended periods of operation • Notify off-duty coordinator for logistics planning
Consider the need or potential for school function relocation and resource movements: • Review business continuity and alternate relocation plans and procedures • Identify resource and logistics support needs • What and where is it required? • How to achieve it • Prompt CMT leader
Incident termination: • Consider when it is appropriate to declare the incident over and return to normal business • Full stand-down • Partial stand-down • Ensure that all teams are informed of the stand-down irrespective of the level • Create a master file of all data and pass to legal for retention — Ensure all log sheets are collected before the team leaves the room — Arrange for copies of all e-mail traffic and incident files to be transferred to a CD Rom and stored — All notebooks to be copied and/or originals to be retained • Lead team debrief on stand-down or soon after. Capture lessons learned/actions. • If there is deemed to be any value in photographing the room layout and key display boards, ensure this is done before the room is tidied • Ensure that a clear action is initiated to have CMT room reset/restocked immediately after stand-down

Action Checklist: External Affairs Leader (Executive Director, Communications/Delegated Authority)

Key roles: 1. Protect district reputation and development of External Affairs Strategy for response 2. Maintain oversight and provide advice for External Affairs issues 3. Liaise with third-party External Affairs representatives to ensure common and consistent messages are transmitted to stakeholders 4. Identify all stakeholders and development of stakeholder management strategy 5. Liaise with city/county counterparts
Action
Register arrival at Crisis Center.
Work with CMT leader and evaluate status of emergency: • Agree to priorities • Agree to the media strategy • Assist with the preparation of press releases using confirmed information only • Ensure that the Media Response Team is properly briefed on handling media inquiries
Relocate to External Affairs Room between CMT update meetings: • Draft press statements • Liaise with government, partners, media teams, and so forth, as required by CMT leader • Brief switchboard operator on transfer of calls to Consumer Affairs • Brief Telephone Response Team (TRT) • Activate TRT direct-dial numbers once TRT is established and fully manned • Monitor/manage TRT (under TRT supervisor) • Liaise with Logistics and Human Resources • Monitor/record all media broadcasts
Ensure that CMT are kept up to date and are fully informed about incident: • Prompt CMT leader for update meetings as required • Ensure a rapid flow of the latest factual and releasable information • Advise on media and government relations • Brief on current media tone and themes • Advise on the current or likely effects of the emergency on SS's image or reputation
Ensure appropriate resources are available. Consider need for the following: • Mobilization of the External Affairs Group • External Affairs support at scene • Call out of the External Affairs Team • Liaise with Human Resources Team
Confirm with Human Resources Team leader what information on casualties can be given to media.

Ensure assistance to CMT: • In the development of Q&As for news conferences • In the consistency of the messages being prepared by SS • In the production and documentation of all statements made to the media • In preparation of SS management who may be speaking • In recorded playback of news broadcasts about the incident
Ensure that copies of press releases are as follows: • Consistent • Scrutinized by legal counsel prior to release • Being given wide, internal distribution • Being given external distribution as per list of companies/authorities/agents/stakeholders, and so on • Agreed to coordinate/coordinated with safety/law enforcement/vested business partners/contractors
Ensure that the district is established as the single authoritative source of information: • Press conferences arranged • Media requests for additional information discussed with CMT leader • Are you up to date on facts? • Is another press release due now? • Next steps?
Provide update of number and nature of calls received at regular intervals.

Action Checklist: Legal Representative (General Counsel/Delegated Authority)

Key roles: 1. Assess legal liabilities, criminal and civil actions 2. Provide legal advice to CMT leadership 3. Review press releases
Action
Register arrival with Crisis Center.
Contact other disciplines and seek advice/mobilize them if required: • Risk • Legal • Finance
Work with CMT leader and evaluate status of emergency relating to the following: • Communication of operational incident details within SS • Students, employees, and families • Partners • Government agencies • Safety and law enforcement agencies • Neighbors • Special interest groups • Suppliers • General • Unions
Determine and advise CMT: • Potential criminal actions • Legal liabilities • Preparation of testimony and counseling of SS representatives at government and legal hearings • Legal advice on communication with the media • Where actions could be interpreted as admission of guilt • Giving clarity to affected parties' legal responsibilities within contractual context • Provide input to situation reports
Ensure CMT responses are recorded for use during postincident investigations: • Recording tapes for emergency lines • Copy of all notes/records/e-mails • CMT notebooks • Message sheets • TRT records
Brief CMT: • On decisions taken • Actions and concerns of the CMT • On legal/finance skills and resources required
Provide input/recommendations for the form and scope of the postincident investigation.

Strategic issues for consideration:

- What insurance coverage is applicable?
- Cost of repair/replacement
- Cost of response to emergency
- Cost of clearance/recovery/cleanup/own resources/other resources
- Cost of liabilities, claims, fines, and compensation. Significant claims to be expected?
- Penalties/fines/damages to be expected? Are actions likely?
- Legal situation on continuing operations. Any injunctions possible?
- Legal costs
- Compensation payment necessary?
- SS professional liability? Partner costs?
- Casualties/medical costs/legal liabilities
- Cash flow implications
- Impact on SS?
- Any other financial impacts
- How will legislators react?

Action Checklist: Security Officer (Director Student Services and Security/Delegated Authority)

Key roles: 1. Initial incident assessment and notification 2. Prime interface with safety incident commander 3. Security control and support to response operations 4. Security guidance to CMT 5. Take charge of security delegation
Action
Make initial assessment (in consultation with CMT leader and on-scene SS incident commander): • What has happened? • Why has it happened? • What is the potential for escalation? Is incident contained? • How has it happened? • Who is injured? • When did it happen? • Where has it happened?
Liaise with safety (police, fire, and emergency medical) and building security: • Threat assessment: real, false alarm, hoax, or uncertain • Determine/agree appropriate immediate response action • Appoint security representative that will interact with safety incident commander • Agree to communications protocol and schedule
Notify/activate security staff, as appropriate: • SS • Contract services
Provide access control and enhanced physical security as deemed necessary: • Buildings • Areas • Surrounding areas • Emergency operations center • Staging areas • Medical care receiving areas (hospitals and clinics) • Remote sites
Coordinate and communicate security buildup with CMT, building security, and with safety organizations.
Liaise with on-site safety incident commander (fire, police, and FBI): • Conduct threat assessment • Review, agree, and modify response actions under way as directed • Advise incident commander of SS response organization and location(s)
Confirm extent of safety and law enforcement involvement:

• Determine/confirm command and control relationships between SS and federal and state law enforcement agencies
Advise CMT of safety incident commander directives.
Attend initial briefing with CMT: • Brainstorm—identify issues, assign and track actions • Advise CMT on security-related directives, issues, concerns, and actions • Establish/agree on objectives developed by CMT
Agree on media strategy with External Affairs: • Confirm prime district spokesperson • Determine whether/when safety desires to issue press releases, make press statement, conduct media interviews, or establish a Joint Information Center • Facilitate/liaise between External Affairs and safety spokespersons(s)
Keep CMT informed and current on security-related matters.
Ensure a Stakeholder Management Plan addresses all relevant law enforcement and security-related constituencies: • Ensure statutory notifications are made and updates provided
Ensure security arrangements are made to support execution of school function relocation plans, as appropriate: • Headquarters building • Schools • Personnel/operations in transit • Receiving sites • Other locations
Maintain records to ensure an auditable trail of information.
Ask yourself—How am I holding up?
Continually review all of the above.
Strategic Issues for consideration: • Is SS effectively positioning itself for a successful outcome? • New/changing threats? • Need to expand/modify liaisons with external safety and law enforcement agencies • Maintaining effective and current information flow to CMT to support decision making

Action Checklist: Environmental, Health, and Safety Representative (Manager, Energy, Environmental, Health, and Safety/Delegated Authority)

Key roles: 1. Responsible for management of environmental-, health-, and safety-related issues
Action
Determine the following: • Nature of incident • Personnel status (injured/missing/deaths, and so forth) • Response operations status • Environmental, health, and safety (EHS) concerns
Work with the CMT leader and evaluate status of emergency: • Agree on priorities — Take care of people — Protect environment — Protect assets — Ensure business continuity
Ensure that CMT has appropriate systems and resources in place for managing EHS issues: • Contribute to the CMT tactical and general deliberations • Advise the CMT on the EHS issues arising from the emergency • Ensure that systems and resources are in place for managing issues related to EHS matters
Ensure that appropriate resources are available: • Consider need for on-site EHS support
Coordinate EHS advice between CMT and the site.
Advise the CMT on whether actual or potential EHS hazards exist and on options to minimize their effect: • Take action, where appropriate, to implement decisions made by the CMT on EHS issues • Keep the CMT leader advised of any pressing safety issues • Ensure safety-related information is accurate and up to date
Consider need for, propose, and initiate appropriate investigations into and reports on the incident.
Provide advice to the CMT on the availability of additional technical advice.
Ensure written situation reports are provided by the CRT, and draft situation reports for the CMT: • Use e-mail if practicable
Brief CMT: • On decisions taken • On actions and concerns of the CRT
Provide EHS input into recommendations for the form and scope of postincident investigation.
Other strategic issues for consideration: • What district assets are affected? • What are likely to be the effects on the EHS? • Short term—implement short-term response • Longer term—decide on longer term responses • How will regulatory agencies react? • How will pressure groups react? • What do our employees need? • What EHS considerations are immediate?

Action Checklist: Risk Representative (Treasurer/Delegated Authority)

Key roles: 1. Liaise with insurance providers 2. Provide general risk-related advice on the response 3. Record and manage all claims
Action
Register arrival with Crisis Center.
Contact other disciplines and seek advice; mobilize them if required: • Risk • Legal • Finance
Work with CMT leader and evaluate status of emergency relating to the following: • Communication of operational incident details within SS • Students, employees, and families • Other SS facilities/sites • Partners • Government • Site neighbors • Special interest groups • Suppliers • General • Unions
Liaise with insurance providers: • Keep insurers informed • Liaise with Finance and develop the optimum cost-recording mechanism to capture insurance-related costs to maximize recovery • Provide risk- and insurance-related advice on the response strategies • Record of all claims (set up a claims hotline) • Strategy for managing claims • Provide advice on recoverable/nonrecoverable costs
Ensure appropriate controls, systems, and audit procedures are implemented with respect to emergency expenditure: • Make arrangements for cash and currency when necessary • Set up a claims hotline as soon as possible, if required • Estimate potential cost of response operations/impacts as requested • Advise the CMT on financial liabilities concerning the incident and ways to minimize them
Ensure CMT responses are recorded for use at any postincident investigation: • Recording tapes for emergency lines from emergency operations center • Copy of all notes/records/e-mails • CMT notebooks • Message sheets • TRT records
Provide input/recommendations for the form and scope of the postincident investigation.
Strategic issues for consideration: • What insurance coverage is applicable? • Cost of repair/replacement

• Cost of response to emergency • Cost of clearance/recovery/damage restoration/cleanup/own resources/other resources • Cost of liabilities, claims, fines, compensation. Significant claims to be expected? • Penalties/fines/damages to be expected? Are actions likely? • Legal situation on continuing operations. Any injunctions possible? • Legal costs • Compensation payment necessary? • SS professional liability? Partner costs? • Casualties/medical costs/legal liabilities • Cash flow implications • Impact on SS? • Any other financial impacts • How will legislators react?

Action Checklist: Planning Representative (Director Strategic Planning/Grants/Delegated Authority)

Key roles: 1. Look ahead, identify issues, consider worst-case scenarios, and develop contingency plans to ensure continuity of response and to minimize impacts 2. Draft emergency response objectives; recommend strategies and priorities for the CMT leader and to assist with stakeholder management 3. Work closely with CMT staff to coordinate production of incident briefing documentation, situation reports, and so forth
Action
Register arrival with the Crisis Center.
Establish the facts, evaluate status of emergency: • Review immediate priorities — Take care of people — Protect environment — Protect assets — Ensure business continuity as efficiently and safely as possible
Support the CMT coordinator in managing the response room and information displays: • Manage the operation and recorder • Identify key issues and ensure that they are tracked on the issues/actions board
Assess the incident using the Crisis Triggers and Assessment checklists: • Agree on the assessment with the CMT leader and present to the CMT at the next update meeting (consider whether the CMT needs to sit full time/part time/partially manned/fully manned/not at all) • Ensure the assessment board is marked up by the log keeper • Ensure that the incident facts are marked on the incident status board • Reassess at regular intervals and update boards
Maintain a high-level view of the incident and make recommendations to the CMT leader as appropriate.
Help External Affairs to develop the Stakeholder Management Plan: • Ensure appropriate internal and external notifications are made
Identify worst-case/best-case/most-likely case scenarios: • Think "outside the box" • Liaise with external organizations as appropriate • Chair planning meetings • Brief CMT • Recommend the creation of project teams as appropriate
Produce an Incident Briefing Document for transmission to senior management as appropriate: • Update regularly
Ensure that CMT has appropriate systems and resources in place for managing EHS and community issues: • Review Environment, Security, and Community Representative Checklists • Mobilize personnel as required

Develop plans for maintaining the response throughout future operational periods. • Focus on the business issues • Liaise with CMT members to ensure that operational issues have been covered. Support as required. • Develop plans for CMT Operations to execute tomorrow/the next day/next week/and so on. For lengthy incidents, consider the use of Incident Action Plans and turning the emergency into a project.
Strategic issues for consideration: • Has SS management been briefed and kept up to date? • Have all stakeholders been identified, prioritized, and engaged? Is there a Stakeholder Management Plan? • Is the CMT planning for the worst-case scenario? • Is the chosen strategy working? • What SS assets are affected? • What is the impact on the SS/local community? • Short term—implement short-term response • Longer term—decide on longer term response • How will regulatory/oversight agencies react? • How will law enforcement agencies and other government/political entities react? • How will parents/community react? • How will special interest groups react?

Action Checklist: Community Representative (Executive Director, Communications/Delegated Authority)

Key roles: 1. Support Planning representative by identifying community issues arising from incident 2. Develop community response strategy that reflects the district's core values 3. Assist External Affairs in development of Stakeholder Management Plan
Action
Register arrival with the Crisis Center.
Start and maintain an Incident Log: • Provide regular updates to the log keeper so that relevant information can be captured in the action/event log
Work with Planning and evaluate the status of the emergency: • Review priorities • Review media strategy • Provide input for press holding statements/press releases • Provide input for the creation of incident objectives
Confirm relevant authorities have been informed: • Identify immediate issues
Assess the impact of the incident on the community: • Type and scope of impact • Magnitude and duration of impact • Areas of community affected • Casualties • Local involvement of safety, government entities, media, and so forth • Other
Seek specialist advice as appropriate: • Internal • External
Consider implications of escalation and need for SS representation on the ground to liaise with community leaders: • Who should represent the district? • What level of management? • Consider safety and security issues
Assist the local response team in providing immediate support: • Confirm the initial response is in accordance with SS core values
Identify status of current district-sponsored community projects and resources.

Work with External Affairs and identify the community stakeholders and the best way of communicating with the community: • Assist in the development of the Stakeholder Management Plan — Media — Authorities — Third party — "Town Hall" Meetings — Flyers — Door to door — Hotlines — Spokesperson(s) — Other
Capture feedback from the community and identify key issues: • Share key issues with CMT, especially External Affairs and security. Sources include the following: — Claims lines — Counseling support — Personal contact — Meetings — Feedback from responders/authorities, and so on
Develop a sustainable strategy for supporting the community: • Ensure strategy is achievable • Try to include some "quick wins" at the start • Don't make promises that can't be kept
Provide regular updates to Planning for input for the Assessment Meeting and/or attend if required.
Provide input to Planning for the Incident Briefing Document.
Assist Finance to estimate costs.

Action Checklist: Log Keeper Representative (District Secretary/Delegated Authority)

Key roles: 1. To keep the incident display boards up to date 2. Assist the Planning representative and recorder in their tasks
Action
Ensure that equipment and boards are properly set up in the Crisis Center: • See Crisis Center Activation Checklist, section 3.2 • Write up boards during briefing • Request the CMT leader provide a recorder, if necessary • Pass logs to recorder for transcription and retention • Print action/event log when complete and pass to recorder for transcription into a word-processing document • Work with Logistics—identify and track resources
Brief recorder on arrival.
Establish and maintain the official CMT Log of Events: • Mark the information on the Log of Events and the relevant CMT wallboards
Ensure all boards and displays are kept up to date: • Request a time out/update from CMT leader if unsure of situation or if information thought to be missing • During updates, stop briefing if you need clarification
Supervise the recorder and ensure the log is typed accurately: • Answer any queries the recorder might have
Ensure that the Log of Events is as follows: • Printed off • Copied and distributed to CMT personnel • Copied to keep an extra copy of the Log of Events for the CMT leader as a permanent record • Used for update messages (e-mails) when appropriate

Action Checklist: Documentation/Recorder Representative (District Secretary/Delegated Authority)

Key roles: 1. Work with the log keeper and maintain full records of the response
Action
• Recognize significant events and ensure they are properly recorded • Maintain a log of all activities and information • Document CMT decisions—liaise with coordinator if any questions or if clarification required • Maintain a record of contacts made with government and other external authorities/agencies made by the CMT • File all incoming and outgoing documents and ensure their traceability • Assist the coordinator in the preparation of status reports • Preserve all documentation that may be required for future insurance and legal claims • Recover documents and notes from all CMT members (on a regular basis) • Keep the minutes of CMT updates • Monitor the accuracy of visual displays • Manage the e-mail interface between the CMT and other response team elements and involved parties

Action Checklist: Logistics Representative (Coordinator, Business Support Services/Delegated Authority)

Key roles: 1. Provide logistics support to the CMT and other response elements 2. Resource appropriate transportation for teams 3. Identify additional resources required 4. Supply contract management 5. Track and account for all personnel members
Action
Register arrival with Crisis Center.
Initiate and maintain contact with CMT, remote team coordinators, and other response team elements as appropriate: • Get an update/establish facts • What does the CMT need? • Ensure response is appropriate and fully resourced • Agree that Logistics call schedules with teams (if required)
Work with CMT and evaluate status of emergency: • Agree on priorities • Take care of people • Protect environment • Protect assets • Business continuity • Casualty evacuation to a safe place for medical treatment • Personnel evacuation /shelter-in-place/student/personnel care and custody issues
Ensure appropriate resources are available: • Consider need for transportation for: — CMT and support staff — Students — Employees — Equipment, supplies, food, sanitation, and other materials
Ensure assistance to personnel evacuated: • Staging area reception (Consider need for postincident debriefing counseling support) — Medical treatment — Cold/inclement weather protection — Cash — Food/water — Accommodations — Others
Control the logistics of technical support sent by the CMT: • Identify/maintain/record the location and movements of people and other support resources dispatched • Maintain a record of all assets available in the field to provide support
Assist CMT in procuring resource drawings, plans, charts, photographs, and other technical support.
Provide advice to the CMT on the availability of additional contracts/logistics advice.
Establish requirements for emergency equipment supplies and other logistical support resources.

Action Checklist: Human Resources Representative (Director, Human Resources/Delegated Authority)

Key roles: 1. Identify and track all personnel members involved in incident 2. Coordinate and provide feedback to employers 3. Ensure consistent messages are conveyed to all concerned 4. Provide advice on union issues 5. Provide student, employee, and family support counseling services where required
Action
Register arrival with Crisis Center.
Work with CMT to establish the facts: • Casualties/numbers/severity of injury/location, and so forth
Mobilize SS counselors (and/or contract counselors) as appropriate: • Brief supervisor(s) • Keep supervisor(s) updated at regular intervals • Consider appropriate staffing levels
Track casualties: • Ensure Human Resources Casualty Board is kept up to date • Check names of casualties against students/employee/visitor lists to ensure that names are correct • Verify discrepancies
Ensure appropriate resources are available. Consider the following: • Alerting all employers of non-SS personnel working at the site and/or within the operation • Arrangements for handling/assisting employees, contractors, next of kin, and those affected by the emergency • Appropriate systems in place for managing Human Resource issues • Telephone Response Team briefed to direct relatives' calls to Human Resource–designated representatives • Liaise across all SS operations, as appropriate • Grief/support counseling available where needed
Communicate factual student/personnel information on the incident to counselors/team: • Share all press releases as soon as possible • Identify other agencies/stakeholders and ensure close liaison with counselors/team • Verify and maintain a record of location and status of persons involved in the incident • Verify details of fatalities, casualties, missing people, and survivors • Determine the nature, tone, and format of communication with students, employees, and families • Conduct postincident debriefing as soon as possible • Transportation for families • Accommodation • Consider need for cash/clothing/communications with families, and so on • Establish call schedules as appropriate
Facilitate visits to hospitals by relatives.

Provide support for families. Consider the following:

- Transportation to hospital/point of arrival
- Child care
- Accommodation
- Passport/visa issues
- Financial support
- Counseling support
- Reassurance as appropriate

Provide advice to CMT:

- Student, family, employee, and union relations
- SS Human Resources Policy

Monitor stress levels of the CMT and support teams.

Ensure adequate debriefing of all responders and assist in recommending the form and scope of the postincident investigation:

- Arrange postincident support debriefing at first opportunity
- Remember families

Strategic issues for consideration:

- What are impacts on
  - Students/employees? Provide compassionate assistance to injured persons and bereaved families, counseling/travel assistance/accommodation/practical help/loans/cash
  - SS contractors and other key third parties?
- Concern at loss of employment, guilt—keep all employees informed/reaffirm positive aspects of SS
- What effect on morale of district personnel? Reinforce district morale
- What are the likely effects of the incident on the workforce and its relationship with the district?
- Stress—who might be affected?
- Monitor for stress at site and in teams
- Response/telephone teams—arrange relief where signs of serious stress are detected
- Students/families/third-party witnesses—put in place a stress counseling program
- Look for and build on positive aspects of incident/response

Action Checklist: Finance Representative (Treasurer/Delegated Authority)

Key roles: 1. May not be an initial core team member but should be mobilized as soon as practicable 2. Tracking and forecasting of costs 3. Identification of financial impacts on stakeholders
Action
Work with CMT leader and evaluate impact of emergency relating to • Students, employees, families, local community, and the general public • Partners • Government • Special interest groups • Unions • Others
Contact appropriate financial advisor. Ensure appropriate resources are available for • Tracking financial liabilities associated with the emergency • Advising SS representatives for impact on SS District Operations • Checking the status of reporting to external agencies • Arranging the calculation of financial impacts
Liaise with SS financial and insurance departments.
Ensure appropriate controls, systems, and audit procedures are implemented with respect to emergency expenditures: • Make arrangements for cash and currency when necessary • Work with Risk to set up a claims hotline as soon as possible • Estimate increases in cost of working • Estimate costs of reinstatement • Advise the CMT on financial liabilities concerning the incident and ways to minimize them
For major incidents: • Determine payment strategy for any casual labor (day rates, and so forth)
Ensure appropriate resources are available for • Cash/credit cards at scene if required • Tracking financial liabilities associated with the emergency • Checking the status of reporting to external agencies • Arranging the calculation of financial impacts
Liaise with CMT Risk and Legal representatives: • Establish what SS insurance policies are relevant. Who are the insurers? • Determine what format/information will be required to substantiate SS claims and gather financial information in that format if possible • Estimate potential financial impact of any shutdown and report to CMT
Ensure appropriate controls, systems, and audit procedures are implemented in respect of emergency expenditure: • Make arrangements for cash and currency when necessary • Estimate potential cost of response operations/impacts as requested • Advise the CMT on financial liabilities concerning the incident and ways to minimize them
Brief CMT:

• Issues • Current and planned tasks • Actions and concerns
Strategic issues for consideration: • What insurance cover is applicable and how should costs be tracked to maximize returns? • Cost of repair/replacement • Cost of response to emergency • Cost of clearance/recovery. Installation/spill contamination cleanup/own resources/other resources • Cost of liabilities, claims, fines, and compensation. Significant claims to be expected? • Penalties/fines/damages to be expected? Are actions likely? • Cash flow implications • Impact on SS? • Any other financial impacts? • How will legislators/agencies and the community react? • How will students, employees, families, unions, and others react?

Action Checklist: CMT Update Briefing

Action
Preparation: • CMT coordinator to give ten-minute warning • CMT coordinator to advise CMT members; ensure log keeper and recorder are in place • Conduct roll call prior to commencing brief from site
CMT Brief: • Review of any outstanding action items from previous meeting • Identify new issues and actions from update call. Delegate responsibility and track actions • CMT members to present progress and provide short status report including, where appropriate, — Planning for worst-case scenarios — Triggers to assist in the proactive recognition of escalation toward the worst-case scenario — Indicators of improvement — How to make best of situation, and what are positives for SS? — Identification of stakeholders and issues — Strategies for dealing with stakeholders and issues • Confirm date and time of next meeting and note on board
Postmeeting actions: • Log keeper captures information • Update complete • CMT leader to update executive team and other senior management, as appropriate

APPENDIX 2

CRISIS RESPONSE TEAM JOB DESCRIPTIONS

The job descriptions included in this section are provided as job aids to assist SS Crisis Response Team members that may be mobilized to a specific school or SS facility to participate in onsite emergency response operations. These procedures conform to protocols established within the framework of the National Incidental Management System and are based upon the Incident Command System.

These protocols describe roles and responsibilities for various responders who may comprise the onsite Crisis Response Team. Safety (fire or police) will assume command of operations during initial stages of the incident response. The extent of activation of the team and specific actions will depend upon the type and magnitude of the incident and will be decided and agreed to between safety and the SS District Crisis Management Team.

Various positions during initial stages of response may be supported and/or staffed by safety agency personnel.

COMMAND SECTION

1. Incident Commander
2. Safety Officer
3. Public Information Officer
4. Liaison Officer

1. Incident Commander

Responsibilities

The Incident Commander is solely responsible for emergency/disaster operations and shall remain at the Command Post to observe and direct all operations.

Ensure the safety of students, staff, and others on campus. Lead by example: your behavior sets tone for staff and students.

Start-up actions

- Obtain your personal safety equipment (i.e., hard hat, vest, clipboard with job description sheet).
- Assess the type and scope of emergency.
- Determine the threat to human life and structures.
- Implement the emergency plan and hazard-specific procedures.
- Develop and communicate an incident action plan with objectives and a timeframe to meet those objectives.
- Activate functions and assign positions as needed.
- Fill in the Incident Assignments form. Refer to appendix 4 for blank copies.
- Appoint a backup or alternate Incident Commander.

Ongoing operational duties

- Continue to monitor and assess the total school situation:
    - View the site map periodically for search and rescue progress and damage assessment information.
    - Check with chiefs for periodic updates.
    - Reassign personnel as needed.
- Report (through Communications) to the school district on the status of students, staff, and facility, as needed (Site Status Report).
- Develop and communicate revised incident action plans as needed.
- Begin student release when appropriate.

NOTE: No student should be released until student accounting is complete. Never send students home before the end of the regular school day unless directed by the superintendent, except at the request of parent/guardian.

- Authorize the release of information.
- Utilize your backup; plan and take regular breaks (5–10 minutes per hour). During break periods, relocate away from the Command Post.
- Plan regular breaks for all staff and volunteers. Take care of your caregivers!
- Release teachers as appropriate and per district guidelines.
- Remain on and in charge of your campus until redirected or released by the superintendent.

Closing down

- Authorize deactivation of sections, branches, or units when they are no longer required.
- At the direction of the superintendent, deactivate the entire emergency response. If the fire department or other outside agency calls an "all clear," contact the district before taking any further action.
- Ensure that any open actions not yet completed will be taken care of after deactivation.
- Ensure the return of all equipment and reusable supplies to Logistics.
- Close out all logs. Ensure that all logs, reports, and other relevant documents are completed and provided to the Documentation Unit.
- Announce the termination of the emergency and proceed with recovery operations if necessary.

Command post equipment/supplies

- Campus map
- Master keys
- Staff and student rosters
- Disaster response forms (refer to appendix 4)
- Emergency plan
- Duplicate rosters (two sets)
- Tables and chairs (if Command Post is outdoors)
- Vests (if available)
- Job description clipboards
- Command Post trays (pens, etc.)
- Two-way radios
- AM/FM radio (battery)

2. Safety Officer

Responsibilities

The Safety Officer ensures that all activities are conducted in as safe a manner as possible under the existing circumstances.

Start-up actions

- Check in with the Incident Commander for a situation briefing.
- Obtain necessary equipment and supplies from Logistics.

- Put on a position identifier, such as a vest, if available.
- Open and maintain a position log (refer to appendix 4 for blank copies). Maintain all required records and documentation to support the history of the emergency or disaster.

Document

- Messages received
- Action taken
- Decision justification and documentation
- Requests filled

Operational duties

- Monitor drills, exercises, and emergency response activities for safety.
- Identify and mitigate safety hazards and situations.
- Stop or modify all unsafe operations.
- Ensure that responders use appropriate safety equipment.
- Think ahead and anticipate situations and problems before they occur.
- Anticipate situation changes, such as cascading events, in all planning.
- Keep the Incident Commander advised of your status and activity and on any problem areas that now need or will require solutions.

Closing down

- When authorized by the Incident Commander, deactivate the unit and close out all logs. Provide logs and other relevant documents to the Documentation Unit.
- Return equipment and reusable supplies to Logistics.

Equipment/supplies

- Vest or position identifier, if available
- Hard hat, if available
- Clipboard, paper, pens
  - Two-way radio, if available

3. Public Information Officer (PIO)

Personnel

- Available staff with assistance from available volunteers

Policy

- The PIO has the right and need to know important information related to an emergency/disaster at the school site as soon as it is available.
- The PIO acts as the official spokesperson for the school site in an emergency situation. If a school district PIO is available, he/she will be the official spokesperson. A school site-based PIO should be used only if the media is on campus and the district PIO is not available.

- News media can play a key role in assisting the school in getting emergency/disaster-related information to students, parents, and other stakeholders.
- Information released must be consistent, accurate, and timely.

Start-up actions

- Determine a possible "news center" site as a media reception area (located away from the Command Post and students). Get approval from the Incident Commander.
- Identify yourself as the PIO (by vest, visor, sign, etc.)
- Consult with the district PIO to coordinate information release.
- Assess the situation and obtain a statement from the Incident Commander. Tape record it if possible.
- Advise arriving media that the site is preparing a press release and the approximate time of its issue.
- Open and maintain a position log (refer to appendix 4 for blank copies) of your actions and all communications. If possible, tape media briefings. Keep all documentation to support the history of the event.

Operational duties

- Keep up to date on the situation.
- Statements must be approved by the Incident Commander and should reflect the following:
  - Reassurance (EGBOK—"Everything's going to be OK.")
  - Incident or disaster cause and time of origin.
  - Size and scope of the incident.
  - Current situation—condition of school site, evacuation progress, care being given, injuries, student release location, and so on. Do not release any names.
  - Resources in use.
  - Best routes to the school, if known and if appropriate.
  - Any information the school wishes to be released to the parents.
- Read statements if possible.
- When answering questions, be complete and truthful, always consider confidentiality and emotional impact. Avoid speculation, bluffing, lying, talking "off the record," arguing, and so on. Avoid using the phrase "no comment."
- Remind school staff and volunteers to refer all questions from the media or waiting parents to the PIO.
- Update information periodically with the Incident Commander.
- Ensure that announcements and other information are translated into other languages as needed.
- Monitor news broadcasts about the incident. Correct any misinformation heard.

Closing down

- At the Incident Commander's direction, release PIO staff when they are no longer needed. Direct staff members to sign out through Timekeeping.
- Return equipment and reusable supplies to Logistics.
- Close out all logs. Provide logs and other relevant documents to the Documentation Unit.

Equipment/supplies

- ID vest
- Battery-operated AM/FM radio
- Paper/pencils/marking pens
- Notebook
- Forms:
  - Sample Information Release
  - School Profile Information

4. Liaison Officer

Responsibilities

The Liaison Officer serves as the point of contact for agency representatives from assisting organizations and agencies outside the school district and assists in coordinating the efforts of these outside agencies by ensuring the proper flow of information.

Start-up actions

- Check in with the Incident Commander for a situation briefing.
- Determine your personal operating location and set it up as necessary.
- Obtain the necessary equipment and supplies from Logistics.
- Put on a position identifier, such as a vest, if available.
- Open and maintain a position log (refer to appendix 4 for blank copies). Maintain all required records and documentation to support the history of the emergency or disaster.

Operational duties

- Brief agency representatives on the current situation, priorities, and incident action plan.
- Ensure coordination of efforts by keeping the Incident Commander informed of agencies' action plans.
- Provide periodic update briefings to agency representatives as necessary.

Closing down

- At the Incident Commander's direction, deactivate the Liaison Officer position and release staff no longer needed. Direct staff members to sign out through Timekeeping.
- Return equipment and reusable supplies to Logistics.
- Close out all logs. Provide logs and other relevant documents to the Documentation Unit.

Equipment/supplies

- Vest or position identifier, if available
- Two-way radio, if available
- Clipboard, paper, pens

OPERATIONS SECTION

1. Operations Section Chief
2. Site Faculty Check/Security
3. Student Care
4. Student Release

1. Operations Section Chief

Responsibilities

The Operations Chief manages the direct response to the disaster, which can include the following:

- Site facility check/security
- Search and rescue
- Medical
- Student care
- Student release.

Start-up actions

- Check in with the Incident Commander for a situation briefing.
- Obtain necessary equipment and supplies from Logistics.
- Put on a position identifier, such as a vest, if available.

Operational duties

- Assume the duties of all operations positions until staff are available and assigned.
- As staff members are assigned, brief them on the situation, and supervise their activities, using the position checklists.
- If additional supplies or staff are needed for the Operations Section, notify Logistics. When additional staff arrive, brief them on the situation, and assign them as needed.
- Coordinate search and rescue operations if it is safe to do so. Appoint a Search and Rescue Team Leader to direct operations, if necessary.
- As information is received from operations staff, pass it on to situation analysis and/or the Incident Commander.
- Inform the Planning Section Chief of operations tasks and priorities.
- Make sure that operations staff are following standard procedures, using appropriate safety gear, and documenting their activities.
- Schedule breaks and reassign staff within the section as needed.

Closing down

- At the Incident Commander's direction, release Operations staff no longer needed. Direct staff members to sign out through Timekeeping.
- Return equipment and reusable supplies to Logistics.
- When authorized by the Incident Commander, deactivate the section and close out all logs. Provide logs and other relevant documents to the Documentation Unit.

Equipment/supplies

- Vest or position identifier, if available
- Search and rescue equipment
- Two-way radio
- Job description clipboard, paper, pens
- MSS:
  - Search and rescue MSS
  - Large campus map

2. Site Facility Check/Security

Personnel

Staff as assigned. Work in pairs.

Responsibilities

Take no action that will endanger yourself.

Start-up actions

- Wear hard hat and identification vest, if available.
- Take appropriate tools, job description clipboard, and radio.
- Put batteries in flashlight, if necessary.

Operational duties

As you complete the following tasks, observe the campus and report any damage by radio to the Command Post.

- Remember: If you are not acknowledged, you have not been heard. Repeat your transmission, being aware of other simultaneous transmissions.
- Lock gates and major external doors.
- Locate, control, and extinguish small fires as necessary.
- Check gas meter and, if gas is leaking, shut down the gas supply.
- Shut down electricity only if building has clear structural damage or if advised to do so by Command Post.
- Post yellow caution tape around damaged or hazardous areas.
- Verify that the campus is "locked down" and report the same to the Command Post.
- Advise the Command Post of all actions taken for information and proper logging.
- Be sure that the entire campus has been checked for safety hazards and damage.
- No damage should be repaired before full documentation, such as photographs and video evidence, is complete unless the repairs are essential to immediate life safety.
- Route fire, rescue, and police, as appropriate.
- Direct all requests for information to the Information Officer.

Closing down

- Return equipment and reusable supplies to Logistics.
- When authorized by the Incident Commander, close out all logs. Provide logs and other relevant documents to the Documentation Unit.

Equipment/supplies

- Vest, hard hat, work gloves, and whistle.
- Two-way radio, master keys, and clipboard with job description.
- Bucket or duffel bag with goggles, flashlight, dust masks, yellow caution tape, and shutoff tools—for gas and water (crescent wrench).

3. Student Care

Classroom teachers, substitute teachers, and staff as assigned.

Responsibilities

Ensure the care and safety of all students on campus.

Start-up actions

- Wear an identification vest, if available.
- Take a job description clipboard and radio.
- Check in with the Operations Section Chief for a situation briefing.
- Make personnel assignments as needed.
- If evacuating:
  - Verify that the assembly area and routes to it are safe.
  - Count or observe the classrooms as they exit, to make sure that all classes evacuate.
  - Initiate the setting up of portable toilet facilities and hand-washing stations.

Operational duties

- Monitor the safety and well-being of the students and staff in the assembly area.
- Administer minor first aid as needed.
- Support the Student Release process by releasing students with the appropriate paperwork.
- When necessary, provide water and food to students and staff.
- Make arrangements for portable toilets if necessary, ensuring that students and staff wash their hands thoroughly to prevent disease.
- Make arrangements to provide shelter for students and staff.
- Arrange activities and keep students reassured.
- Update records of the number of students and staff in the assembly area (or in the buildings).
- Direct all requests for information to the PIO.

Closing down

- Return equipment and reusable supplies to Logistics.

- When authorized by the Incident Commander, close out all logs. Provide logs and other relevant documents to the Documentation Unit.

Equipment/supplies

- Vest
- Clipboard with job description
- First aid kit
- Student activities: books, games, coloring books, and so on
- Forms:
  - Student Accounting
  - Notice of First Aid Care
- Two-way radio
- Water, food, sanitation supplies

4. Student Release

School secretary, available staff, and disaster volunteers. Use a buddy system. The Student Release process is supported by designated runners.

Responsibilities

Assure the reunification of students with their parents or authorized adult through separate Request and Release Gates. Refer to Section 8.13 Student/Family Reunification for details.

Start-up actions

- Obtain and wear a vest or position identifier, if available.
- Check with the Operations Section Chief for assignment to the Request Gate or Release Gate.
- Obtain necessary equipment and forms from Logistics.
- Secure the area against unauthorized access. Mark the gates with signs.
- Set up the Request Gate at the main student access gate. Use alphabetical grouping signs to organize parent requests.
- Have Student Release Forms available for parents outside of the fence at the Request Gate. Assign volunteers to assist.
- Set up the Release Gate some distance from the Request Gate.

Operational duties

Follow the procedures outlined below to ensure the safe reunification of students with their parents or guardians:

- Refer all requests for information to the POI. Do not spread rumors!
- If volunteers arrive to help, send those with disaster volunteer badges with photo ID to Logistics. If they are not registered (i.e., do not have badges), direct them to the Operations Section Chief.

Procedures

- The requesting adult fills out a Student Release Form, gives it to a staff member, and shows identification.
- The staff member verifies the identification, pulls the Emergency Card from the file, and verifies that the requester is listed on the card.
- The staff member instructs the requester to proceed to the Release Gate.
- If there are two copies of the Emergency Cards (one at each gate), staff files the Emergency Card in the out box. If there is only one copy, a runner takes the card with the Student Release Form, and staff files a blank card with the student's name on it in the out box.
- The designated runner takes the form(s) to the designated classroom.

NOTE: If a parent refuses to wait in line, don't argue. Note the time with appropriate comments on the Emergency Card and place it in the out box.

If the student is with the class:

- Designated runner shows the Student Release Form to the teacher.
- The teacher marks the box, "Sent with Runner."
- If appropriate, the teacher sends the parent copy of the First Aid Form with the runner.
- The runner walks the student(s) to the Release Gate.
- The runner hands the paperwork to release personnel.
- Release staff, match the student to the requester, verify proof of identification, ask the requester to fill out and sign the lower portion of Student Release Form, and release the student. Parents are given the Notice of First Aid Care Given, if applicable.

If the student is not with the class:

- The teacher makes the appropriate notation on the Student Release Form:
  - "Absent" if the student was never in school that day.
  - "First Aid" if the student is in the Medical Treatment area.
  - "Missing" if the student was in school but now cannot be located.
- The runner takes Student Release Form to the Command Post.
- The Command Post verifies the student's location, if known, and directs the runner accordingly.
- If the designated runner is retrieving multiple students and one or more are missing, the runner walks the available students to the Release Gate before returning forms to the Command Post for verification.
- The parent should be notified of the missing student's status and escorted to a crisis counselor.
- If the student is in First Aid, the parent should be escorted to the safety emergency medical personnel.
- If the student was marked absent, the parent will be notified by the principal or designated staff member.

Closing down

- At the direction of the Operations Section Chief, return equipment and unused supplies to Logistics.
- Complete all paperwork and turn it in to the Documentation Unit.

Equipment/supplies

- Job description clipboards
- Pens, stapler
- Box(es) of Emergency Cards
- Signs to mark Request Gate and Release Gate
- Signs for alphabetical grouping to organize the parents (A–F, etc.)
- Empty file boxes to use as out boxes
- Student Release Form (copies for every student)

PLANNING SECTION

1. Planning Section Chief
2. Documentation
3. Situation Analysis

1. Planning Section Chief

Responsibilities

This section is responsible for the collection, evaluation, documentation, and use of information about the development of the incident and the status of resources. Maintain accurate records and site map. Provide ongoing analysis of situation and resource status.

Start-up actions

- Check in with the Incident Commander for a situation briefing.
- Obtain necessary equipment and supplies from Logistics.
- Put on a position identifier, such as a vest, if available.

Operational duties

- Assume the duties of all Planning Section positions until staff is available and assigned.
- As (or if) staff is assigned, brief them on the situation and supervise their activities, utilizing the position checklists.
- Assist the Incident Commander in writing action plans.

Closing down

- At the Incident Commander's direction, deactivate the section and close out all logs.

- Verify that the closing tasks of all Planning Section positions have been accomplished.
- Return equipment and reusable supplies to Logistics.

Equipment/supplies

- Two-way radio
- File box(es)
- Dry-erase pens
- Large laminated site map of campus
- Forms:
  - Emergency Time/Situation Report
  - Sample Log
  - Student Accounting Form
- Paper, pens
- Job description clipboard
- Tissues

2. Documentation

Responsibilities

This section is responsible for the collection, evaluation, documentation, and use of information about the development of the incident and the status of resources.

Start-up actions

- Check in with the Planning Section Chief for a situation briefing.
- Obtain necessary equipment and supplies from Logistics.
- Put on a position identifier, such as a vest, if available.
- Determine whether there will be a Finance/Administration Section. If there is none, the Documentation Clerk will be responsible for maintaining all records of any expenditures as well as all personnel timekeeping records.

Operational duties

Records:

- Maintain a time log of the incident, noting all actions and reports. (See the sample log in appendix 4.)
- Record content of all radio communication with the district EOC.
- Record verbal communication for basic content.
- Log in all written reports.
- File all reports for reference (file box).
- Important: A permanent log may be typed or rewritten at a later time for clarity and better understanding. Keep all original notes and records—they are legal documents.

Student and Staff Accounting:

- Receive, record, and analyze Student Accounting forms. (Refer to appendix 4 for blank copies.)

- Check off staff roster. Compute the number of students, staff, and others on campus for situation analysis. Update periodically.
- Report missing persons and site damage to the Command Post.
- Report first aid needs to the medical team leader.
- File forms for reference.

Closing down

- Collect and file all paperwork and documentation from deactivating sections.
- Securely package and store these documents for future use.
- Return equipment and reusable supplies to Logistics.

Equipment/supplies

- Two-way radio
- File box(es)
- Forms:
  - Emergency Time/Situation Report
  - Student Accounting Form
  - Sample Log
- Paper, pens
- Job description clipboard

3. Situation Analysis

Responsibilities

This section is responsible for the collection, evaluation, documentation, and use of information about the development of the incident and the status of resources. Maintain accurate site map. Provide ongoing analysis of situation and resource status.

Start-up actions

- Check in with Planning Section Chief for a situation briefing.
- Obtain necessary equipment and supplies from Logistics.
- Put on a position identifier, such as a vest, if available.

Operational duties

Situation Status (Map):

- Collect, organize, and analyze situation information.
- Mark the site map appropriately as related reports are received, including but not limited to search and rescue reports and damage updates, giving a concise picture of the status of the campus.
- Preserve the map as a legal document until it is photographed.
- Use an area-wide map to record information on major incidents, road closures, utility outages, and so on. (This information may be useful to staff for planning routes home, etc.)

Situation Analysis:

- Provide current situation assessments based on analysis of information received.
- Develop situation reports for the Command Post to support the action planning process.
- Think ahead and anticipate situations and problems before they occur.
- Report only to Command Post personnel. Refer all other requests to the PIO.

Closing down

- Close out all logs and turn all documents in to Documentation.
- Return equipment and reusable supplies to Logistics.

Equipment/supplies

- Two-way radio
- Paper, pens, dry-erase pens, tissues
- Job description clipboards
- Large laminated site map of campus
- File box(es)
- Map of county or local area

LOGISTICS SECTION

1. Logistics Section Chief
2. Supplies/Facilities
3. Staffing
4. Communications

1. Logistics Section Chief

Responsibilities

The Logistics Section is responsible for providing facilities, services, personnel, equipment, and materials in support of the incident.

Start-up actions

- Check in with the Incident Commander for a situation briefing.
- Open the supplies container or other storage facility.
- Put on position identifier, such as a vest, if available.
- Begin distribution of supplies and equipment as needed.
- Ensure that the Command Post and other facilities are set up as needed.

Operational duties

- Assume the duties of all Logistics positions until staff is available and assigned.

- As (or if) staff is assigned, brief them on the situation and supervise their activities, utilizing the position checklists.
- Coordinate supplies, equipment, and personnel needs with the Incident Commander.
- Maintain security of the cargo container, supplies, and equipment.

Closing down

- At the Incident Commander's direction, deactivate the section and close out all logs.
- Verify that closing tasks of all Logistics positions have been accomplished. Secure all equipment and supplies.

Equipment/supplies

- Two-way radio
- Job description clipboard
- Paper, pens
- Cargo container or other storage facility and all emergency supplies stored on campus
- Clipboards with volunteer sign-in sheets
- Forms:
  - Inventory of emergency supplies on campus
  - Site Status Report
  - Communications Log
  - Message forms

2. Supplies/Facilities

Responsibilities

This unit is responsible for providing facilities, equipment, supplies, and materials in support of the incident.

Start-up actions

- Check in with the Logistics Section Chief for a situation briefing.
- Open the supplies container or other storage facility if necessary.
- Put on a position identifier, such as a vest, if available.
- Begin distribution of supplies and equipment as needed.
- Set up the Command Post.

Operational duties

- Maintain security of the cargo container, supplies, and equipment.
- Distribute supplies and equipment as needed.
- Assist team members in locating appropriate supplies and equipment.
- Set up the staging area, sanitation area, feeding area, and other facilities as needed.

Closing down

- At the Logistic Chief's direction, receive all equipment and unused supplies as they are returned.
- Secure all equipment and supplies.

Equipment/supplies

- Two-way radio
- Job description clipboard
- Paper, pens
- Cargo container or other storage facility and all emergency supplies stored on campus
- Form: Inventory of emergency supplies at school/facility

3. Staffing

Responsibilities

This unit is responsible for coordinating the assignment of personnel (staff, students, disaster volunteers) in support of the incident.

Start-up actions

- Check in with the Logistics Section Chief for a situation briefing.
- Put on a position identifier, such as a vest, if available.
- Open logs to list staff, volunteers, and students who are awaiting assignment.

Operational duties

- Deploy personnel as requested by the Incident Commander.
- Sign in volunteers, making sure volunteers are wearing their ID badges and are on the site disaster volunteer list. Unregistered volunteers should be sent to the Operations Section Chief for instructions.

Closing down

- Ask volunteers to sign out.
- At the Logistic Section Chief's direction, close out all logs and turn them in to Documentation.
- Return all equipment and supplies.

Equipment/supplies

- Two-way radio
- Job description clipboard
- Paper, pens
- Cargo container or other storage facility and all emergency supplies stored on campus

- Clipboards with volunteer sign-in sheets
- Forms:
  - Inventory of emergency supplies at school/facility
  - List of registered disaster volunteers

4. Communications

Responsibilities

This unit is responsible for establishing, coordinating, and directing verbal and written communications within the school disaster site and with the school district. If the school district cannot be contacted, communications may be made with outside agencies when necessary.

Personnel

- A school staff member with a campus two-way radio, supported by disaster volunteer runners.

Start-up actions

- Set up the Communications station in a quiet location with access to the Command Post.
- Turn on radios and advise the Command Post when ready to accept traffic.

Operational duties

- Communicate with the district EOC per district procedure. At the direction of the Incident Commander, report the status of students, staff, and campus, using the Site Status Report form (refer to appendix 4 for blank copies).
- Receive and write down all communications from the district EOC.
- Use designated runners to deliver messages to the Incident Commander with copies to the Planning Section Chief.
- Maintain the Communications Log: date/time/originator/recipient.
- Follow communications protocol. Do not contact the city directly if the district EOC is available.
- Direct the media to the PIO.
- Monitor local emergency news:
- Radio:
  - ________
  - ________
  - ________
  - ________
  - ________
  - ________
- TV:
  - ________
  - ________

- ◦ ________
- ◦ ________
- ◦ ________
- Internet:
  - ◦ www.xxx.com

Closing down

- Close out all logs, message forms, and so on, and turn them over to Documentation.
- Return all equipment and unused supplies to Logistics.

Equipment/supplies

- Two-way radios with spare batteries for each
- Job description clipboard
- Paper, pens
- Table and chairs
- AM/FM radio
- File boxes, tote tray for outgoing messages
- Forms:
  - Site Status Report
  - Message forms

FINANCE/ADMINISTRATION SECTION

1. Finance/Administration Chief
2. Timekeeping
3. Purchasing

1. Finance/Administration Section Chief

Responsibilities

The Finance/Administration Section is responsible for financial tracking, procurement, and cost analysis related to the disaster or emergency. It maintains financial records and tracks and records staff hours.

Start-up actions

- Check in with the Incident Commander for a situation briefing.
- Put on a position identifier, such as a vest, if available.
- Locate and set up work space.
- Check in with the Documentation Clerk to collect records and information that relate to personnel timekeeping and/or purchasing.

Operational duties

- Assume the duties of all Finance/Administration positions until staff is available and assigned.

- As (or if) staff is assigned, brief them on the situation and supervise their activities, utilizing the position checklists.

Closing down

- At the Incident Commander's direction, deactivate the section and close out all logs.
- Verify that the closing tasks of all Finance/Administration positions have been accomplished. Secure all documents and records.

Equipment/supplies

- Job description clipboard.
- Paper, pens.
- Form: Staff Duty Log.

2. Timekeeping

Responsibilities

This unit is responsible for maintaining accurate and complete records of staff hours.

Start-up actions

- Check in with the Finance/Administration Section Chief for a situation briefing.
- Put on a position identifier, such as a vest, if available.
- Locate and set up work space.
- Check in with the Documentation Clerk to collect records and information that relates to personnel timekeeping.

Operational duties

- Meet with the Finance/Administration Section Chief to determine the process for tracking regular and overtime hours of staff.
- Ensure that accurate records are kept of all staff members, indicating the hours worked.
- If district personnel not normally assigned to the site are working, be sure that records of their hours are kept.

Closing down

- Close out all logs.
- Secure all documents and records.

Equipment/supplies

- Job description clipboard.
- Staff Duty Log.

3. Purchasing

Responsibilities

This unit is responsible for maintaining accurate and complete records of purchases. Most purchases will be made at the district level; however, in emergency situations, it may be necessary for school sites to acquire certain items quickly.

Start-up actions

- Check in with the Finance/Administration Section Chief for a situation briefing.
- Put on a position identifier, such as a vest, if available.
- Locate and set up work space.
- Check in with the Documentation Clerk to collect records and information that relate to purchasing.

Operational duties

- Meet with the Finance/Administration Section Chief to determine the process for tracking purchases.
- Support Logistics in making any purchases which have been approved by the Incident Commander.

Closing down

- Close out all logs.
- Secure all documents and records.

Equipment/supplies

- Job description clipboard.
- Paper, pens.

APPENDIX 3

EMERGENCY RESPONSE FLIP CHARTS

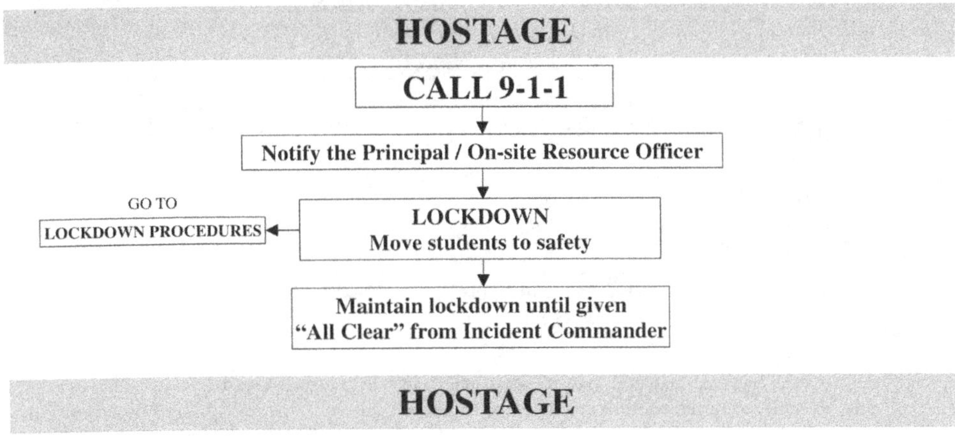

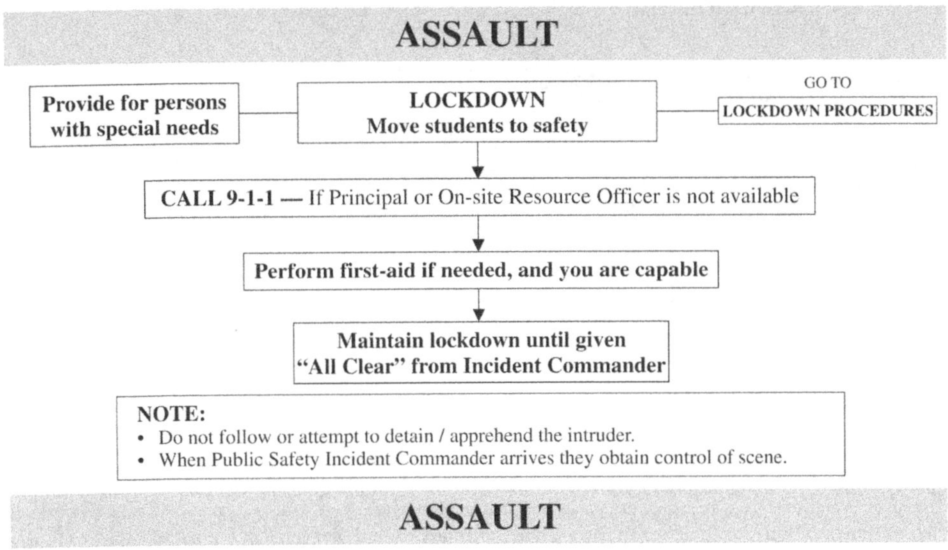

REVERSE EVACUATION

Reverse Evacuation is the process of moving all students and staff inside the building because there is a threat outside the building.

Notify Principal / On-site Resource Officer

↓

CALL 9-1-1 — If Principal or On-site Resource Officer is not available

↓

Move students inside to safety — **Provide for persons with special needs**

↓

Stay inside until given "All Clear" by Incident Commander

NOTE:

- Reverse Evacuation may soon be followed by lockdown or shelter-in-place.
- Make sure threat does not come in with others.
- When Public Safety Incident Commander arrives they obtain control of scene.

REVERSE EVACUATION

FIGHTING

OUT OF CONTROL

Notify Principal / On-site Resource Officer

↓

CALL 9-1-1 — if Principal or On-site Resource Officer is not available

↓

LOCKDOWN (if needed) move students to safety — GO TO LOCKDOWN PROCEDURES — **Provide for persons with special needs**

↓

Maintain lockdown until given "All Clear" from Incident Commander

CONTROLLABLE

Notify Principal / On-site Resource Officer

↓

If there is no Principal / On-site Resource Officer — identify yourself to the fighters and tell them to stop fighting

↓

Get medical attention if necessary and perform first-aid if needed, and you are capable

NOTE:

- If there is no Principal or On-site Resource Officer and you are capable, calmly take charge of the situation, and disperse any crowd.
- Separate the combatants while avoiding physical force.
- Remove participants to the office.
- If safe, direct rescue personnel to injured and give any required assistance.
- Attempt to determine the number of victims and identify witnesses.
- When Public Safety Incident Commander arrives they obtain control of scene.

FIGHTING

WEAPONS

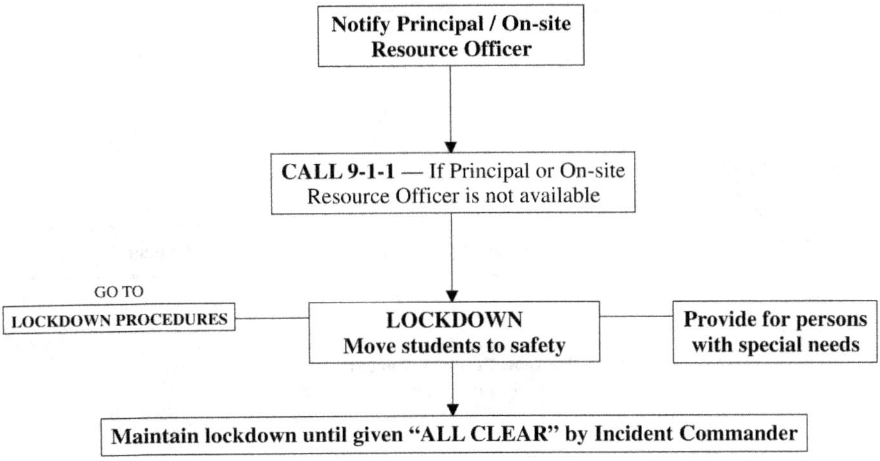

NOTE:

- Remain Calm — do not raise voice and speak clearly.
- Gather as much information as possible to relay to authorities.
- Isolate individual or suspect if safe to do so.
- If interaction with the individual is imminent, do not use force.
- When Public Safety Incident Commander arrives they obtain control of scene.

WEAPONS

SHOOTING

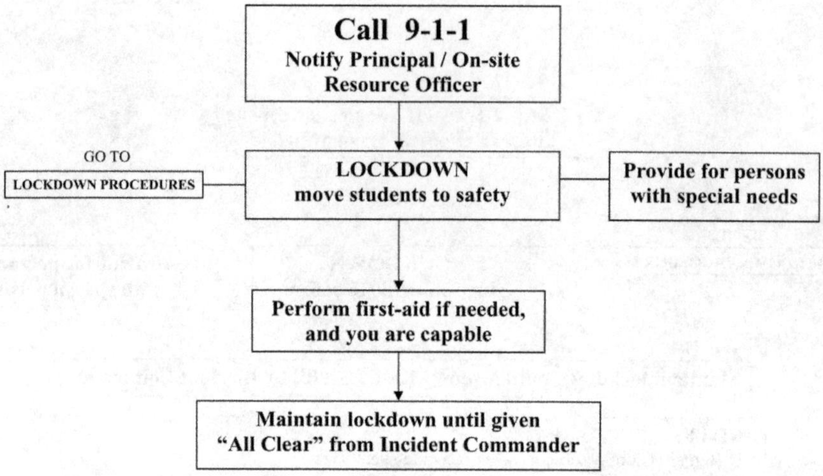

NOTE:

- If safe, direct rescue personnel to injured and give any required assistance.
- When Public Safety Incident Commander arrives they obtain control of scene.
- Keep students away from doors and windows.

SHOOTING

GAS LEAK / CHEMICAL / HAZMAT SPILL

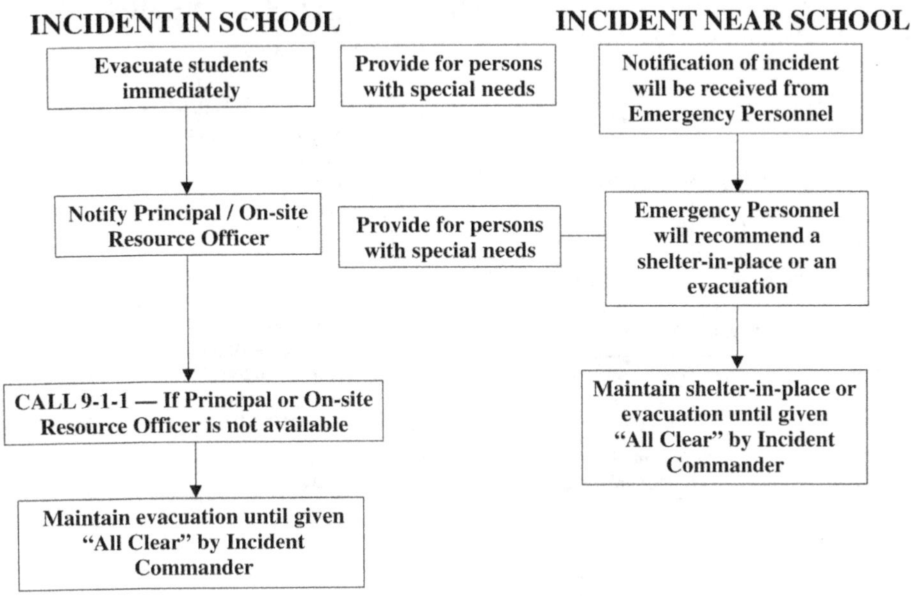

NOTE:
- Keep students and staff from going near the area of contamination.
- If evacuation is necessary take attendance regularly.
- When Public Safety Incident Commander arrives they obtain control of scene.

GAS LEAK / CHEMICAL / HAZMAT SPILL

FIRE / EVACUATION

Evacuation should occur when there is some kind of a threat that can directly affect those in the school facility. Direction should be given as to how far evacuees need to go.

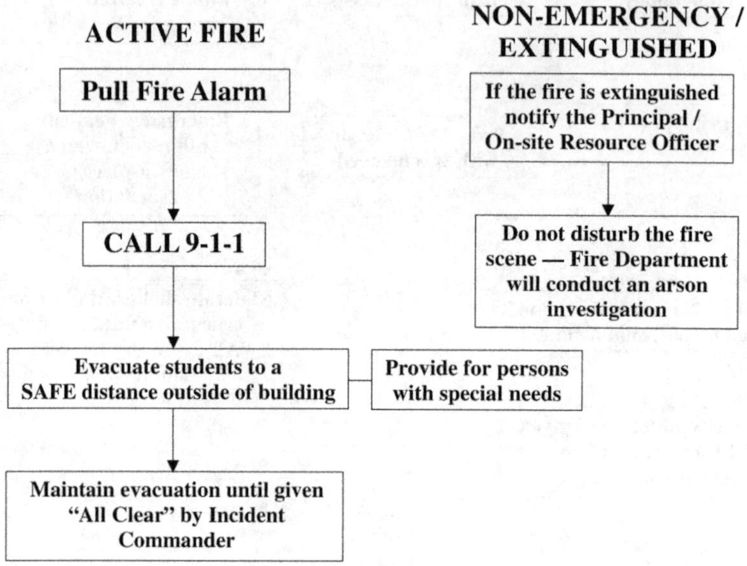

NOTE:

- If advised by Emergency Personnel, or there is inclement weather, students may need to be moved to another location.
- Follow normal fire drill route. Follow alternate route if normal route is too dangerous.
- When Public Safety Incident Commander arrives they obtain control of scene.
- False alarms should be called in to Student Services / Security.
- If the principal is not available, non-emergency fires that have been extinquished should be reported to the Fire Department at xxx-xxx-xxxx.

FIRE / EVACUATION

SHELTER-IN-PLACE

Shelter-in-Place provides refuge for students, staff and public within a school building during an emergency. Shelters are located in areas that maximize the safety of inhabitants. Safe areas may change depending on emergency.

During a Shelter-in-Place have students and staff stay in the rooms they are currently in. You may be asked to move to a designated area such as a tornado shelter. Close and lock all windows and keep everyone quiet until given further instruction.

When advised over the PA system "Shelter-in-Place"

Follow directions for "Shelter-in-Place". You may remain in classroom or you may move to tornado shelter positions

Shelter-In-Place

Have students move under desks and keep them away from doors and windows

Maintain "Shelter-in-Place" until informed by Incident Commander that it is okay to resume normal status

Provide for persons with special needs

Shelter In Tornado Area

Quickly, but safely, move students to tornado shelter area

Maintain "Shelter-in-Place" until informed by Incident Commander that it is okay to resume normal status

NOTE:

- Keep everyone calm.
- Keep away from doors and windows.
- When Public Safety Incident Commander arrives they obtain control of scene.
- Sheltering-in-Place may need to happen in many situations and often should be done if deemed necessary before the announcement is given.

SHELTER-IN-PLACE

LOCKDOWN PROCEDURES

Lockdown - those conditions requiring complete separation and protection of school staff and students from any situation regarding an existing internal or external situation that could directly threaten their safety.

During a lockdown doors should be locked and no one is to enter or leave the room. Windows should be covered and shut, if safe to do so. It is safest to move away from the door. Make sure everyone is quiet and remains in lockdown until given all clear by Incident Commander.

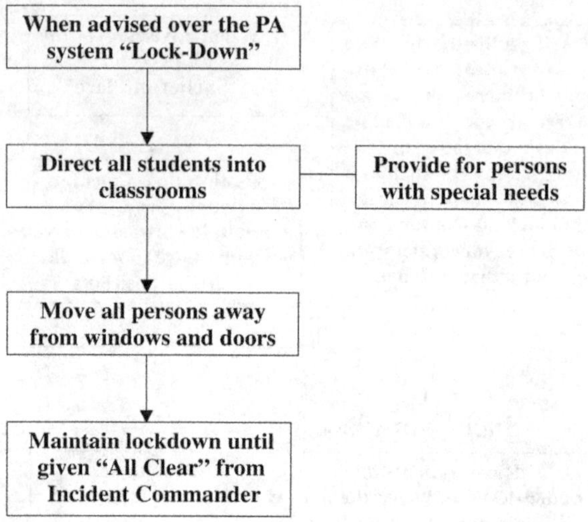

NOTE:

- PA announcement will be in plain language and provide as much information about the incident as possible.
- Under lockdown, conditions in a specific classroom can be communicated using color coded cards. A card, either posted in the window, or slipped under the door, can alert emergency responders to the status of students in individual classrooms:
  - Green Card — NO injuries
  - Yellow Card — Injuries have occurred but are minor.
  - Red Card — Injuries have occurred and medical assistance is needed immediately.
- When Public Safety Incident Commander arrives they obtain control of scene.

LOCKDOWN PROCEDURES

MEDICAL EMERGENCIES

Assess the Situation

SERIOUS

Call 9-1-1
Notify Principal / On-site Resource Officer
(if necessary)

Make student / employee comfortable

Follow medical protocol for student / employee if on file
(may need to send someone to get medication kit)

Administer first-aid if trained, and you are able to
(wear appropriate Personal Protective Equipment)

Monitor respiration and circulation until Emergency Medical Personnel arrive

NOT SERIOUS

Make student / employee comfortable

Follow medical protocol for student / employee if on file
(may need to send someone to get medication kit)

Notify Principal
(if necessary)

Blood / Bodily Fluid Precautions

When exposed to Blood or other Bodily Fluids it is very important to use extreme caution. Personal Protective Equipment such as gloves should be worn at all times and disposed of in a biohazard materials bag. If Personal Protective Equipment is not available, avoid coming in direct contact with blood or other bodily fluids. When you are finished administering medical assistance wash your hands and any other areas that may have come in contact with the possibly infectious material with soap and warm running water for at least 15 seconds.

NOTE:

- Stay Calm!
- **ONLY** move the person if there is a question of their, or your, safety.
- If poisoning suspected, call the Poison Control Center, 1-800-222-1222 and follow medical direction.
- Record time and site of insect sting, ingestion of food, exposure to other allergen; and name of medicine, dosage and time; if appropriate.
- When Public Safety Incident Commander arrives they obtain control of scene.
- For clean up of bodily fluids refer to the guidelines regarding clean-up of potentially infectious materials.

MEDICAL EMERGENCIES

SEVERE WEATHER — TORNADO

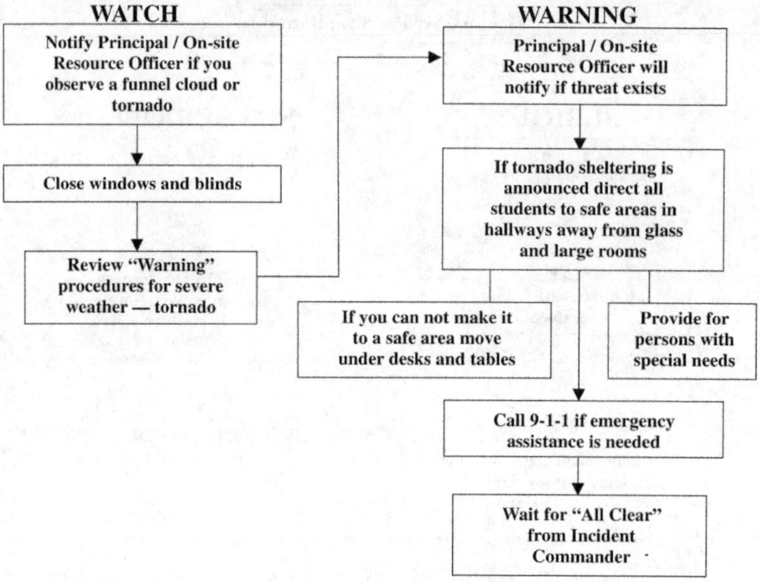

Definitions of a Tornado Watch and Warning as provided by the National Weather Service:

Tornado Watch

This is issued by the National Weather Service when conditions are favorable for the development of tornadoes in and close to the watch area. Their size can vary depending on the weather situation. They are usually issued for a duration of 4 to 8 hours. They normally are issued well in advance of the actual occurrence of sever weather. During the watch, people should review tornado safety rules and be prepared to move to a place of safety if threatening weather approaches.

Tornado Warning

This is issued when a tornado is indicated or sighted by spotters; therefore, people in the affected area should seek safe shelter immediately. They can be issued without a Tornado Watch being already in effect. They are usually issued for a duration of 30 minutes.

NOTE:
- Stay Calm!
- Remind teachers to take class rosters.
- Remain in safe areas until warning expires or until incident commander has issued an "All Clear" signal.
- Keep everyone away from doors, windows and rooms with high ceilings.
- When Public Safety Incident Commander arrives they obtain control of scene.

SEVERE WEATHER — TORNADO

DISTURBANCES / DEMONSTRATIONS

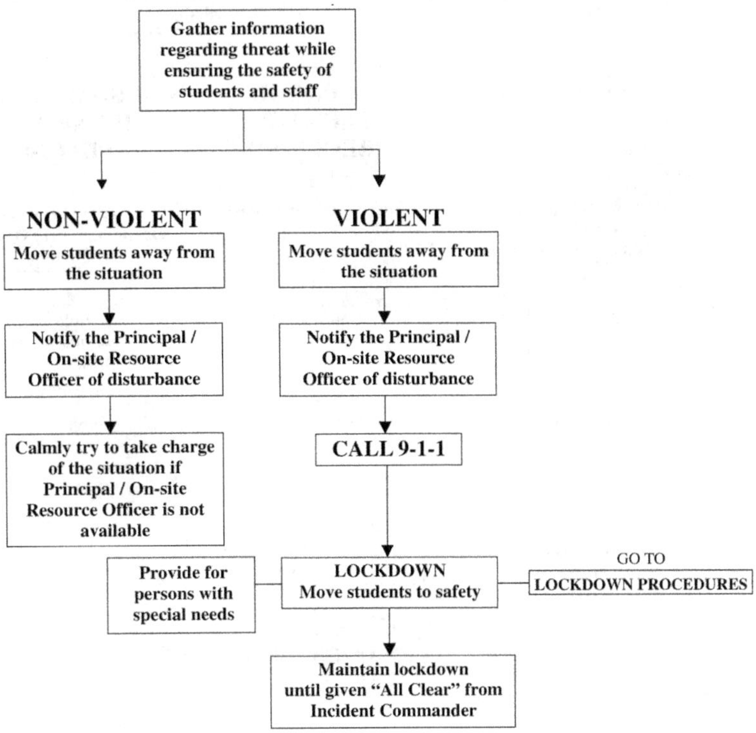

NOTE:

- If safe, direct rescue personnel to injured and give any required assistance.
- When Public Safety Incident Commander arrives they obtain control of scene.
- Take class attendance to see who is accounted for and who is not.

DISTURBANCES / DEMONSTRATIONS

BIOLOGICAL / FOOD POISONING

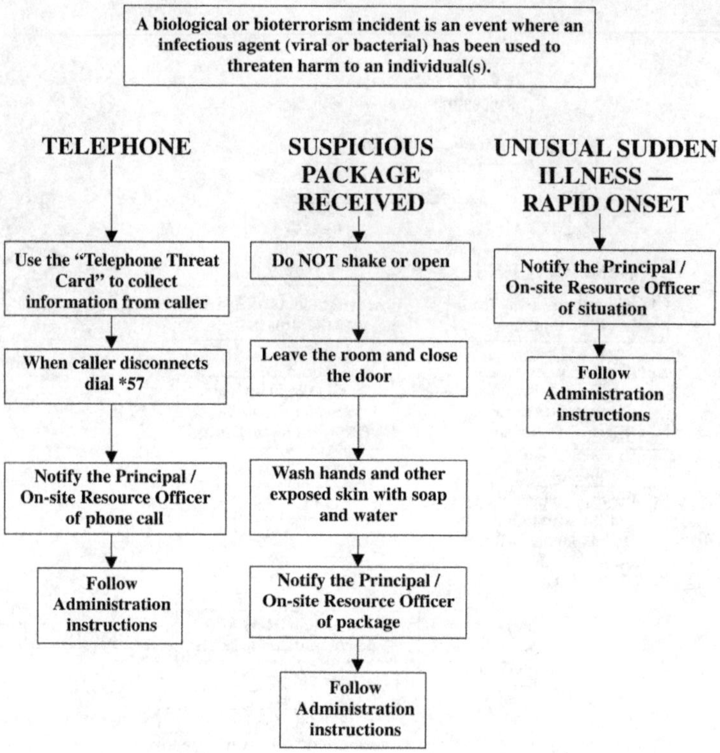

NOTE:
- Isolate the area.
- Do not touch or move the object.
- Do not attempt to cover or hide the device.
- Avoid using any electrical devices around the object—as they could trigger the device.
- When Public Safety Incident Commander arrives they obtain control of scene.

BIOLOGICAL / FOOD POISONING

DEATH / SUICIDE — ATTEMPT, THREAT

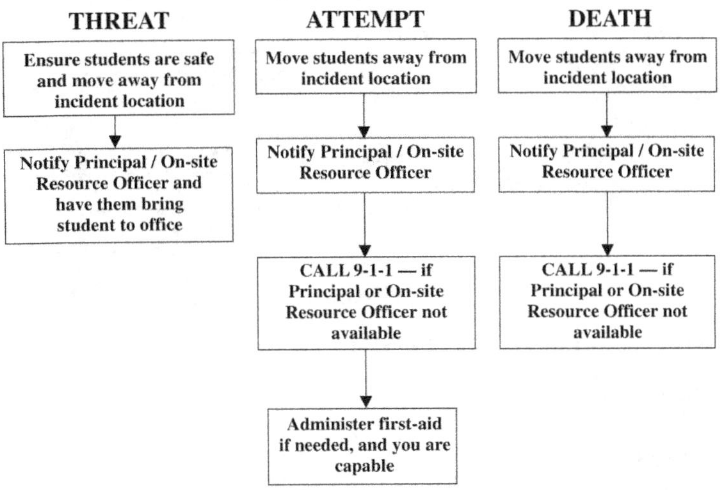

THREAT

NOTE:
- Keep student isolated from other students until counselor or psychologist arrives.

ATTEMPT

NOTE:
- DO NOT divulge the fact that death has/may have occurred — This is the role of Public/Hospital medical officials.
- When Public Safety Incident Commander arrives they obtain control of scene.

DEATH

NOTE:
- Move students and staff away from victim and close off the area.
- DO NOT touch the victim or anything at the scene.
- When Public Safety Incident Commander arrives they obtain control of scene.

DEATH / SUICIDE — ATTEMPT, THREAT

DISGRUNTLED / UNAUTHORIZED PERSONS

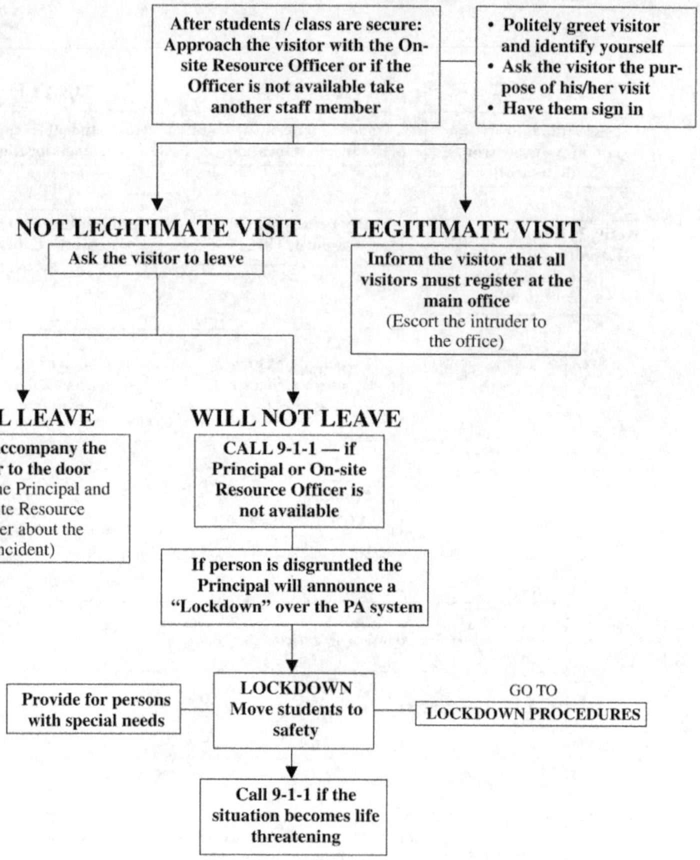

NOTE:
- If Disgruntled Person is outdoors immediately call 9-1-1.
- Remember to remain calm and be courteous and confident.
- Do not touch the person and keep at a reasonable distance.
- Make sure to be attentive to what the person has to say.
- When Public Safety Incident Commander arrives they obtain control of scene.

DISGRUNTLED / UNAUTHORIZED PERSONS

BOMB / BOMB THREAT

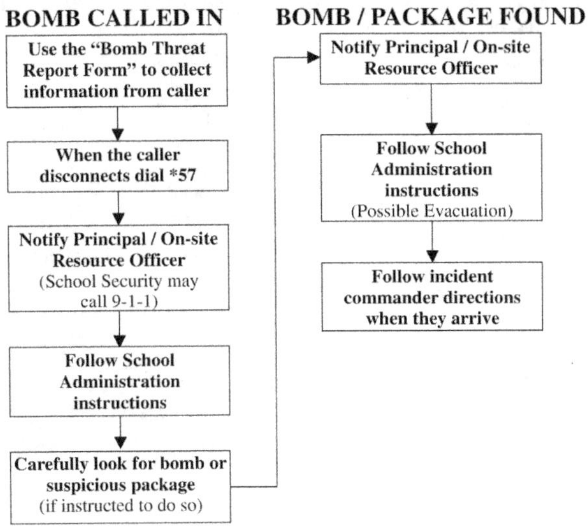

NOTE:

- Isolate area.
- Do not touch or move the object.
- Do not attempt to cover or hide the device.
- Avoid using any electrical devices around the object — as they could trigger the device.
- When Public Safety Incident Commander arrives they obtain control of scene.

BOMB THREAT REPORT FORM

Number at which call is received:
( ) ______________________________

Time: ______________ pm / am Date: __/__/__

Sex of Caller: M / F Race: __________

Length of Call: min. ______ sec. _____ Age: __________

Exact Wording of the Threat:
(Place on another sheet)

QUESTIONS TO ASK?

When is the bomb going to explode?

Where is it right now?

What does it look like?

What kind of bomb is it?

What will cause it to explode?

Did you place the bomb?

Why?

What is your address?

What is your name?

CALLER'S VOICE

_____ Calm
_____ Angry
_____ Excited
_____ Slow
_____ Rapid
_____ Soft
_____ Loud
_____ Laughter
_____ Crying
_____ Normal
_____ Distinct
_____ Slurred
_____ Nasal
_____ Stutter
_____ Lisp
_____ Raspy
_____ Deep
_____ Ragged
_____ Clearing throat
_____ Deep breathing
_____ Cracking voice
_____ Disguised
_____ Accent
_____ Familiar

If voice is familiar, whom did it sound like?
____
____

Background sounds:

_____ Street noises
_____ Animal noises
_____ Voices
_____ PA System
_____ Music
_____ Motor
_____ Machinery
_____ Clear
_____ Static
_____ Local
_____ House

Other Noises:
____
____
____

Threat Language:

_____ Foul
_____ Irrational
_____ Message read by threat maker
_____ Well spoken (educated)
_____ Taped
_____ Incoherent

Remarks: ______________________________
____
____

Report call immediately to:
Principal
(Principal will begin Notification Chart)
____

Date: __/__/__ Phone: ( ) __________
Name: ______________________________

NOTE: WHEN CALL IS OVER—HANG UP AND IMMEDIATELY DIAL *57

BOMB / BOMB THREAT

APPENDIX 4

FORMS

Staff Skills Survey and Inventory

Name and School ________________/______________________ Room ________

During any disaster situation, it is important to be able to draw from all available resources. The special skills, training and capabilities of the staff will play a vital role in coping with the effects of any disaster incident, and they will be of paramount importance during and after a major or catastrophic disaster. The purpose of this survey/inventory is to pinpoint those staff members with equipment and the special skills that might be needed. Please indicate the areas that apply to you and return this survey to your administrator.

PLEASE CHECK ANY OF THE FOLLOWING IN WHICH YOU HAVE EXPERTISE AND TRAINING.

CIRCLE YES OR NO WHERE APPROPRIATE.

____ First Aid (current card yes/no) ____ CPR (current yes/no) ____ Triage ____ Firefighting

____ Construction (electrical, plumbing, carpentry, etc.) ____ Running/Jogging

____ Emergency Planning ____ Emergency Management ____ Search and Rescue

____ Law Enforcement Bi/Multilingual (what language[s]) ____________________

____ Mechanical Ability ____ Structural Engineering ____ Bus/Truck Driver (class 1 or 2 license yes/no)

____ Shelter Management ____ Survival Training and Techniques ____ Food Preparation

____ Ham Radio Operator ____ CB Radio ____ Journalism

____ Camping ____ Waste Disposal ____ Recreational Leader

DO YOU KEEP A PERSONAL EMERGENCY KIT? ______

In your car? ______ In your room? ______

DO YOU HAVE MATERIALS IN YOUR ROOM THAT WOULD BE OF USE DURING AN EMERGENCY?

(i.e., athletic bibs, traffic cones, carpet squares) ______ Yes ______ No

DO YOU HAVE EQUIPMENT OR ACCESS TO EQUIPMENT OR MATERIALS AT YOUR SCHOOL SITE THAT COULD BE USED AN IN EMERGENCY? ______ Yes ______ No

PLEASE LIST EQUIPMENT AND MATERIALS.

____

____

COMMENTS ____________________________________

____

WHAT WOULD MAKE YOU FEEL MORE PREPARED SHOULD A DISASTER STRIKE WHILE YOU WERE AT SCHOOL?

____

____

EMERGENCY INFORMATION SHEET

Room Number__________________

Student's Name ____

Grade_________________________

Address ____

Home Phone____

Parent/Guardian Names

____

Mother/Guardian Work Address

____

Father/Guardian Work Address

____

Mother/Guardian Work Phone ____

Father/Guardian Work Phone____

In an emergency or major disaster during school hours, my child may be released to the following persons: (THESE NAMES SHOULD BE THE SAME AS ON YOUR CHILD'S GREEN EMERGENCY CARD)

Name_______________ Address ________________ Phone ________________

Name_______________ Address ________________ Phone ________________

Name_______________ Address ________________ Phone ________________

Name_______________ Address ________________ Phone ________________

Name_______________ Address ________________ Phone ________________

My child needs to take the following medication: ____________________________

Medication__________________________ Dosage_______________________
Time________________

Have you provided the school office with a supply of this medicine?

____

My child may__________or may not______________ have Tylenol if needed during an emergency.

List allergies to medicine or foods

____

Doctor__________________ Address________________ Phone_____________

Family out-of-state telephone contact to be used in an emergency:

Name_____________________________ Phone___________________________

Date ________ Parent/Guardian Signature_________________________________

Student Released to: ____

____

Date:_____________Time:_____________Destination:____________________

STUDENT ACCOUNTING FORM

Room No. _______________________ Date _______________________

Enrolled per Register ______________ Reported by ___________________

Not in School Today _______________ Received by ___________________

Present Now _____________________

1. Students or classroom volunteers elsewhere (off campus, left in room, other location, etc.)

Name	Location	Problem

2. Students on playground needing more first aid than you can handle:

Name	Location	Problem

____
____
____
____

Additional comments: (report fire, gas/water leaks, blocked exits, structural damage, etc.)

NOTICE OF FIRST AID CARE

DATE: ____________________
SCHOOL: ____________________

Dear Parent:

________________________________ was injured at school and has been given first aid. If you feel further care is necessary please consult your family physician.

Destination: (if not presently onsite)

____

Transporting Agency: (if not presently onsite)

____

Time: ____________________

Remarks:

____

____

Please sign and return one copy to school. Retain a copy for your records.

____________________ ____________________

PARENT'S SIGNATURE SCHOOL REPRESENTATIVE'S SIGNATURE

Note: One copy goes home with student
One copy stays with teacher or medical treatment team records

Date:

SITE STATUS REPORT

TO: ___________ FROM: (name) ___________ LOCATION: _______________
DATE: _______ TIME: ______ PERSON IN CHARGE AT SITE: _____________
Message via: Two-way Radio _____ Telephone ______ Messenger ________________

EMPLOYEE/STUDENT STATUS

	Absent	Injured	# Sent to hosp./med	Dead	Missing	Unaccounted for (away from site)	# Released to parents	# Being super-vised
Students								
Site Staff								
Others								

STRUCTURAL DAMAGE: Check damage/problem and indicate location(s).

Check ✓	Damage/problem	Location(s)
	Gas leak	
	Water leak	
	Fire	
	Electrical	
	Communications	
	Heating/cooling	
	Other (list):	

MESSAGE: (include kind of immediate assistance required; can you hold out without assistance/how long?; overall condition of campus, neighborhood, and street conditions; outside agencies on campus and actions; names of injured, dead, missing, and accounted for ASAP)

Update Report

Name __________________________________ Time ________________________________
_________ # children remaining at school
_________ # staff members remaining to care for children
______ Assistance required: ______ water _____ food ______ blankets ______ # people to help

STUDENT RELEASE FORM

To be taken by Runner

Please Print

Student's Name ____

Teacher______________________________ Grade__________________

Requested By ____

****

To be filled in by Request Gate staff

Proof of ID __________ Name on Emergency Card ______________________
(yes) (no)

****

Student's Status
To be filled in by teacher

Sent with Runner_______ Absent__________ First Aid ________ Missing ______

****

To be filled in by Request Gate staff

Proof of ID __________ Name on Emergency Card ______________________
(yes) (no)

****

To be filled in by Requester
<u>at Release Gate</u>

Requester Signature ____

Destination: ____

Date: ____________________

Time: ____________________

****

SEARCH AND RESCUE TEAMS

SEARCH AND RESCUE TEAM LEADER

Note: Number of teams will vary depending on size of campus.

	NAMES	Radio	Keys	Hard Hat	Goggles	Bucket	Vest	Clipboard	Backpack
SEARCH AND RESCUE TEAM #1 NOTES:	1. 2.								
SEARCH AND RESCUE TEAM #2 NOTES:	1. 2.								
SEARCH AND RESCUE TEAM #3 NOTES:	1. 2.								
SEARCH AND RESCUE TEAM #4 NOTES:	1. 2.								
SEARCH AND RESCUE TEAM #5 NOTES:	1. 2.								

S&R Team Leader

- Assign teams based on available manpower; minimum two persons. Attempt to place one experienced person on each team.
- Perform visual check of outfitted team leaving Command Post; include radio check. Advise teams of known injuries.

- Remain at Command Post table.
- Be attentive to all S&R-related communications.
- Utilize boxes above to record location of injured students. For example: report of two injured students in Room 20 would be recorded as “S/2 = RM 20” in box under team #3.
- Utilize manpower pool to aid S&R (i.e., request for backboard and carryout or request for rescue equipment).

Sample
Search & Rescue Map
Search & Rescue #2

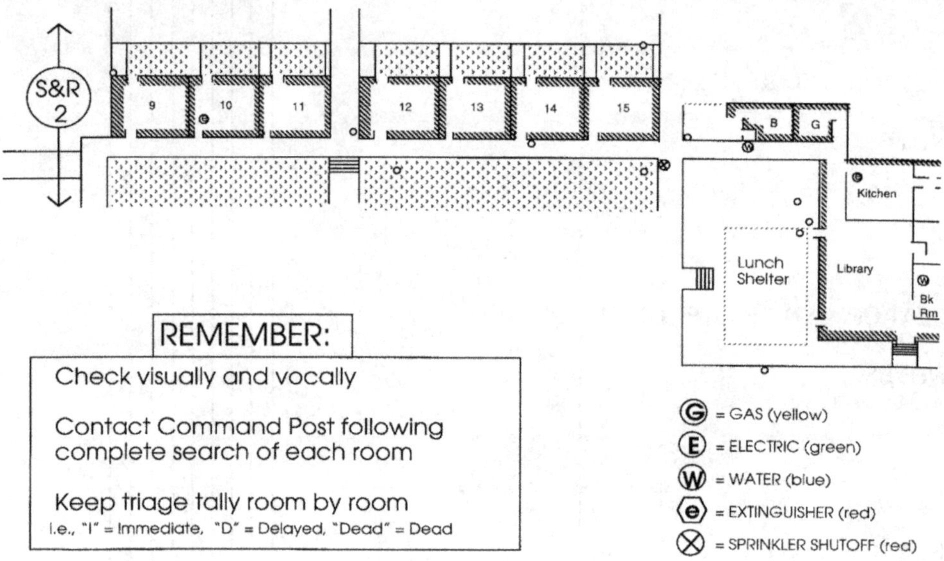

PUBLIC INFORMATION RELEASE

Check (_) as appropriate: District/District-Wide __________ School __________

Date: ______________________ Time: _________________________________

NOTE: If this is used as a script, read only those items checked. Make no other comments.

(Check off, fill in, and cross off as appropriate.)

______________________ has just experienced a(n) ______________________

- The (students/employees) ([are being] or [have been]) accounted for.
- No further information is available at this time.
- Emergency medical services ([are here] or [are on the way] or [are not available to us]).

- Police ([are here] or [are on the way] or [are not available to us]).
- Fire Dept./paramedics ([are here] or [are on the way] or [are not available to us]).
- ______________________ ([are here] or [are on the way] or [are not available to us]).
- Communication center(s) for parents (is/are) being set up at ______________________ to answer questions about individual students.
- Communication center(s) for families (is/are) being set up at ______________________
- to answer questions about individual employees.
- Injuries have been reported at __________ and are being treated at the site by
- (staff/professional medical responders). (#)_____ reported injured.
- Students have been taken to a safe area, ______ , and are with [(classroom
- teachers/staff) or (__________)].
- (#) Students have been taken to the local emergency room for treatment of serious injury.
- Parents of injured students should go to the emergency room at __________
- (#) Confirmed deaths have been reported at ______________________ ____
- Names cannot be released until families have been notified.
- Structural damage has been reported at the following sites: ______________

Release restrictions _____ No ____Yes
If yes, what?
Released to the public as Public Information Release #__________________
(Date/Time: ______________)

Emergency Time/Situation/Response Report

TIME	SITUATION	RESPONSE	INITIAL

Sample Log

Date ___________

Important: A permanent log may be typed or rewritten later for clarity and better understanding. Keep all original notes and records. **They are legal documents!**

Time	**Recorded Action(s)**

WE ARE
SHELTERING IN PLACE

DO NOT ENTER

SHELTER-IN-PLACE DRILL CHECKLIST

Good	Need to Improve How?	
_______	_______________	1. Parents were informed about the drill.
_______	_______________	2. Scenario reviewed with staff prior to event.
_______	_______________	3. Students/staff went inside, closed the doors, and closed and locked all windows.
_______	_______________	4. Hang signs on doors and office to indicate "Sheltering In Place."
_______	_______________	5. Roll call of students and staff.
_______	_______________	6. Place wet towels across the bottom of doors to the outside. (Simulate during drill.)
_______	_______________	7. Tape up any vents that can't be closed. (Simulate during drill.)
_______	_______________	8. Tape around window if air is leaking in. (Simulate during drill.)
_______	_______________	**9.** Turn off air conditioning and exhaust fan in kitchen. (Simulate during drill.)
_______	_______________	10. Close drapes and curtains.
_______	_______________	11. Turn off pilot lights, but electricity should remain on. (Simulate during drill.)
_______	_______________	12. Designate room for people who come to school during drill.
_______	_______________	13. Post signs indicating location of visitor's room.
_______	_______________	14. Alternate restroom facilities available in each room.
_______	_______________	15. Alternate source of water for rooms without sinks.
_______	_______________	16. Are there provisions for students needing medication during shelter-in-place?
_______	_______________	**17.** Call 911 and the School District to tell them you are sheltering-in-place. (Simulate during drill.)
_______	_______________	**18.** No one goes out during this time.
_______	_______________	19. Do you have a method of communicating with the office during a drill?

Please use the reverse side of this checklist for your suggestions and candid comments. Please return this form within five days to the principal. This form should be completed by all adult participants and observers at school site.

Name: ___________________ **Date:** ____________ **Site:**________________

DISTRICT-WIDE EMERGENCY DRILL

Site Observer Checklist

Duck/Cover/Hold:

- The teacher led duck/cover/hold on by example.
- The students knew the proper procedure:
- The students □ ducked under cover □ covered their eyes □ held on
- The teacher gave instructions and reassurances.
- The teacher checked self and evaluated situation.
- The teacher asked the students to check themselves and others.
- The teacher evaluated the situation and waited for class composure before asking for an evacuation.
- The teacher checked with the buddy teacher.

Special situations: If there were special situations, either planned or unplanned, how were they handled?

Evacuation and Student Accounting:

- Evacuation was orderly.
- If there were no serious "injuries," the teacher evacuated with the buddy teacher, one at the front of the line, one at the rear. If a teacher was incapacitated, buddy teacher evacuated both classes. If a student was injured and could not be moved, one teacher remained with the injured, while the buddy teacher evacuated both classes to the assembly area.
- At the assembly area, the students sat down while the teacher took roll.
- A Student Accounting Form was sent to the Command Post

Special situations: If there were special situations, either planned or unplanned, how were they handled?

Comments:

CHECKLIST TO CLOSE SCHOOL

- The School Director is notified of a situation that might require the closing of a school or site, either by local authorities (police, fire, HAZMAT, civil defense, etc.) or by status report(s) to the district EOC from the sites.
- Gather information (weather forecasts, road conditions, damage assessment of facilities, injuries, staff available, utilities and sanitation capabilities, local authority recommendations/directives) in the EOC.
- Only the district superintendent is authorized to close schools. The recommendation to close schools will be made by the school director.
- All requests/recommendations by local civil authorities and/or agencies (police, fire, HAZMAT, Red Cross) shall be directed to the emergency manager.
- If school is to be closed the following day, this decision will be made, if possible, before sending children and staff home.
- Affected sites are notified.
- Civil authorities are notified.
- County is notified to begin the Emergency Alert System.
- The superintendent will put the message on the telephone information line. Updates made periodically.
- Transportation is notified when and where buses are needed.
- The superintendent notifies the school board.
- The district public Information Officer notifies community relations director, who will furnish the media with up-to-date information. Updates will be made periodically.

For reopening:

- Decision to reopen site is made. Date/Time ________________________________
- District Public Information Officer Sandra Wilkins notifies the media as to the reopening.
- Information message is updated.
- Personnel is notified of site reopening.

CLOSING OF SCHOOL ACTION CHECKLIST

This checklist delineates the action to be taken if school, schools, or the entire district must be closed due to an emergency situation. It is imperative that all staff members be aware of these procedures and be prepared to carry them out if the principal is not available to make the determinations required. If specific persons are designated to perform these tasks, their names should be noted as appropriate.

A copy of this checklist must be forwarded to the School Director and the District Emergency Manager following the incident.

Site:__________________________________ Date:__________________________

Type of Incident:______________________________ Time: __________________

Check off completed tasks as appropriate to the circumstances of the situation.

- The site requested/directed by civil authority to close school.
  Immediately _______ Next day ______

  Request made by (name and agency) ____________________________________
- School director advised of civil authority's directive to close school.
- Superintendent makes final decision. Decision: _______________________________
  ____
- Decision announced to site. By ____________________________ Time: _______
- Telephone calling trees began. By __________________________ Time: _______

Reopening of site:

- School is notified to reopen. By ____
  Date: __________________________ Time: ________________________
- Employees are notified. By: ____
  Date: __________________________ Time: ________________________
- Students are notified. Calling trees began by: ________________________________
  Date: __________________________ Time: ________________________

UNSAFE BUILDING EVACUATION PLAN WORKSHEET		
PERSON IN CHARGE OF SCHOOL	**ALTERNATE**	**ALTERNATE**
STAFF ASSIGNMENTS		
DUTY	**STAFF NAME**	**ALTERNATE**
All teachers are to stay with classroom groups or assigned neighborhood groups.		
Notify the following from an off-site location Emergency assistance: 911 or ______________ Security: xxx-xxxx Superintendent: xxx-xxxx or xxx-xxxx Transportation: xxx-xxxx or xxx-xxxx		
Search halls, rest rooms, common areas, and kitchen.		
Replacements for missing or injured teachers.		
Provide medical assistance.		
Shut off power, gas, and water.		
Notify as appropriate PTA Evacuation Leader: Name________________ Phone ____________ PTA Evacuation Alternate: Name________________ Phone ____________ Emergency message: xxx-xxxx (general public) Emergency message: xxx-xxxx (employees) Activate CALLING TREE: Phone ____________		
Post premade signs and direct the public.		
Lead students and teachers to alternate site (attach map and agreement). Site 1 ______________________________ Site 2 ______________________________		
Emergency Equipment		
___ Air horn or bell ___ Radio (battery or solar) ___ First aid kit ____ Lights/lanterns ___ Answering machine ____ Unlisted phone line ___ Generator ____ Megaphone ___ Evacuation plan/ master lists ____ Paper and marker pens ___ Evacuation boxes ____ Stretcher		
Other		

SAFE BUILDING EVACUATION PLAN WORKSHEET		
PERSON IN CHARGE OF SCHOOL	**ALTERNATE**	**ALTERNATE**
STAFF ASSIGNMENTS		
DUTY	**STAFF NAME**	**ALTERNATE**
Notify Superintendent: xxx-xxxx or xxx-xxxx		
Notify as appropriate PTA Evacuation Leader: Name_________________ Phone ___________ PTA Evacuation Alternate: Name_________________ Phone ___________ Emergency message: xxx-xxxx (general public) Emergency message: xxx-xxxx (employees) Activate CALLING TREE: Phone _____________		
Call students and teachers to preassigned inside locations.		
Post premade signs and direct the public.		
Sign out students to responsible adults.		
Emergency Equipment		
___ Air horn or bell ___ Radio (battery/solar) ___ First aid kit ___ Lights/lanterns ___ Answering machine ___ Unlisted phone line ___ Generator ___ Megaphone ___ Evacuation plan/master lists ___ Paper and marker pens ___ Evacuation boxes ___ Stretcher		
IF A LONG DURATION STAY IS REQUIRED		
Trauma, shock, medical care		
Provide light and warmth		
Food arrangements		
Sleeping accommodations		
Internal communication/morale		
Other		

CRISIS RESPONSE TEAM REPORT FORMAT

Report to document crisis management activities; to be submitted to Executive Director, Business Affairs within five days of the incident:

Crises Response Team Report

Date of Report____________

School ____

Student'(s) Name(s):	Parent/Guardian:
______________________	______________________
______________________	______________________
______________________	______________________

Description of incident (include date, time, place):

Immediate intervention by crisis response team:

Follow-up procedures (with student, with student body [if appropriate], with faculty):

Follow-up with parent(s)/guardian(s) of student(s) involved:

Case Manager: ______________________________________

Reviewed by Principal: _______________________________

Date: ________________________

EVALUATION OF CRISIS MANAGEMENT TEAM INTERVENTION

The School System Crisis Management Team (CMT) seeks your input to help us assess the effectiveness of the intervention during the recent crisis at your school. We would appreciate your taking a few minutes of your time to complete this form as soon as possible and return it to the executive director, Business Affairs within five days of the incident.

Please indicate with an "X" the response that most closely reflects your evaluation of the following:

1. Speed of CMT's response to the crisis:

______	______	______	______	______
Very delayed		Adequate		Very timely

2. Comprehensiveness of CMT's response:

______	______	______	______	______
Very limited		Adequate		Exceptional

3. Effectiveness of team's response in meeting students' needs:

______	______	______	______	______
Not helpful		Adequate		Very helpful

4. Assistance to teachers in the classroom:

______	______	______	______	______
Not supportive		Adequate		Very supportive

5. Quality of communication with the school faculty and staff:

______	______	______	______	______
Not informative		Adequate		Very explanatory

6. Support and guidance to individual teachers and staff:

______	______	______	______	______
Ineffectual		Adequate		Very empathetic

7. Assistance to the administrative staff:

______	______	______	______	______
Limited			Adequate	Very comprehensive

8. Amount of time allocated for intervention:

______	______	______	______	______
Insufficient		Adequate		Very sufficient

9. Support to families in need:

___	___	___	___	___
Lacking		Adequate		Outstanding

10. Communication of information to parents:

___	___	___	___	___
Minimal		Adequate		Thorough

11. Please circle the adjectives that best describe the students' reactions to the CMT's intervention:

ambivalent	angry	apathetic	grateful	hostile
negative	positive	receptive	relieved	satisfied

12. Please describe any significant reactions the students had to the CMT intervention that should be considered in future interventions.

13. Please comment on any aspects of the intervention you found particularly helpful or areas that you think should be modified.

School:

____

Signature (Optional):

____

Please return to Executive Director, Business Affairs
(within five days of the incident)

AUTHORIZATION TO RELEASE CHILDREN IN AN EMERGENCY

Our school has developed an emergency plan in case of any disaster that might occur. The emergency plan is devoted to the welfare and safety of your child during school hours. The plan is available for inspection in the school office.

We are requesting your assistance at this time:

Should there be an emergency, such as a major fire, tornado, explosion, and so on, your child may be required to remain in the care of the school until it is deemed safe by an Emergency Services authority that the child can be released. At that point, children may be released only to properly authorized parents and/or designees. Therefore, please list as many names *(with local telephone numbers and addresses)* as possible, of those persons to whom you would allow your child's release in the event of an emergency. Be sure to notify those persons listed that you have authorized their supervision in case of emergency.

In the event that you should be unable to come to school, it is essential that others be designated to care for your child. No child will be released to the care of unauthorized persons.

We appreciate your cooperation in this important matter.

Child: ______________ **Teacher:** ________________ **School Year:** __________

Please release my child to any of the persons listed below:

Name	Relationship	Address	Phone
			Home
			Work
			Cell
			Home
			Work
			Cell
			Home
			Work
			Cell
			Home
			Work
			Cell

Parent/Guardian: ________________________________ **Date:** ____________

Signature

Home Phone: _____________ **Work Phone:** __________ **Cell Phone:** _________

SITE CRISIS RESPONSE TEAM ASSIGNMENT FORM

POSITION	ORGANIZATION	NAME
Command Section:		
Incident Commander		
Safety Officer		
Public Information Officer		
Liaison Officer		
Operations Section:		
Operations Section Chief		
Site Facility Check/Security		
Student Care		
Student Release		
Planning Section:		
Planning Section Chief		
Documentation		
Situation Analysis		
Logistics Section:		
Logistics Section Chief		
Supplies/Facilities		
Staffing		
Communications		
Finance/Administration Section:		
Finance/Administration Chief		
Timekeeping		
Purchasing		

General Message Form	
Incident Name/Location:	**Date/Time of Message:**
To:	
From:	
Message:	
Reply:	
Signature (person replying):	**Date/Time of Reply:**

APPENDIX 5

SAMPLE COMMUNICATIONS AND LETTERS

Tips for Communicating with Parents

In the event of an emergency, parents have very specific information needs.

First, parents want to know their child is safe; then, parents want to know the details of the emergency situation, to know how it was handled, and to know that their children will be safe in the future.

The first reactions are likely to involve fear. Upon learning of an incident at the school, parents are likely to descend upon the school in search of their child or to telephone, frantically seeking information. Establishing a system for responding quickly to parent needs for information is an important part of preplanning.

Anger is another common reaction of parents, particularly in the case of senseless acts of violence. In the event of a crisis or disaster:

1. Tell parents exactly what is known to have happened. Do not embellish or speculate.
2. Implement the plan to manage phone calls and parents who arrive at school.
3. Schedule and attend an open question-and-answer meeting for parents as soon after the incident as possible. The meeting is an opportunity for school officials to listen and respond to parent concerns (which are helpful in combating rumors and other misinformation) and to work on restoring parental trust in the school.
4. In the event of an incident that involved damage or destruction, an open house for parents and other members of the community to see the school restored to its "normal" state helps everyone get beyond the crisis.

Sample Annual Letter to Parents/Guardians:

Description of School District Emergency Program and Parent-Guardian Expectations

[SCHOOL LETTERHEAD]

Dear Parents/Guardians:

We want you to be aware that this school has made many preparations to deal effectively with emergency situations that could occur in or around the school, both during the school day and during after-hours activities. While we hope that a natural disaster or other serious incident never occurs, our goal is to be prepared for any potential emergency. At all times, our first priority is to protect all students, staff, and guests from harm.

In order for our emergency response plans to be effective, we depend on the cooperation and assistance of many people, such as the police and the fire department. We also depend on you, as parents and guardians, to support our disaster preparation and response efforts. Your cooperation is vital to helping us protect the safety and welfare of all children and school employees.

Therefore, we ask parents to observe the following procedures in the event of a school emergency:

1. Do not telephone the school. We understand and respect your concern, but it is essential that the telephone system is available for emergency communications.
2. Make sure that we have emergency contact information for each of your children at all times. We must be able to contact you or your designated representative in an emergency.
3. Tune your radio to WAKR—AM 1590, WTOU—AM 1350, WTAM—AM 1100, WHLO—AM 640, WNIR—FM 100.1 WKDD—FM 98.1, WQMX—FM 94.9, WONE—FM 97.5, or WCPN—FM 90.3 for emergency announcements and status reports. You also will receive instructions on where you should go and how/when you may be able to pick up your children. Our school emergency plan includes evacuation procedures with several alternative destinations. When appropriate and safe, students may be released to their parents/guardians from these shelter locations. Under those circumstances, we will be prepared to implement procedures for confirming the identity of individuals who arrive to pick up each child. When arriving to pick up your children, please make sure that you have with you your driver's license or government issued picture identification.
4. Do not come to the school or alternate destination until instructed to do so. It may be necessary to keep the streets and parking lot clear for emergency vehicles. If evacuation is required, students may be transported to a location away from school. You will be notified of this through the media bulletins.
5. Talk to your children and emphasize how important it is for them to follow instructions from their teachers and school officials during any emergency.
6. Parents and other adults must stay calm and focused in an emergency, mindful that their actions and comments will be the example that, to a great extent, determines the children's response.

7. Carefully read all information you receive from the school. You may receive updates about our safety procedures from time to time.
8. When your child is at home following an emergency, try to keep your child away from news being broadcast over the various media. Have an emergency plan at home. Decide on a telephone number to call outside our community or a place to rendezvous if separated. Keep a "disaster supplies kit" containing drinking water, nonperishable food, batteries, flashlight, radio, medication, toothbrushes, and so on.

We are proud that *School Name* is a safe school, and we are doing everything possible to keep it that way. We appreciate your cooperation and support. If you have any questions about this letter or other aspects of our safety procedures, please contact me at _____________.

Sincerely,

Principal

Sample Letters to Parents/Guardians Notice of Bus Accident

[SCHOOL LETTERHEAD]

Dear Parents/Guardians,

This morning, prior to school, there was an accident involving a school bus and an automobile. There were known injuries to the passengers of the car. The children on Bus #_ witnessed the aftermath of the accident but were not involved in it.

The children from the bus involved in the accident were taken to the library by the guidance counselors and administration. The children were asked if they were injured in any way and their parents were then contacted. Your child, because of being on Bus # _____, may show delayed reaction to the accident. Please be alert over the next several days to symptoms of delayed reaction, including:

- a desire to be alone, unusually quiet
- loss of appetite
- problems with sleeping, nightmares
- difficulty with concentration
- crying
- angry outburst, short temper
- headaches, upset stomach
- depressed, sad

Your child may also exhibit some physical complaints. Please contact *(principal's name)* to fill out an accident report. The school will be offered support services for students needing help dealing with the accident. We will also provide counseling services to parents in helping their children to cope. Please don't hesitate to call if you have any questions or concerns.

Sincerely,

Principal

Sample Joint APS/Community Letter to Parents/Guardians about Crisis

(i.e., shooting or other violent acts)

[SCHOOL LETTERHEAD]

Dear Parents:

The tragic event of *(date)* has hurt/affected the entire *(school)* community. This unfortunate incident shows us that we must band together as parents and seek positive solutions to problems that victimize us.

It is for this reason I'm asking ALL parents to join me in a special meeting *(day, date, time)* in the school auditorium. We will learn the steps the administration is taking to ensure the welfare and safety of our children; we will discuss what we, as parents and community, can do to assist our children and school staff.

There will be representatives from the school staff, schools administration, School Police Department, guidance office, and the representatives of the School Mayor's Office to address ALL our concerns.

Bus transportation will be provided for the *(area)* families. *(Here include the place and time of bus pickups.)*

Child care will be provided at the school for your convenience.

Refreshments will be served at the end of the meeting.

If you would like further information about this meeting, contact *(name/telephone)*.

Cordially,

____________________ ____________________

Principal Mayor's Office

Sample Thank You Letter to Faculty and Staff Members

[SCHOOL LETTERHEAD]

Dear Faculty and Staff Members:

We would like to thank you for your support during the recent crisis at our school. Your professionalism and dedication were evident as we all worked to quiet and soothe scared students and allay their fears while still tending to instructional responsibilities.

We know that this has been an extremely difficult time for you as well as the students. Without your courage and concern, our school could not possibly have come through this crisis as well as we did.

Thank you once again. Your expertise and commitment have enabled all of us to work together as a team and overcome this tragic situation.

Sincerely,

Principal/Superintendent

Sample Letter to Students and Parents: Student Sudden Death

[SCHOOL LETTERHEAD]

Dear Students and Parents:

On Friday, John Doe, a fifth-grade student at *(name of school)* was in a terrible accident. Apparently, John was hit by a car that was speeding and had crossed over the median strip to the wrong side of the road. John died instantly.

John's death is a tragic, emotional loss for the entire school family. I am sure all of us will make every effort to comfort and support John's family as they attempt to deal with this traumatic loss. There are no adequate words to express our sense of grief and sympathy for the family.

Because John's death is felt so deeply by so many, on Monday and thereafter as needed, we will bring in our crisis team to discuss this accident and loss with students and faculty.

We encourage each of you to discuss this loss with your child. In order to help you do so, we are holding a parent meeting on *(date and time)*, in the school cafeteria. The topic will be ways to help children cope with loss and will be presented by a local mental health professional. John's family, friends, and the school are suffering deeply. Please join us in supporting John's family.

Sincerely,

Principal

Sample Letter to Families Following an Assault/Sexual Assault

[SCHOOL LETTERHEAD]

Dear Parents/Caregivers:

This letter is to inform you that a student assault has occurred. *(Insert information here regarding whether the assault took place on or off campus and any other information that can be provided without violating a victim's confidentiality.)*

All of us will be feeling a variety of emotions, including shock, sadness, and anger. I want you to know that we, the *(school name)* staff, care about the feelings our students may be experiencing.

The Crisis Management Team has made plans to respond to the emotional needs of the students. *(Clarify what is being done: grief counseling, classroom debriefing, and referrals.)*

Any time violence touches us, it is extremely stressful. Please inform your child's *(teacher or counselor)* if there is additional information the school should be aware of so we can provide the support your child needs.

Please feel free to call me at *(school telephone number)* if you have any questions or concerns.

Sincerely,

Principal

Sample Letter to Parents, Students and Staff: Following a Bomb Threat

[SCHOOL LETTERHEAD]

Dear Parents, Students, and Staff,

Somewhere between 7:30 a.m. and 7:45 a.m. *(add day/date)*, a note was left in the drop box in the attendance office making reference to a bomb going off at school. There was no specific reference to a location *(school name)*, a date, or a time. The attendance clerk contacted the school administration immediately. The police were notified along with District Office personnel. School Police responded to the site within minutes. Law Enforcement officers and site administrators evacuated classes at 9 a.m. and faculty and the deputies conducted an organized and thorough room-by-room search. At approximately 10:15 a.m. law enforcement officials determined that the school/facility was safe and gave us permission to reenter the building. After a short break, classes resumed as regularly scheduled.

The administration of *(school name)* and local law enforcement agencies are taking this event very seriously. If anyone has information regarding the person responsible for leaving the note, please contact the police immediately.

School will be open tomorrow; classes will be conducted as usual. Increased security will be in place. We will continue to communicate with you via the auto dialer and written messages. Counselors and administrative staff are available to assist students who wish to talk about their concerns or fears regarding this or other issues that are going on nationally. Please feel free to contact us for further information and remember the anonymous tip line is available for your use as needed by dialing *(phone number)*.

Sincerely,

Principal

Sample Letter to Families Following a Death

[SCHOOL LETTERHEAD]

Dear Parents /Guardians:

I am sorry to inform you that a *(staff person/student/friend), (name)* at *(school name),* has died *(day/date). (Insert what information can be shared about the cause and circumstances of the death.)*

Death can be difficult for us to understand, especially when it is sudden. All of us will be feeling a variety of emotions: shock, sadness, and confusion. What is important is that we care for and support each other.

The Crisis Management Team has made plans to respond to the emotional needs of the students. *(Spell out what is being done: grief counseling, classroom debriefing, and referrals.)*

If your family has experienced a death or similar loss recently, the death of *(name)* may bring up feelings about that death. This is a normal experience. Please let your child's *(teacher or counselor)* know if there is any additional information the school should be aware of so we can provide the support your child needs.

Any time death touches us, it is extremely stressful. This sudden death may be disturbing to you as well as to your child. It is for this reason, we especially want you to know of our caring and support.

Sincerely,

Principal

Sample Letter to Staff Following a Death

[SCHOOL LETTERHEAD]

Dear Staff,

There are times when it is necessary to communicate news that is painful for all of us. During those times we must be prepared to support each other as we deal with the many feelings that we begin to experience. It is with great sorrow that I inform you that *(name of person)* at *(school name)* has died.

Death can be difficult for us to understand, especially when it is sudden. We will all begin to feel different emotions: shock, sadness, confusion even some anger. What is most important is that we care for and support each other.

Although rare, sometimes students may be so overwhelmed by the death of someone important to them, that they may express suicidal thoughts or actions. Please contact an appropriate staff person or counselor if you notice a student who appears to be having more difficulty with his/her feelings than might be expected. Information regarding youth suicide is included in this letter.

In memory of *(name), (indicate here what activity or activities the school is planning).* Administration will keep you updated as more information is given to us at the school.

Sincerely,

Principal

Sample Letter to Students Following a Death

[SCHOOL LETTERHEAD]

Dear Students,

I have asked your teacher to read this letter to you because I want to make sure that all students receive the same information about the recent tragedy at our school. It gives me a great sorrow to inform you that *(name),* a *(teacher, student/friend)* at *(school name)*, has died *(day/date). (Insert what information can be shared about the cause and circumstances of the death.)*

Death can be difficult for us to understand, especially when it is sudden. Many of us may be confronted with a variety of emotions which might include shock, sadness, and confusion. I want to assure you that we, the *(school name)* staff, care about you and the feelings you may be experiencing.

Please know that we want to support you during this time. The Crisis Management Team will be available to meet with you in *(place)* to assist you in dealing with any feeling you may be having. You might wish to share memories you have of *(name).* Crisis Management Team members will also be available at any time during the day to help you if you feel a more urgent need to talk with someone. *(Insert here specific information on how students can access Team members for support).* I want to encourage those students who may be particularly upset, perhaps even struggling with a death in the family or of a friend, to talk with Crisis Management Team members. They will be available all day to meet with you.

Anytime death touches us, it is extremely stressful. This sudden death may be quite shocking to you and confuse you. For these reasons, we especially want you to know of our caring and support.

Sincerely,

Principal

Sample Letter to Families Following Tragedy

[SCHOOL LETTERHEAD]

Dear Parents/Guardians,

I would like to share with you the response of our school district to today's tragedy. Although these events were sudden and unexpected, students acted appropriately, and staff responded professionally. Many teachers discussed the events with students. Students who were uncomfortable or worried were allowed to go to the office for assistance.

I would like to offer some guidelines for supporting your children through this challenging time, as follows:

- Take the lead from your child. If your child wants to talk about the events, do so. Focus on the facts. Listen to their thoughts and feelings. Listening is more important than talking, but do share your own feelings as a way to help them accept and understand their feelings.
- Students who were emotionally fragile prior to these events may have a bigger response than others. Some children's reactions may be delayed by several days.
- Children may need assurance that actions are being taken to keep them safe and secure.
- Especially for younger children, using art to express feelings may be helpful. Provide paper and drawing implements to give children an opportunity for this expression.
- An opportunity to write about the events of the day may also be a good release.
- Phone calls or visits with close friends and extended family members may create a sense of security and closeness.

We will continue to assess the needs of our students and respond to those needs. Please let your child's teacher or school administrator know if you have a particular concern, if you think your child needs some extra attention and support, or if your family has suffered a loss during these events.

Pending any action by state officials, all of our schools will remain in session. However, if you feel that you want to keep your child at home, please do so. As usual, call the office to report the absence.

If you need further information, contact the Office of the Superintendent at *(phone number)*, or the Office of Instruction at *(phone number)*.

Sincerely,

Principal

Sample Letter to Media Following a Trauma

[SCHOOL LETTERHEAD]

TO: *(Community/media group)*
RE: *(Specific event)*

This letter is to inform you that a traumatic event has occurred on our campus. I am writing this letter because I feel it is important that the community receives the same information about the recent tragedy. *(Insert what information can be shared about the cause and circumstance of the event without violating confidentiality.)*

All of us will be feeling a variety of emotions, including shock, sadness, and anger, I want you to know that we, the *(school name)* staff care about the feelings our students and community may be experiencing.

The Crisis Management Team has made plans to respond to the emotional needs of the students and their families. *(Clarify what is being done: grief counseling, classroom debriefing, and referrals.)*

Anytime violence touches us it is extremely stressful. This recent event may trigger feelings about personal trauma and loss. This is a normal experience. We hope all affected will receive the support needed.

Sincerely,

Principal

Sample Announcement: Student Suicide
(To be read to the students by the classroom teacher.)

[SCHOOL LETTERHEAD]

TO: School Faculty
FROM: Principal
SUBJECT: *(Crisis)*

John Doe committed suicide early Saturday morning. As a faculty we extend our sympathy to John's family and friends.

We encourage all students to consider the tragic nature of this death and to realize that death is final. John's death is a reminder to us all that the act of taking one's life is not an appropriate solution to any of life's problems nor is it an act of courage. Please let your teachers know if you would like to talk to a counselor or other staff member.

Funeral services for John will be held in ___________ and there will/will not be a memorial service in this area. Expressions of sympathy may be sent to *(name and address)*.

ACRONYMS AND ABBREVIATIONS

A

AA&E Arms Ammunition and Explosives
AAR After Action Report
ACL Access Control List
ACP Access control point
ACS Access Control System
ADA Americans with Disabilities Act
ADAAG Americans with Disabilities Act Accessibility Guidelines
AECS Automated Entry Control System
AFJMAN Air Force Joint Manual, also may be known as AFMAN (I) for Air Force Manual
AFMAN Air Force Manual
ALERT Automated Local Evaluation in real time
AMS Aerial Measuring System
ANFO Ammonium Nitrate and Fuel Oil
ANS Alert and Notification System
ANSI American National Standards Institute
ANSIR Awareness of National Security Issues and Response Program
AOR Area of Responsibility
AP Armor piercing
APHL Agency for Public Health Laboratories
ARAC Atmospheric Release Advisory Capability

ARC American Red Cross
ARG Accident Response Group
ARS Agriculture Research Service
ASCE American Society of Civil Engineers
ASHRAE American Society of Heating, Refrigerating, and Air-Conditioning Engineers
ASTHO Association for State and Territorial Health Officials
ASTM American Society for Testing and Materials
ASZM-TED A copper-silver-zinc-molybdenum-triethylenediamine
AT Antiterrorism
ATC Air Traffic Control
ATF Bureau of Alcohol, Tobacco, and Firearms
ATSD(CS) Assistant to the Secretary of Defense for Civil Support
ATSDR Agency for Toxic Substances and Disease Registry
AWG American wire gauge

B

BCA Benefit/Cost Analysis
BCC Backup Control Center
BCP Business Continuity Plan
BDC Bomb Data Center
BLASTOP Blast-Resistant Window Program
BMS Balanced magnetic switch
BW Biological warfare

C

CAMEO Computer-Aided Management of Emergency Operations
CB Citizens Band
CBIAC Chemical and Biological Defense Information and Analysis Center
CBR Chemical, biological, or radiological
CBRNE Chemical, biological, radiological, nuclear, or explosive
CCTV Closed circuit television
CDC Centers for Disease Control and Prevention
CDR Call Detail Report
CDRG Catastrophic Disaster Response Group
CEO Chief Executive Officer
CEPPO Chemical Emergency Preparedness and Prevention Office
CERCLA Comprehensive Environmental Response, Compensation, and Liability Act
CERT Community Emergency Response Team
CFD Computational Fluid Dynamics
CFO Chief Financial Officer

CFR Code of Federal Regulations
CHEMTREC Chemical Manufacturers' Association Chemical Transportation Emergency Center
CHPPM Center for Health Promotion and Preventive Medicine
CIAO Chief Infrastructure Assurance Office
CIAO Critical Infrastructure Assurance Officer
CICG Critical Infrastructure Coordination Group
CIO Chief Information Officer
CIP Critical Infrastructure Protection
CIRG Crisis Incident Response Group
CIT Crisis Intervention Team
CJCS Chairman of the Joint Chiefs of Staff
CM Consequence Management
CM Crisis Management
CMS Call Management System
CMT Crisis Management Teams
CMU Concrete masonry unit
CMU Crisis Management Unit (CIRG)
COB Continuity of Business
COBIT™ Control Objectives for Information Technology
CO/DO Central Office/Direct Outdial
CONEX Container Express
CONOPS Concept of Operations
COO Chief Operating Officer
COOP Continuity of Operations Plan
COR Class of Restriction
COS Class of Service
CPG Civil Preparedness Guide
CPTED Crime Prevention Through Environmental Design
CPX Command Post Exercise
CRU Crisis Response Unit
CSEPP Chemical Stockpile Emergency Preparedness Program
CSI Construction Specifications Institute
CSREES Cooperative State Research, Education, and Extension Service
CST Civil Support Team
CSTE Council of State and Territorial Epidemiologists
CT Counterterrorism
CW/CBD Chemical Warfare/Contraband Detection

D

DBMS Database Management System
DBT Design Basis Threat
DBU Dial backup
DD Data Dictionary
DES Data Encryption Standard

DEST Domestic Emergency Support Team
DFO Disaster Field Office
DHS Department of Homeland Security
DISA Direct Inward System Access
DMA Disaster Mitigation Act of 2000
DMAT Disaster Medical Assistance Team
DMCR Disaster Management Central Resource
DMORT Disaster Mortuary Operational Response Team
DOC Department of Commerce
DoD Department of Defense
DOE Department of Energy
DOJ Department of Justice
DOS Department of State
DOT Department of Transportation
DPP Domestic Preparedness Program
DRC Disaster Recovery Center
DTCTPS Domestic Terrorism/Counterterrorism Planning Section (FBI HQ)
DTIC Defense Technical Information Center
DTM Data-transmission media
DWI Disaster Welfare Information

E

EAS Emergency Alert System
ECL Emergency Classification Level
EECS Electronic Entry Control System
EFR Emergency First Responder
EM Emergency Management
EMAC Emergency Medical Assistance Compact
EMI Emergency Management Institute
EMP Electromagnetic pulse
EMS Emergency Medical Services
EOC Emergency Operations Center
EOD Explosive Ordnance Disposal
EOP Emergency Operating Plan
EOP Emergency Operations Plan
EPA Environmental Protection Agency
EPCRA Emergency Planning and Community Right-to-Know Act
EPG Emergency Planning Guide
EPI Emergency Public Information
EP&R Directorate of Emergency Preparedness and Response (DHS)
EPZ Emergency Planning Zone
ERP Emergency Response Plan
ERT Emergency Response Team
ERT-A Emergency Response Team Advance Element
ERT-N Emergency Response Team National

ERTU Evidence Response Team Unit
ESC Expandable Shelter Container
ESF Emergency Support Function
ESS Electronic Security System
EST Emergency Support Team
ETL Engineering Technical Letter
EU Explosives unit

F

FAsT Field Assessment Team
FASRO Florida Association of School Resources Officers
FBI Federal Bureau of Investigation
FCC Federal Communications Commission
FCC Fire Control Center
FCO Federal Coordinating Officer
FEM Finite element
FEMA Federal Emergency Management Agency
FEST Foreign Emergency Support Team
FHBM Flood Hazard Boundary Map
FIA Federal Insurance Administration
FIPS Federal Information Processing Standard
FIRM Flood Insurance Rate Map
FIS Flood Insurance Study
FISCAM Federal Information Systems Control Audit Manual
FMFIA Federal Manager's Financial Integrity Act
FNS Food and Nutrition Service
FOIA Freedom of Information Act
FOUO For Official Use Only
FPEIS Final Programmatic Environmental Impact Statement
FRERP Federal Radiological Emergency Response Plan
FRF Fragment retention film
FRL Facility Restriction Level
FRMAC Federal Radiological Monitoring and Assessment Center
FRP Federal Response Plan
FS Forest Service
FSTFS Frame-Supported Tensioned Fabric Structure
FTP File Transfer Protocol
FTX Functional Training Exercise

G

GAO General Accounting Office
GAR Governor's Authorized Representative

GC/MS Gas chromatograph/mass spectrometer
GIS Geographic Information System
GP General Purpose
GPS Global Positioning System
GSA General Services Administration

H

HazMat Hazardous material
HAZUS Hazards US
HEPA High efficiency particulate air
HEU Highly enriched uranium
HF High frequency
HHS Department of Health and Human Services
HIRA Hazard Identification and Risk Assessment
HMRU Hazardous Materials Response Unit
HQ Headquarters
HRCQ Highway Route Controlled Quantity
HRT Hostage Rescue Team (CIRG)
HSOs Homeland Security Offices
HTIS Hazardous Technical Information Services (DoD)
HVAC Heating, ventilation, and air conditioning

I

IAQ Indoor Air Quality
IC Incident Commander
ICDDC Interstate Civil Defense and Disaster Compact
ICP Incident Command Post
ICS Incident Command System
ID Identification
IDS Intrusion Detection System
IED Improvised Explosive Device
IEMS Integrated Emergency Management System
IESNA Illuminating Engineering Society of North America
IID Improvised Incendiary Device
IMSion Mobility spectrometry
IND Improvised Nuclear Device
IPL Initial Program Load
IR Infrared
IRZ Immediate Response Zone
IS Information System
ISACF Information Systems Audit and Control Foundation
ISC Interagency Security Committee

ISO International Organization for Standardization
ISP Internet Service Provider
IT Information Technology

J

JIC Joint Information Center
JIISE Joint Interagency Intelligence Support Element
JIS Joint Information System
JNACC Joint Nuclear Accident Coordinating Center
JOC Joint Operations Center
JSMG Joint Service Materiel Group
JTF-CS Joint Task Force for Civil Support
JTTF Joint Terrorism Task Force
JTWG Joint Terrorism Working Group

K

kHz Kilohertz
kPa kilo Pascal

L

LAN Local Area Network
LAW Light Antitank Weapon
LBNL Lawrence Berkley National Lab
LCM Life-cycle Management
LED Light-emitting diode
LEED Leadership in Energy and Environmental Design
LEPC Local Emergency Planning Committee
LF Low frequency
LFA Lead Federal Agency
LLNL Lawrence Livermore National Laboratory
LOP Level of protection
LOS Line of sight
LPHA Local Public Health Agency
LPHS Local Public Health System

M

MAC Moves, Adds, Changes
MEDCOM Medical Command

MEI Minimum Essential Infrastructure
M/E/P Mechanical/Electrical/Plumbing
MEP Mission Essential Process
MERV Minimum efficiency reporting value
MMRS Metropolitan Medical Response System
MOU/A Memorandum of Understanding/Agreement
mph Miles per hour
MPOP Minimum-Points-of-Presence
Ms Millisecond
MSCA Military Support to Civil Authorities
MSDS Material Safety Data Sheet
MSS Medium Shelter System
MW Medium wave

N

NACCHO National Association for County and City Health Officials
NAP Nuclear Assessment Program
NAVFAC Naval Facilities Command
NBC Nuclear, biological, and chemical
NCJ National Criminal Justice
NCP National Contingency Plan (aka National Oil and Hazardous Substances Pollution Contingency Plan)
NDA National Defense Area
NDMS National Disaster Medical System
NDPO National Domestic Preparedness Office
NECA National Electrical Contractors Association
NEST Nuclear Emergency Search Team
NETC National Emergency Training Center
NFA National Fire Academy
NFIP National Flood Insurance Program
NFPA National Fire Protection Association
NFPC National Fire Protection Code
NIJ National Institute of Justice
NIMS National Incident Management System
NIOSH National Institute for Occupational Safety and Health
NMRT National Medical Response Team
NMS Network Management System
NOAA National Oceanic and Atmospheric Administration
NPP Nuclear Power Plant
NRC National Response Center
NRC Nuclear Regulatory Commission
NRT National Response Team
NSC National Security Council
NTIS National Technical Information Service

NUREG Nuclear Regulation
NWS National Weather Service

O

OCC Operational Control Center
ODP Office of Disaster Preparedness
OEP Office of Emergency Preparedness
OES Office of Emergency Services
OFCM Office of the Federal Coordinator for Meteorology
OHS Office of Homeland Security
OJP Office of Justice Programs
O&M Operations and maintenance
OMB Office of Management and Budget
OPA Oil Pollution Act
OSC On-scene Coordinator
OSD Office of Secretary of Defense
OSHA Occupational Safety and Health Administration
OSLDPS Office for State and Local Domestic Preparedness Support

P

Pa Pascal
PA Public address
PAZ Protective Action Zone
PBCC Palm Beach Community College
PBX Public Branch Exchange
PC Personal computer
PCC Policy Coordinating Committee
PCCIP President's Commission on Critical Infrastructure Protection
PCM Procedures Control Manual
PDA Personal data assistant
PDA Preliminary Damage Assessment
PDD Presidential Decision Directive
PHS Public Health Service
PIN Personal Identification Number
PIO Public Information Officer
PL Public Law
POC Point of Contact
POD Probability of detection
POI Probability of intrusion
POL Petroleum, Oils, and Lubricants
POV Privately owned vehicle
PPA Performance Partnership Agreement
Ppm Parts per million

PSE Particle size efficiency
Psi Pounds per square inch
PT Preparedness, Training, and Exercises Directorate (FEMA)
PTE Potential Threat Element
PTZ Pan-tilt-zoom (camera)
PVB Polyvinyl butyral
PZ Precautionary Zone

R

RACES Radio Amateur Civil Emergency Service
RAP Radiological Assistance Program
RCRA Research Conservation and Recovery Act
RDD Radiological Dispersal Device
RDT&E Research, Development, Test, and Evaluation
REACT Radio Emergency Associated Communications Team
REAC/TS Radiation Emergency Assistance Center/Training Site
REM Roentgen Man Equivalent
REP Radiological Emergency Preparedness Program
RF Radio frequency
ROC Regional Operations Center
ROD Record of Decision
RPG Rocket Propelled Grenade
RRIS Rapid Response Information System (FEMA)
RRP Regional Response Plan
RRT Regional Response Team

S

S&R Search and Rescue
SAA State Administrative Agency
SAC Special Agent in Charge (FBI)
SAFEVU Safety Viewport Analysis Code
SAME Specific Area Message Encoder
SARA Superfund Amendments and Reauthorization Act
SATCOM Satellite communications
SAW Surface acoustic wave
SBCCOM Soldier and Biological Chemical Command (US Army)
SCADA Supervisory, Control, and Data Acquisition
SCBA Self-Contained Breathing Apparatus
SCC Security Control Center
SCO State Coordinating Officer
SDO Single-degree-of-freedom
SEA Southeast Asia
SEB State Emergency Board

SEL Standardized Equipment List
SEMA State Emergency Management Agency
SERC State Emergency Response Commission
SESIR School Environment Safety Incident Reporting System
SFO Senior FEMA Official
SIOC Strategic Information and Operations Center (FBI HQ)
SLA Service Level Agreement
SLG State and Local Guide
SNM Special Nuclear Material
SpNS Special Needs Shelters
SOP Standard Operating Procedure
SPCA Society for the Prevention of Cruelty to Animals
SPSA Super Power Small Arms
SSS Small Shelter System
STC Sound Transmission Class
SWAT Special Weapons and Tactics

T

TAC Trunk Access Codes
TDR Transferable development right
TEA Threat Environment Assessment
TEMPER Tent, Extendable, Modular, Personnel
TERC Tribal Emergency Response Commission
TIA Terrorist Incident Appendix
TIM toxic industrial material
TM Technical Manual
TNT Trinitrotoluene
TRIS Toxic Release Inventory System
TSC Triple-standard concertina
TSO Time Share Option
TTG Thermally tempered glass

U

UC Unified Command
UCS Unified Command System
UFAS Uniform Federal Accessibility Standards
UFC Unified Facilities Criteria
UL Underwriters Laboratories
ULPA Ultra low penetration air
UPS Uninterrupted power supply
URV UVGI Rating Values
U.S United States

USA United States Army
USAF United States Air Force
USC US Code
USDA US Department of Agriculture
USFA US Fire Administration
USGBC US Green Building Council
USGS US Geological Survey
US&R Urban Search and Rescue
UV Ultraviolet
UVGI Ultraviolet germicidal irradiation

V

VA Department of Veterans Affairs
VAP Vulnerability Assessment Plan
VAV Variable Air Volume
VDN Vector Directory Number
VHF Very high frequency
VRU Voice Response Unit
VSS Video Surveillance System

W

WAN Wide Area Network
wg Water gauge
WINGARD Window Glazing Analysis Response and Design
WINLAC Window Lite Analysis Code
WMD Weapons of Mass Destruction
WMD-CST WMD Civil Support Tea

GLOSSARY

General Glossary

A

Access group. A software configuration of an Access Control System that groups together access points or authorized users for easier arrangement and maintenance of the system.

Access road. Any roadway, such as a maintenance, delivery, service, emergency, or other special limited-use road, that is necessary for the operation of a building or structure.

Accountability. The explicit assignment of responsibilities for oversight of areas of control to executives, managers, staff, owners, providers, and users of minimum essential infrastructure resource elements.

Acoustic eavesdropping. The use of listening devices to monitor voice communications or other audibly transmitted information with the objective to compromise information.

Active vehicle barrier. An impediment placed at an access control point that may be manually or automatically deployed in response to detection of a threat.

Aerosol. Fine liquid or solid particles suspended in a gas (e.g., fog or smoke).

Aggressor. Any person seeking to compromise a function or structure.

Airborne contamination. Chemical or biological agents introduced into and fouling the source of supply of breathing or conditioning air.

Airlock. A building entry configuration with which airflow from the outside can be prevented from entering a toxic-free area. An airlock uses two doors, only one of which can be opened at a time, and a blower system to maintain positive air pressures and purge contaminated air from the airlock before the second door is opened.

Alarm assessment. Verification and evaluation of an alarm alert through the use of closed circuit television or human observation. Systems used for alarm assessment are designed to respond rapidly, automatically, and predictably to the receipt of alarms at the security center.

Alarm printers. Alarm printers provide a hard-copy of all alarm events and system activity, as well as limited backup in case the visual display fails.

Alarm priority. A hierarchy of alarms by order of importance. This is often used in larger systems to give priority to alarms with greater importance.

Annunciation. A visual, audible, or other indication by a security system of a condition.

Antiterrorism (AT). Defensive measures used to reduce the vulnerability of individuals, forces, and property to terrorist acts.

Area Commander. A military commander with authority in a specific geographical area or military installation.

Area lighting. Lighting that illuminates a large exterior area.

Areas of potential compromise. Categories where losses can occur that will impact either a department's or an agency's minimum essential infrastructure and its ability to conduct core functions and activities.

Assessment. The evaluation and interpretation of measurements and other information to provide a basis for decision-making.

Assessment System elements. Detection measures used to assist guards in visual verification of Intrusion Detection System Alarms and Access Control System functions and to assist in visual detection by guards. Assessment System elements include closed circuit television and protective lighting.

Asset. A resource of value requiring protection. An asset can be tangible (e.g., people, buildings, facilities, equipment, activities, operations, and information) or intangible (e.g., processes or a company's information and reputation).

Asset protection. Security program designed to protect personnel, facilities, and equipment, in all locations and situations, accomplished through planned and integrated application of combating terrorism, physical security, operations security, and personal protective services, and supported by intelligence, counterintelligence, and other security programs.

Asset value. The degree of debilitating impact that would be caused by the incapacity or destruction of an asset.

Attack. A hostile action resulting in the destruction, injury, or death to the civilian population, or damage or destruction to public and private property.

Audible alarm device. An alarm device that produces an audible announcement (e.g., bell, horn, and siren) of an alarm condition.

B

Balanced magnetic switch. A door position switch utilizing a reed switch held in a balanced or center position by interacting magnetic fields when not in alarm condition.

Ballistics attack. An attack in which small arms (e.g., pistols, submachine guns, shotguns, and rifles) are fired from a distance and rely on the flight of the projectile to damage the target.

Barbed tape or concertina. A coiled tape or coil of wires with wire barbs or blades deployed as an obstacle to human trespass or entry into an area.

Barbed wire. A double strand of wire with four-point barbs equally spaced along the wire deployed as an obstacle to human trespass or entry into an area.

Barcode. A black bar printed on white paper or tape that can be easily read with an optical scanner.

Biological agents. Living organisms or the materials derived from them that cause disease in or harm to humans, animals, or plants or cause deterioration of material. Biological agents may be used as liquid droplets, aerosols, or dry powders.

Biometric reader. A device that gathers and analyzes biometric features.

Biometrics. The use of physical characteristics of the human body as a unique identification method.

Blast curtains. Heavy curtains made of blast-resistant materials that could protect the occupants of a room from flying debris.

Blast-resistant glazing. Window opening glazing that is resistant to blast effects because of the interrelated function of the frame and glazing material properties frequently dependent upon tempered glass, polycarbonate, or laminated glazing.

Blast vulnerability envelope. The geographical area in which an explosive device will cause damage to assets.

Bollard. A vehicle barrier consisting of a cylinder, usually made of steel and sometimes filled with concrete, placed on end in the ground and spaced about 3 feet apart to prevent vehicles from passing, but allowing entrance of pedestrians and bicycles.

Boundary penetration sensor. An interior intrusion detection sensor that detects attempts by individuals to penetrate or enter a building.

Building hardening. Enhanced construction that reduces vulnerability to external blast and ballistic attacks.

Building separation. The distance between closest points on the exterior walls of adjacent buildings or structures.

Business Continuity Program. An ongoing process supported by senior management and funded to ensure that the necessary steps are taken to identify the impact of potential losses, maintain viable recovery strategies and recovery plans, and ensure continuity services through personnel training, plan testing, and maintenance.

C

Cable barrier. Cable or wire rope anchored to and suspended off the ground or attached to chain-link fence to act as a barrier to moving vehicles.

Capacitance sensor. A device that detects an intruder approaching or touching a metal object by sensing a change in capacitance between the object and the ground.

Card reader. A device that gathers or reads information when a card is presented as an identification method.

Chemical agent. A chemical substance that is intended to kill, seriously injure, or incapacitate people through physiological effects. Generally separated by severity of effect (e.g., lethal, blister, and incapacitating).

Chimney effect. Air movement in a building between floors caused by differential air temperature (differences in density), between the air inside and outside the building. It occurs in vertical shafts, such as elevators, stairwells, and conduit/wiring/piping chases. Hotter air inside the building will rise and be replaced by infiltration with colder outside air through the lower portions of the building. Conversely, reversing the temperature will reverse the flow (down the chimney). Also known as stack effect.

Clear zone. An area that is clear of visual obstructions and landscape materials that could conceal a threat or perpetrator.

Closed circuit television (CCTV). An electronic system of cameras, control equipment, recorders, and related apparatus used for surveillance or alarm assessment.

CCTV pan-tilt-zoom camera. A CCTV camera that can move side to side, up and down, and zoom in or out.

CCTV pan-tilt-zoom control. The method of controlling the PTZ functions of a camera.

CCTV pan-tilt-zoom controller. The operator interface for performing PTZ control.

CCTV switcher. A piece of equipment capable of presenting multiple video images to various monitors, recorders, and so on.

Collateral damage. Injury or damage to assets that are not the primary target of an attack.

Combating terrorism. The full range of federal programs and activities applied against terrorism, domestically and abroad, regardless of the source or motive.

Community. A political entity that has the authority to adopt and enforce laws and ordinances for the area under its jurisdiction. In most cases, the community is an incorporated town, city, township, village, or unincorporated area of a county; however, each state defines its own political subdivisions and forms of government.

Components and cladding. Elements of the building envelope that do not qualify as part of the main wind-force resisting system.

Confidentiality. The protection of sensitive information against unauthorized disclosure and sensitive facilities from physical, technical, or electronic penetration or exploitation.

Consequence management. Measures to protect public health and safety, restore essential government services, and provide emergency relief to governments, businesses, and individuals affected by the consequences of terrorism. State and local governments exercise the primary authority to respond to the consequences of terrorism.

Contamination. The undesirable deposition of a chemical, biological, or radiological material on the surface of structures, areas, objects, or people.

Continuity of services and operations. Controls to ensure that, when unexpected events occur, departmental/agency minimum essential infrastructure services and operations, including computer operations, continue without interruption or are promptly resumed, and that critical and sensitive data are protected through adequate contingency and business recovery plans and exercises.

Control center. A centrally located room or facility staffed by personnel charged with the oversight of specific situations and/or equipment.

Controlled area. An area into which access is controlled or limited. It is that portion of a restricted area usually near or surrounding a limited or exclusion area. Correlates with exclusion zone.

Controlled lighting. Illumination of specific areas or sections.

Controlled perimeter. A physical boundary at which vehicle and personnel access is controlled at the perimeter of a site. Access control at a controlled perimeter should demonstrate the capability to search individuals and vehicles.

Conventional construction. Building construction that is not specifically designed to resist weapons, explosives, or chemical, biological, and radiological effects. Conventional construction is designed only to resist common loadings and environmental effects, such as wind, seismic, and snow loads.

Coordinate. To advance systematically an exchange of information among principals who have or may have a need to know certain information in order to carry out their roles in a response.

Counterintelligence. Information gathered and activities conducted to protect against espionage, other intelligence activities, sabotage, or assassinations conducted for or on behalf of foreign powers, organizations, or persons; or international terrorist activities, excluding personnel, physical, document, and communications security programs.

Counterterrorism (CT). Offensive measures taken to prevent, deter, and respond to terrorism.

Covert entry. Attempts to enter a facility by using false credentials or stealth.

Crash bar. A mechanical egress device located on the interior side of a door that unlocks the door when pressure is applied in the direction of egress.

Crime Prevention through Environmental Design. A crime prevention strategy based on evidence that the design and form of the built environment can influence human behavior. CPTED usually involves the use of three principles: namely, natural surveillance (by placing physical features, activities, and people to maximize visibility); natural access control (through the judicial placement of entrances, exits, fencing, landscaping, and lighting); and territorial reinforcement (using buildings, fences, pavement, signs, and landscaping to express ownership).

Crisis Management. The measures taken to identify, acquire, and plan the use of resources needed to anticipate, prevent, and/or resolve a threat or act of terrorism.

Critical assets. Those assets essential to the minimum operations of the organization, and to ensure the health and safety of the general public.

Critical infrastructure. Primary infrastructure systems (e.g., utilities, telecommunications, and transportation) whose incapacity would have a debilitating impact on the organization's ability to function.

D

Damage assessment. The process used to appraise or determine the number of injuries and deaths, damage to public and private property, and the status of key facilities and services (e.g., hospitals and other health care facilities, fire and police stations, communications networks, water and sanitation systems, utilities, and transportation networks) resulting from a manmade or natural disaster.

Data-gathering panel. A local processing unit that retrieves, processes, stores, and/or acts on information in the field.

Data transmission equipment. A path for transmitting data between two or more components (e.g., a sensor and alarm reporting system, a card reader and controller, a CCTV camera and monitor, or a transmitter and receiver).

Decontamination. The reduction or removal of a chemical, biological, or radiological material from the surface of a structure, area, object, or person.

Defense layer. Building design or exterior perimeter barriers intended to delay attempted forced entry.

Defensive measures. Protective measures that delay or prevent attack on an asset or that shield the asset from weapons, explosives, and CBR effects. Defensive measures include site work and building design.

Delay rating. A measure of the effectiveness of penetration protection of a defense layer.

Design Basis Threat. The threat (e.g., tactics and associated weapons, tools, or explosives) against which assets within a building must be protected and upon which the security engineering design of the building is based.

Design constraint. Anything that restricts the design options for a protective system or that creates additional problems for which the design must compensate.

Design opportunity. Anything that enhances protection, reduces requirements for protective measures, or solves a design problem.

Design team. A group of individuals from various engineering and architectural disciplines responsible for the protective system design.

Detection layer. A ring of intrusion detection sensors located on or adjacent to a defensive layer or between two defensive layers.

Detection measures. Protective measures that detect intruders, weapons, or explosives; assist in assessing the validity of detection; control access to protected areas;

and communicate the appropriate information to the response force. Detection measures include Detection Systems, Assessment Systems, and Access Control System elements.

Detection system elements. Detection measures that detect the presence of intruders, weapons, or explosives. Detection system elements include Intrusion Detection Systems, weapons and explosives detectors, and guards.

Disaster. An occurrence of a natural catastrophe, technological accident, or human-caused event that has resulted in severe property damage, deaths, and/or multiple injuries.

Disaster Field Office. The office established in or near the designated area of a presidentially declared major disaster to support federal and state response and recovery operations.

Disaster Recovery Center. Places established in the area of a presidentially declared major disaster, as soon as practicable, to provide victims the opportunity to apply in person for assistance and/or obtain information relating to that assistance.

Domestic terrorism. The unlawful use, or threatened use, of force or violence by a group or individual based and operating entirely within the United States or Puerto Rico without foreign direction committed against persons or property to intimidate or coerce a government, the civilian population, or any segment thereof in furtherance of political or social objectives.

Door position switch. A switch that changes state based on whether or not a door is closed. Typically, a switch mounted in a frame that is actuated by a magnet in a door.

Door strike, electronic. An electromechanical lock that releases a door plunger to unlock the door. Typically, an electronic door strike is mounted in place of or near a normal door strike plate.

Dose rate (radiation). A general term indicating the quantity (total or accumulated) of ionizing radiation or energy absorbed by a person or animal, per unit of time.

Dosimeter. An instrument for measuring and registering total accumulated exposure to ionizing radiation.

Dual technology sensor. A sensor that combines two different technologies in one unit.

Duress alarm devices. Also known as panic buttons, these devices are designated specifically to initiate a panic alarm.

E

Effective stand-off distance. A stand-off distance at which the required LOP can be shown to be achieved through analysis or can be achieved through building hardening or other mitigating construction or retrofit.

Electromagnetic pulse (EMP). A sharp pulse of energy radiated instantaneously by a nuclear detonation that may affect or damage electronic components and equipment. EMP can also be generated in lesser intensity by nonnuclear means in specific frequency ranges to perform the same disruptive function.

Electronic emanations. Electromagnetic emissions from computers, communications, electronics, wiring, and related equipment.

Electronic-emanations eavesdropping. Use of electronic-emanation surveillance equipment from outside a facility or its restricted area to monitor electronic emanations from computers, communications, and related equipment.

Electronic Entry Control Systems. Electronic devices that automatically verify authorization for a person to enter or exit a controlled area.

Electronic Security System. An integrated system that encompasses interior and exterior sensors, closed circuit television systems for assessment of alarm conditions, EECSs, data-transmission media, and alarm reporting systems for monitoring, control, and display of various alarm and system information.

Emergency. Any natural or human-caused situation that results in or may result in substantial injury or harm to the population or substantial damage to or loss of property.

Emergency Alert System. A communications system of broadcast stations and interconnecting facilities authorized by the FCC. The system provides the president and other national, state, and local officials the means to broadcast emergency information to the public before, during, and after disasters.

Emergency Environmental Health Services. Services required to correct or improve damaging environmental health effects on humans, including inspection for food contamination, inspection for water contamination, and vector control; providing for sewage and solid waste inspection and disposal; cleanup and disposal of hazardous materials; and sanitation inspection for emergency shelter facilities.

Emergency Medical Services. Services including personnel, facilities, and equipment required to ensure proper medical care for the sick and injured from the time of injury to the time of final disposition, including medical disposition within a hospital, temporary medical facility, or special care facility; release from the site; or declared dead. Further, EMS specifically include those services immediately required to ensure proper medical care and specialized treatment for patients in a hospital and coordination of related hospital services.

Emergency Mortuary Services. Services required to ensure adequate death investigation, identification, and disposition of bodies; removal, temporary storage, and transportation of bodies to temporary morgue facilities; notification of next of kin; and coordination of mortuary services and burial of unclaimed bodies.

Emergency Operations Center. The protected site from which state and local civil government officials coordinate, monitor, and direct emergency response activities during an emergency.

Emergency Operations Plan. A document that describes how people and property will be protected in disasters and disaster threat situations; details who is responsible for carrying out specific actions; identifies the personnel, equipment, facilities, supplies, and other resources available for use in the disaster; and outlines how all actions will be coordinated.

Emergency Planning Zones (EPZ). Areas around a facility for which planning is needed to ensure prompt and effective actions are taken to protect the health and safety of the public if an accident or disaster occurs. In the Radiological Emergency Preparedness Program, the two EPZs are **Plume Exposure Pathway** (10-mile EPZ)—a circular geographic zone (with a 10-mile radius centered at the nuclear power plant) for which plans are developed to protect the public against exposure to radiation emanating from a radioactive plume caused as a result of an accident at the nuclear power plant. **Ingestion Pathway** (50-mile EPZ)—a circular geographic zone (with a 50-mile radius centered at the nuclear power plant) for which plans are developed to protect the public from the ingestion of water or food contaminated as a result of a nuclear power plant accident. In the CSEPP, the EPZ is divided into three concentric circular zones: **Immediate Response Zone** (IRZ)—a circular zone ranging from 10 to 15 kilometers (6–9 miles) from the potential chemical event source, depending on the stockpile location on-post. Emergency response plans developed for the IRZ must provide for the most rapid and effective protective actions possible, because the IRZ will have the highest concentration of agent and the least amount of warning time. **Protective Action Zone** (PAZ)—an area that extends beyond the IRZ to approximately 16–50 kilometers (10–30 miles) from the stockpile location. The PAZ is that area where public protective actions may still be necessary in case of an accidental release of chemical agent, but where the available warning and response time is such that most people could evacuate. However, other responses (e.g., sheltering) may be appropriate for institutions and special populations that could not evacuate within the available time. **Precautionary Zone** (PZ)—the outermost portion of the EPZ for CSEPP, extending from the PAZ outer boundary to a distance where the risk of adverse impacts to humans is negligible. Because of the increased warning and response time available for implementation of response actions in the PZ, detailed local emergency planning is not required, although Consequence Management planning may be appropriate.

Emergency Public Information. Information that is disseminated primarily in anticipation of an emergency or at the actual time of an emergency and, in addition to providing information frequently directs actions, instructs, and transmits direct orders.

Emergency Response Team. An interagency team, consisting of the lead representative from each federal department or agency assigned primary responsibility for an ESF and key members of the FCO's staff, formed to assist the FCO in carrying out his/her coordination responsibilities.

Emergency Response Team Advance Element (ERT-A). For federal disaster response and recovery activities under the Stafford Act, the portion of the ERT that is first deployed to the field to respond to a disaster incident. The ERT-A is the nucleus of the full ERT.

Emergency Response Team National (ERT-N). An ERT that has been established and rostered for deployment to catastrophic disasters where the resources of the FEMA Region have been, or are expected to be, overwhelmed. Three ERT-Ns have been established.

Emergency Support Function. In the FRP, a functional area of response activity established to facilitate the delivery of federal assistance required during the immediate response phase of a disaster to save lives, protect property and public health, and maintain public safety. ESFs represent those types of federal assistance that the state will most likely need because of the impact of a catastrophic or significant disaster on its own resources and response capabilities, or because of the specialized or unique nature of the assistance required. ESF missions are designed to supplement state and local response efforts.

Emergency Support Team. An interagency group operating from FEMA Headquarters. The EST oversees the national-level response support effort under the FRP and coordinates activities with the ESF primary and support agencies in supporting federal requirements in the field.

Entity-wide security. Planning and management that provides a framework and continuing cycle of activity for managing risk, developing security policies, assigning responsibilities, and monitoring the adequacy of the entity's physical and cyber security controls.

Entry control point. A continuously or intermittently manned station at which entry to sensitive or restricted areas is controlled.

Entry control stations. Entry control stations should be provided at main perimeter entrances where security personnel are present. Entry control stations should be located as close as practical to the perimeter entrance to permit personnel inside the station to maintain constant surveillance over the entrance and its approaches.

Equipment closet. A room where field control equipment, such as data gathering panels and power supplies, are typically located.

Evacuation. Organized, phased, and supervised dispersal of people from dangerous or potentially dangerous areas.

Evacuation, mandatory or directed. This is a warning to persons within the designated area that an imminent threat to life and property exists and individuals MUST evacuate in accordance with the instructions of local officials.

Evacuation, spontaneous. Residents or citizens in the threatened areas observe an emergency event or receive unofficial word of an actual or perceived threat and, without receiving instructions to do so, elect to evacuate the area. Their movement, means, and direction of travel are unorganized and unsupervised.

Evacuation, voluntary. This is a warning to persons within a designated area that a threat to life and property exists or is likely to exist in the immediate future. Individuals issued this type of warning or order are NOT required to evacuate; however, it would be to their advantage to do so.

Evacuees. All persons removed or moving from areas threatened or struck by a disaster.

Exclusion area. A restricted area containing a security interest. Uncontrolled movement permits direct access to the item. See controlled area and limited area.

Exclusion zone. An area around an asset that has controlled entry with highly restrictive access. See controlled area.

Explosives disposal container. A small container into which small quantities of explosives may be placed to contain their blast pressures and fragments if the explosive detonates.

F

Facial recognition. A biometric technology that is based on features of the human face.

Federal Coordinating Officer. The person appointed by the FEMA Director to coordinate federal assistance in a presidentially declared emergency or major disaster.

Federal On-scene Commander. The FBI official designated upon JOC activation to ensure appropriate coordination of the overall US government response with federal, state, and local authorities, until such time as the attorney general transfers the LFA role to FEMA.

Federal Response Plan. The FRP establishes a process and structure for the systematic, coordinated, and effective delivery of federal assistance to address the consequences of any major disaster or emergency.

Fence protection. An intrusion detection technology that detects a person crossing a fence by various methods, such as climbing, crawling, cutting, and so on.

Fence sensor. An exterior intrusion detection sensor that detects aggressors as they attempt to climb over, cut through, or otherwise disturb a fence.

Fiber optics. A method of data transfer by passing bursts of light through a strand of glass or clear plastic.

Field Assessment Team (FAST). A small team of pre-identified technical experts who conduct an assessment of response needs (not a PDA) immediately following a disaster.

Field of view. The visible area in a video picture.

First responder. Local police, fire, and emergency medical personnel who first arrive on the scene of an incident and take action to save lives, protect property, and meet basic human needs.

Flash flood. Follows a situation in which rainfall is so intense and severe and runoff so rapid that it precludes recording and relating it to stream stages and other information in time to forecast a flood condition.

Flood. A general and temporary condition of partial or complete inundation of normally dry land areas from overflow of inland or tidal waters, unusual or rapid accumulation or runoff of surface waters, or mudslides/mudflows caused by accumulation of water.

Forced entry. Entry to a denied area achieved through force to create an opening in fence, walls, doors, and so on, or to overpower guards.

Fragment retention film. A thin, optically clear film applied to glass to minimize the spread of glass fragments when the glass is shattered.

Frame rate. In digital video, a measurement of the rate of change in a series of pictures, often measured in frames per second (fps).

Frangible construction. Building components that are designed to fail to vent blast pressures from an enclosure in a controlled manner and direction.

G

Glare security lighting. Illumination projected from a secure perimeter into the surrounding area, making it possible to see potential intruders at a considerable distance while making it difficult to observe activities within the secure perimeter.

Glass-break detector. An intrusion detection sensor that is designed to detect breaking glass either through vibration or acoustics.

Glazing. A material installed in a sash, ventilator, or panes (e.g., glass and plastic, including material, such as thin granite installed in a curtain wall).

Governor's Authorized Representative. The person empowered by the governor to execute, on behalf of the state, all necessary documents for disaster assistance.

Grid wire sensor An intrusion detection sensor that uses a grid of wires to cover a wall or fence. An alarm is sounded if the wires are cut.

H

Hand geometry. A biometric technology that is based on characteristics of the human hand.

Hazard. A source of potential danger or adverse condition.

Hazard mitigation. Any action taken to reduce or eliminate the long-term risk to human life and property from hazards. The term is sometimes used in a stricter sense to mean cost-effective measures to reduce the potential for damage to a facility or facilities from a disaster event.

Hazardous material (HazMat). Any substance or material that, when involved in an accident and released in sufficient quantities, poses a risk to people's health, safety, and/or property. These substances and materials include explosives, radioactive materials, flammable liquids or solids, combustible liquids or solids, poisons, oxidizers, toxins, and corrosive materials.

High-hazard areas. Geographic locations that, for planning purposes, have been determined through historical experience and vulnerability analysis to be likely to experience the effects of a specific hazard (e.g., hurricane, earthquake, and hazardous materials accident), resulting in vast property damage and loss of life.

High-risk target. Any material resource or facility that, because of mission sensitivity, ease of access, isolation, and symbolic value, may be an especially attractive or accessible terrorist target.

Human-caused hazard. Human-caused hazards are technological hazards and terrorism. They are distinct from natural hazards primarily in that they originate from human activity. Within the military services, the term "threat" is typically used for human-caused hazard. See definitions of technological hazards and terrorism for further information.

Hurricane. A tropical cyclone, formed in the atmosphere over warm ocean areas, in which wind speeds reach 74 miles per hour or more and blow in a large spiral around a relatively calm center or "eye." Circulation is counter-clockwise in the Northern Hemisphere and clockwise in the Southern Hemisphere.

I

Impact analysis. A management level analysis that identifies the impacts of losing the entity's resources. The analysis measures the effect of resource loss and escalating losses over time in order to provide the entity with reliable data upon which to base decisions on hazard mitigation and continuity planning.

Incident Command System. A standardized organizational structure used to command, control, and coordinate the use of resources and personnel who have responded to the scene of an emergency. The concepts and principles for ICS include common terminology, modular organization, integrated communication, unified command structure, consolidated action plan, manageable span of control, designated incident facilities, and comprehensive resource management.

Insider compromise. A person authorized access to a facility (an insider) compromises assets by taking advantage of that accessibility.

Intercom door/gate station. Part of an intercom system where communication is typically initiated, usually located at a door or gate.

Intercom master station. Part of an intercom system that monitors one or more intercom door/gate stations; typically, where initial communication is received.

Intercom switcher. Part of an intercom system that controls the flow of communications between various stations.

Intercom system. An electronic system that allows simplex, half-duplex, or full-duplex audio communications.

International terrorism. Violent acts or acts dangerous to human life that are a violation of the criminal laws of the United States or any state, or that would be a criminal violation if committed within the jurisdiction of the United States or any state. These acts appear to be intended to intimidate or coerce a civilian population, influence the policy of a government by intimidation or coercion, or affect the conduct of a government by assassination or kidnapping. International terrorist acts occur outside the United States, or transcend national boundaries in terms of the means by which they are accomplished, the persons they appear intended to coerce or intimidate, or the locale in which their perpetrators operate or seek asylum.

Intrusion detection sensor. A device that initiates alarm signals by sensing the stimulus, change, or condition for which it was designed.

Intrusion Detection System. The combination of components, including sensors, control units, transmission lines, and monitor units, integrated to operate in a specified manner.

Isolated fenced perimeters. Fenced perimeters with 100 feet or more of space outside the fence that is clear of obstruction, making approach obvious.

J

Jersey barrier. A protective concrete barrier initially and still used as a highway divider that now also functions as an expedient method for traffic speed control at entrance gates and to keep vehicles away from buildings.

Joint Information Center. A central point of contact for all news media near the scene of a large-scale disaster. News media representatives are kept informed of activities and events by Public Information Officers who represent all participating federal, state, and local agencies that are collocated at the JIC.

Joint Information System. Under the FRP, connection of public affairs personnel, decision-makers, and news centers by electronic mail, fax, and telephone when a single federal-state-local JIC is not a viable option.

Joint Interagency Intelligence Support Element. An interagency intelligence component designed to fuse intelligence information from the various agencies participating in a response to a WMD threat or incident within an FBI JOC. The JIISE is an expanded version of the investigative/intelligence component that is part of the standardized FBI Command Post structure. The JIISE manages five functions, including security, collections management, current intelligence, exploitation, and dissemination.

Joint Operations Center. Established by the LFA under the operational control of the federal OSC, as the focal point for management and direction of onsite activities, coordination/establishment of state requirements/priorities, and coordination of the overall federal response.

Jurisdiction. Typically counties and cities within a state, but states may elect to define differently in order to facilitate their assessment process.

L

Laminated glass. A flat lite of uniform thickness consisting of two monolithic glass plies bonded together with an interlayer material as defined in Specification C1172. Many different interlayer materials are used in laminated glass.

Landscaping. The use of plantings (shrubs and trees), with or without landforms and/or large boulders, to act as a perimeter barrier against defined threats.

Laser card. A card technology that uses a laser reflected off a card for uniquely identifying the card.

Layers of protection. A traditional approach in security engineering using concentric circles extending out from an area to be protected as demarcation points for different security strategies.

Lead agency. The federal department or agency assigned lead responsibility under US law to manage and coordinate the federal response in a specific functional area.

Lead Federal Agency. The agency designated by the president to lead and coordinate the overall federal response is referred to as the LFA and is determined by the type of emergency. In general, an LFA establishes operational structures and procedures to assemble and work with agencies providing direct support to the LFA in order to provide an initial assessment of the situation, develop an action plan, monitor and update operational priorities, and ensure each agency exercises its concurrent and distinct authorities under US law and supports the LFA in carrying out the president's relevant policy. Specific responsibilities of an LFA vary, according to the agency's unique statutory authorities.

Level of protection. The degree to which an asset is protected against injury or damage from an attack.

Liaison. An agency official sent to another agency to facilitate interagency communications and coordination.

Limited area. A restricted area within close proximity of a security interest. Uncontrolled movement may permit access to the item. Escorts and other internal restrictions may prevent access to the item. See controlled area and exclusion area.

Line of sight. Direct observation between two points with the naked eye or hand-held optics.

Line-of-sight sensor. A pair of devices used as an intrusion detection sensor that monitor any movement through the field between the sensors.

Line supervision. A data integrity strategy that monitors the communications link for connectivity and tampering. In Intrusion Detection System sensors, line supervision is often referred to as two-state, three-state, or four-state in respect to the number of conditions monitored. The frequency of sampling the link also plays a big part in the supervision of the line.

Local government. Any county, city, village, town, district, or political subdivision of any state, and Indian tribe or authorized tribal organization, or Alaska Native village or organization, including any rural community or unincorporated town or village or any other public entity.

M

Magnetic lock. An electromagnetic lock that unlocks a door when power is removed.

Magnetic stripe. A card technology that uses a magnetic stripe on the card to encode data used for unique identification of the card.

Mail-bomb delivery. Bombs or incendiary devices delivered to the target in letters or packages.

Man-trap. An access control strategy that uses a pair of interlocking doors to prevent tailgating. Only one door can be unlocked at a time.

Mass care. The actions that are taken to protect evacuees and other disaster victims from the effects of the disaster. Activities include providing temporary shelter, food, medical care, clothing, and other essential life support needs to those people who have been displaced from their homes because of a disaster or threatened disaster.

Mass notification. Capability to provide real-time information to all building occupants or personnel in the immediate vicinity of a building during emergency situations.

Microwave motion sensor. An intrusion detection sensor that uses microwave energy to sense movement within the sensor's field of view. These sensors work similar to radar by using the Doppler effect to measure a shift in frequency.

Military installations. Army, Navy, Air Force, and Marine Corps bases, posts, stations, and annexes (both contractor and government operated), hospitals, terminals, and other special mission facilities, as well as those used primarily for military purposes.

Minimum essential infrastructure resource elements. The broad categories of resources, all or portions of which constitute the minimal essential infrastructure necessary for a department, agency, or organization to conduct its core mission(s).

Minimum measures. Protective measures that can be applied to all buildings regardless of the identified threat. These measures offer defense or detection opportunities for minimal cost, facilitate future upgrades, and may deter acts of aggression.

Mitigation. Those actions taken to reduce the exposure to and impact of an attack or disaster.

Motion detector. An intrusion detection sensor that changes state based on movement in the sensor's field of view.

Moving vehicle bomb. An explosive-laden car or truck driven into or near a building and detonated.

Mutual aid agreement. A prearranged agreement developed between two or more entities to render assistance to the parties of the agreement.

N

Natural hazard. Naturally occurring events, such as floods, earthquakes, tornadoes, tsunami, coastal storms, landslides, and wildfires, that strike populated areas. A natural event is a hazard when it has the potential to harm people or property (FEMA 386-2, Understanding Your Risks). The risks of natural hazards may be increased or decreased as a result of human activity; however, they are not inherently human-induced.

Natural protective barriers. Natural protective barriers are mountains and deserts, cliffs and ditches, water obstacles, or other terrain features that are difficult to traverse.

Non-exclusive zone. An area around an asset that has controlled entry, but shared or less restrictive access than an exclusive zone.

Non-persistent agent. An agent that, upon release, loses its ability to cause casualties after 10 to 15 minutes. It has a high evaporation rate, is lighter than air, and will disperse rapidly. It is considered to be a short-term hazard; however, in small, unventilated areas, the agent will be more persistent.

Nuclear, biological, or chemical weapons. Also called Weapons of Mass Destruction (WMD). Weapons that are characterized by their capability to produce mass casualties.

Nuclear detonation. An explosion resulting from fission and/or fusion reactions in nuclear material, such as that from a nuclear weapon.

O

On-Scene Coordinator. The federal official predesignated by the EPA and US Coast Guard to coordinate and direct response and removals under the National Oil and Hazardous Substances Pollution Contingency Plan.

Open systems architecture. A term borrowed from the IT industry to claim that systems are capable of interfacing with other systems from any vendor, which also uses open system architecture. The opposite would be a proprietary system.

Operator interface. The part of a security management system that provides the user interface to humans.

Organizational areas of control. Controls consist of the policies, procedures, practices, and organization structures designed to provide reasonable assurance that business objectives will be achieved and that undesired events will be prevented or detected and corrected.

P

Passive infrared motion sensor. A device that detects a change in the thermal energy pattern caused by a moving intruder and initiates an alarm when the change in energy satisfies the detector's alarm-criteria.

Passive vehicle barrier. A vehicle barrier that is permanently deployed and does not require response to be effective.

Patch panel. A concentrated termination point that separates backbone cabling from devices cabling for easy maintenance and troubleshooting.

Perimeter barrier. A fence, wall, vehicle barrier, landform, or line of vegetation applied along an exterior perimeter used to obscure vision, hinder personnel access, or hinder or prevent vehicle access.

Persistent agent. An agent that, upon release, retains its casualty-producing effects for an extended period of time, usually anywhere from 30 minutes to several days. A persistent agent usually has a low evaporation rate and its vapor is heavier than air; therefore, its vapor cloud tends to hug the ground. It is considered to be a long-term hazard. Although inhalation hazards are still a concern, extreme caution should be taken to avoid skin contact, as well.

Physical security. The part of security concerned with measures/concepts designed to safeguard personnel; to prevent unauthorized access to equipment, installations, materiel, and documents; and to safeguard them against espionage, sabotage, damage, and theft.

Planter barrier. A passive vehicle barrier, usually constructed of concrete and filled with dirt (and flowers for aesthetics). Planters, along with bollards, are the usual street furniture used to keep vehicles away from existing buildings. Overall size and the depth of installation below grade determine the vehicle stopping capability of the individual planter.

Plume. Airborne material spreading from a particular source; the dispersal of particles, gases, vapors, and aerosols into the atmosphere.

Polycarbonate glazing. A plastic glazing material with enhanced resistance to ballistics or blast effects.

Predetonation screen. A fence that causes an anti-tank round to detonate or prevents it from arming before it reaches its target.

Preliminary Damage Assessment. A mechanism used to determine the impact and magnitude of damage and the resulting unmet needs of individuals, businesses, the public sector, and the community as a whole. Information collected is used by the state as a basis for the governor's request for a presidential declaration, and by FEMA to document the recommendation made to the president in response to the governor's request. PDAs are made by at least one state and one federal representative. A local government representative familiar with the extent and location of damage in the community often participates; other state and federal agencies and voluntary relief organizations also may be asked to participate, as needed.

Preparedness. Establishing the plans, training, exercises, and resources necessary to enhance mitigation of and achieve readiness for response to, and recovery from all hazards, disasters, and emergencies, including WMD incidents.

Pressure mat. A mat that generates an alarm when pressure is applied to any part of the mat's surface, such as when someone steps on the mat. Pressure mats can be used to detect an intruder approaching a protected object, or they can be placed by doors and windows to detect entry.

Primary asset. An asset that is the ultimate target for compromise by an aggressor. Primary gathering building. Inhabited buildings routinely occupied by fifty or more personnel. This designation applies to the entire portion of a building that meets the population density requirements for an inhabited building.

Probability of detection. A measure of an intrusion detection sensor's performance in detecting an intruder within its detection zone.

Probability of intercept. The probability that an act of aggression will be detected and that a response force will intercept the aggressor before the asset can be compromised.

Progressive collapse. A chain reaction failure of building members to an extent disproportionate to the original localized damage. Such damage may result in upper floors of a building collapsing onto lower floors.

Protective barriers. Define the physical limits of a site, activity, or area by restricting, channeling, or impeding access and forming a continuous obstacle around the object.

Protective measures. Elements of a protective system that protect an asset against a threat. Protective measures are divided into defensive and detection measures.

Protective system. An integration of all of the protective measures required to protect an asset against the range of threats applicable to the asset.

Proximity sensor. An intrusion detection sensor that changes state based on the close distance or contact of a human to the sensor. These sensors often measure the change in capacitance as a human body enters the measured field.

Public Information Officer. A federal, state, or local government official responsible for preparing and coordinating the dissemination of EPI.

R

Radiation. High-energy particles or gamma rays that are emitted by an atom as the substance undergoes radioactive decay. Particles can be either charged alpha or beta particles or neutral neutron or gamma rays.

Radiation sickness. The symptoms characterizing the sickness known as radiation injury, resulting from excessive exposure of the whole body to ionizing radiation.

Radiological monitoring. The process of locating and measuring radiation by means of survey instruments that can detect and measure (as exposure rates) ionizing radiation.

Recovery. The long-term activities beyond the initial crisis period and emergency response phase of disaster operations that focus on returning all systems in the community to a normal status or to reconstitute these systems to a new condition that is less vulnerable.

Regional Operations Center. The temporary operations facility for the coordination of federal response and recovery activities located at the FEMA Regional Office (or Federal Regional Center) and led by the FEMA Regional Director or Deputy Director until the DFO becomes operational. After the ERT-A is deployed, the ROC performs a support role for federal staff at the disaster scene.

Report printers. A separate, dedicated printer attached to the ESSs used for generating reports utilizing information stored by the central computer.

Request-to-exit device. Passive infrared motion sensors or push buttons that are used to signal an EECS that egress is imminent or to unlock a door.

Resolution. The level to which video details can be determined in a CCTV scene is referred to as resolving ability or resolution.

Resource management. Those actions taken by a government to identify sources and obtain resources needed to support disaster response activities; coordinate the supply, allocation, distribution, and delivery of resources so that they arrive where and when most needed; and maintain accountability for the resources used.

Response. Executing the plan and resources identified to perform those duties and services to preserve and protect life and property, as well as provide services to the surviving population.

Response force. The people who respond to an act of aggression. Depending on the nature of the threat, the response force could consist of guards, special reaction teams, military or civilian police, an explosives ordnance disposal team, or a fire department.

Response time. The length of time from the instant an attack is detected to the instant a security force arrives onsite.

Restricted area. Any area with access controls that is subject to these special restrictions or controls for security reasons. See controlled area, limited area, exclusion area, and exclusion zone.

Retinal pattern. A biometric technology that is based on features of the human eye.

RF data transmission. A communications link using radio frequency to send or receive data.

Risk. The potential for loss of, or damage to, an asset. It is measured based upon the value of the asset in relation to the threats and vulnerabilities associated with it.

Rotating drum or rotating plate vehicle barrier. An active vehicle barrier used at vehicle entrances to controlled areas based on a drum or plate rotating into the path of the vehicle when signaled.

Routinely occupied. For the purposes of these standards, an established or predictable pattern of activity within a building that terrorists could recognize and exploit.

RS-232 data. IEEE Recommended Standard 232; a point-to-point serial data protocol with a maximum effective distance of 50 feet.

RS-422 data. IEEE Recommended Standard 422; a point-to-point serial data protocol with a maximum effective distance of 4,000 feet.

RS-485 data. IEEE Recommended Standard 485; a multi-drop serial data protocol with a maximum effective distance of 4,000 feet.

S

Sacrificial roof or wall. Roofs or walls that can be lost in a blast without damage to the primary asset.

Safe haven. Secure areas within the interior of the facility. A safe haven should be designed such that it requires more time to penetrate by aggressors than it takes for the response force to reach the protected area to rescue the occupants. It may be a haven from a physical attack or air-isolated haven from CBR contamination.

Scramble keypad. A keypad that uses keys on which the numbers change pattern with each use to enhance security by preventing eavesdropping observation of the entered numbers.

Secondary asset. An asset that supports a primary asset and whose compromise would indirectly affect the operation of the primary asset.

Secondary hazard. A threat whose potential would be realized as the result of a triggering event that of itself would constitute an emergency (e.g., dam failure might be a secondary hazard associated with earthquakes).

Secure/access mode. The state of an area monitored by an intrusion detection system in regards to how alarm conditions are reported.

Security analysis. The method of studying the nature of and the relationship between assets, threats, and vulnerabilities.

Security console. Specialized furniture, racking, and related apparatus used to house the security equipment required in a control center.

Security engineering. The process of identifying practical, risk managed short- and long-term solutions to reduce and/or mitigate dynamic manmade hazards by integrating multiple factors, including construction, equipment, manpower, and procedures.

Security engineering design process. The process through which assets requiring protection are identified, the threat to and vulnerability of those assets is determined, and a protective system is designed to protect the assets.

Security Management System database. In a Security Management System, a database that is transferred to various nodes or panels throughout the system for faster data processing and protection against communications link downtime.

Security Management System distributed processing. In a Security Management System, a method of data processing at various nodes or panels throughout the system for faster data processing and protection against communications links downtime.

Segregation of duties. Policies, procedures, and an organizational structure established so that one individual cannot control key aspects of physical and/or computer-related operations and thereby conduct unauthorized actions or gain unauthorized access to minimum essential infrastructure resource elements.

Semi-isolated fenced perimeters. Fence lines where approach areas are clear of obstruction for 60 to 100 feet outside of the fence and where the general public or other personnel seldom have reason to be in the area.

Senior FEMA Official. The official appointed by the Director of FEMA, or his representative, that is responsible for deploying to the JOC to serve as the senior interagency consequence management representative on the Command Group, and to manage and coordinate activities taken by the Consequence Management Group.

Serial interface. An integration strategy for data transfer where components are connected in series.

Shielded wire. Wire with a conductive wrap used to mitigate electromagnetic emanations.

Situational crime prevention. A crime prevention strategy based on reducing the opportunities for crime by increasing the effort required to commit a crime, increasing the risks associated with committing the crime, and reducing the target appeal or vulnerability (whether property or person). This opportunity reduction is achieved by management and use policies, such as procedures and training, as well as physical approaches, such as alteration of the built environment.

Smart card. A newer card technology that allows data to be written, stored, and read on a card typically used for identification and/or access.

Software level integration. An integration strategy that uses software to interface systems. An example of this would be digital video displayed in the same computer application window and linked to events of a security management system.

Specific threat. Known or postulated aggressor activity focused on targeting a particular asset.

Stand-off distance A distance maintained between a building or portion thereof and the potential location for an explosive detonation or other threat.

Stand-off weapons. Weapons, such as anti-tank weapons and mortars, that are launched from a distance at a target.

State Coordinating Officer. The person appointed by the governor to coordinate state, commonwealth, or territorial response and recovery activities with FRP-related activities of the federal government, in cooperation with the FCO.

State Liaison. A FEMA official assigned to a particular state, who handles initial coordination with the state in the early stages of an emergency.

Stationary vehicle bomb. An explosive-laden car or truck stopped or parked near a building.

Storm surge. A dome of sea water created by the strong winds and low barometric pressure in a hurricane that causes severe coastal flooding as the hurricane strikes land.

Strain sensitive cable. Strain sensitive cables are transducers that are uniformly sensitive along their entire length and generate an analog voltage when subjected to mechanical distortions or stress resulting from fence motion. They are typically attached to a chain-link fence about halfway between the bottom and top of the fence fabric with plastic ties.

Structural protective barriers. Manmade devices (e.g., fences, walls, floors, roofs, grills, bars, roadblocks, signs, or other construction) used to restrict, channel, or impede access.

Superstructure. The supporting elements of a building above the foundation.

Supplies-bomb delivery. Bombs or incendiary devices concealed and delivered to supply or material handling points, such as loading docks.

System events. Events that occur normally in the operation of a security management system. Examples include access control operations and changes of state in intrusion detection sensors.

System software. Controls that limit and monitor access to the powerful programs and sensitive files that control the computer hardware and secure applications supported by the system.

T

Tactics. The specific methods of achieving the aggressor's goals to injure personnel, destroy assets, or steal materiel or information.

Tamper switch. Intrusion detection sensor that monitors an equipment enclosure for breach.

Tangle-foot wire Barbed wire or tape suspended on short metal or wooden pickets outside a perimeter fence to create an obstacle to approach.

Taut wire sensor. An intrusion detection sensor utilizing a column of uniformly spaced horizontal wires, securely anchored at each end and stretched taut. Each wire is attached to a sensor to indicate movement of the wire.

Technical assistance. The provisioning of direct assistance to states and local jurisdictions to improve capabilities for program development, planning, and operational performances related to responses to WMD terrorist incidents.

Technological hazards. Incidents that can arise from human activities, such as manufacture, transportation, storage, and use of hazardous materials. For the sake of simplicity, it is assumed that technological emergencies are accidental and that their consequences are unintended.

TEMPEST. An unclassified short name referring to investigations and studies of compromising emanations. It is sometimes used synonymously for the term "compromising emanations" (e.g., TEMPEST tests, TEMPEST inspections).

Terrorism. The unlawful use of force and violence against persons or property to intimidate or coerce a government, the civilian population, or any segment thereof, in furtherance of political or social objectives.

Thermally tempered glass. Glass that is heat-treated to have a higher tensile strength and resistance to blast pressures, although with a greater susceptibility to airborne debris.

Threat. Any indication, circumstance, or event with the potential to cause loss of, or damage to an asset.

Threat analysis. A continual process of compiling and examining all available information concerning potential threats and human-caused hazards. A common method to evaluate terrorist groups is to review the factors of existence, capability, intentions, history, and targeting.

Time/date stamp. Data inserted into a CCTV video signal with the time and date of the video as it was created.

TNT equivalent weight. The weight of TNT (trinitrotoluene) that has an equivalent energetic output to that of a different weight of another explosive compound.

Tornado. A local atmospheric storm, generally of short duration, formed by winds rotating at very high speeds, usually in a counter-clockwise direction. The vortex, up to several hundred yards wide, is visible to the observer as a whirlpool-like column of winds rotating about a hollow cavity or funnel. Winds may reach 300 miles per hour or higher.

Toxic-free area. An area within a facility in which the air supply is free of toxic chemical or biological agents.

Toxicity. A measure of the harmful effects produced by a given amount of a toxin on a living organism.

Triple-standard concertina wire. This type of fence uses three rolls of stacked concertina. One roll will be stacked on top of two other rolls that run parallel to each other while resting on the ground, forming a pyramid.

Tsunami. Sea waves produced by an undersea earthquake. Such sea waves can reach a height of 80 feet and can devastate coastal cities and low-lying coastal areas.

Twisted pair wire. Wire that uses pairs of wires twisted together to mitigate electromagnetic interference.

Two-person rule. A security strategy that requires two people to be present in or gain access to a secured area to prevent unobserved access by any individual.

U

Unobstructed space. Space around an inhabited building without obstruction large enough to conceal explosive devices 150 millimetre (6 inches) or greater in height.

Unshielded wire. Wire that does not have a conductive wrap.

V

Vault. A reinforced room for securing items.

Vertical rod. Typical door hardware often used with a crash bar to lock a door by inserting rods vertically from the door into the doorframe.

Vibration sensor. An intrusion detection sensor that changes state when vibration is present.

Video intercom system. An intercom system that also incorporates a small CCTV system for verification.

Video motion detection. Motion detection technology that looks for changes in the pixels of a video image.

Video multiplexer. A device used to connect multiple video signals to a single location for viewing and/or recording.

Visual displays. A display or monitor used to inform the operator visually of the status of the ESS.

Visual surveillance. The aggressor uses ocular and photographic devices (such as binoculars and cameras with telephoto lenses) to monitor facility or installation operations or to see assets.

Voice recognition. A biometric technology that is based on nuances of the human voice.

Volumetric motion sensor. An interior intrusion detection sensor that is designed to sense aggressor motion within a protected space.

Vulnerability. Any weakness that can be exploited by an aggressor or, in a non-terrorist threat environment, make an asset susceptible to hazard damage.

W

Warning. The alerting of emergency response personnel and the public to the threat of extraordinary danger and the related effects that specific hazards may cause.

Watch. Indication in a defined area that conditions are favorable for the specified type of severe weather (e.g., flash flood watch, severe thunderstorm watch, tornado watch, tropical storm watch).

Waterborne contamination. Chemical, biological, or radiological agent introduced into and fouling a water supply.

Weapons-grade material. Nuclear material considered most suitable for a nuclear weapon. It usually connotes uranium enriched to above 90 percent uranium-235 or plutonium with greater than about 90 percent plutonium-239.

Weapons of Mass Destruction (WMD). Any device, material, or substance used in a manner, in a quantity or type, or under circumstances showing an intent to cause death or serious injury to persons, or significant damage to property. An explosive, incendiary, or poison gas, bomb, grenade, rocket having a propellant charge of more than four ounces, or a missile having an explosive incendiary charge of more than 0.25 ounce, or mine or device similar to the above; poison gas; weapon involving a disease organism; or weapon that is designed to release radiation or radioactivity at a level dangerous to human life.

Weigand protocol. A security industry standard data protocol for card readers.

Z

Zoom. The ability of a CCTV camera to close and focus or open and widen the field of view.

Security Glossary

Access control point. A station at an entrance to a building or a portion of a building where identification is checked and people and hand-carried items are searched.

Access controls. Procedures and controls that limit or detect access to minimum essential infrastructure resource elements (e.g., people, technology, applications, data, and/or facilities), thereby protecting these resources against loss of integrity, confidentiality, accountability, and/or availability.

Access Control System. Also referred to as an EECSs; an electronic system that controls entry and egress from a building or area.

Access Control System elements Detection measures used to control vehicle or personnel entry into a protected area. Access Control System elements include locks, EECSs, and guards.

Accountability. The explicit assignment of responsibilities for oversight of areas of control to executives, managers, staff, owners, providers, and users of minimum essential infrastructure resource elements.

Active vehicle barrier. An impediment placed at an access control point that may be manually or automatically deployed in response to detection of a threat.

Aerosol. Fine liquid or solid particles suspended in a gas (e.g., fog or smoke).

Aggressor. Any person seeking to compromise a function or structure.

Airborne contamination. Chemical or biological agents introduced into and fouling the source of supply of breathing or conditioning air.

Antiterrorism (AT). Defensive measures used to reduce the vulnerability of individuals, forces, and property to terrorist acts.

Area lighting. Lighting that illuminates a large exterior area.

Assessment. The evaluation and interpretation of measurements and other information to provide a basis for decision-making.

Asset. A resource of value requiring protection. An asset can be tangible (e.g., people, buildings, facilities, equipment, activities, operations, and information) or intangible (e.g., processes or a company's information and reputation).

Asset protection. Security program designed to protect personnel, facilities, and equipment, in all locations and situations, accomplished through planned and integrated application of combating terrorism, physical security, operations security, and personal protective services, and supported by intelligence, counterintelligence, and other security programs.

Asset value. The degree of debilitating impact that would be caused by the incapacity or destruction of an asset.

Attack. A hostile action resulting in the destruction, injury, or death to the civilian population, or damage or destruction to public and private property.

B

Balanced magnetic switch. A door position switch utilizing a reed switch held in a balanced or center position by interacting magnetic fields when not in alarm condition.

Ballistics attack. An attack in which small arms (e.g., pistols, submachine guns, shotguns, and rifles) are fired from a distance and rely on the flight of the projectile to damage the target.

Barbed tape or concertina. A coiled tape or coil of wires with wire barbs or blades deployed as an obstacle to human trespass or entry into an area.

Barbed wire. A double strand of wire with four-point barbs equally spaced along the wire deployed as an obstacle to human trespass or entry into an area.

Barcode. A black bar printed on white paper or tape that can be easily read with an optical scanner.

Biological agents. Living organisms or the materials derived from them that cause disease in or harm to humans, animals, or plants or cause deterioration of material. Biological agents may be used as liquid droplets, aerosols, or dry powders.

Blast curtains. Heavy curtains made of blast-resistant materials that could protect the occupants of a room from flying debris.

Blast-resistant glazing. Window opening glazing that is resistant to blast effects because of the interrelated function of the frame and glazing material properties frequently dependent upon tempered glass, polycarbonate, or laminated glazing.

Blast vulnerability envelope. The geographical area in which an explosive device will cause damage to assets.

Bollard. A vehicle barrier consisting of a cylinder, usually made of steel and sometimes filled with concrete, placed on end in the ground and spaced about 3 feet apart to prevent vehicles from passing, but allowing entrance of pedestrians and bicycles.

Building hardening. Enhanced construction that reduces vulnerability to external blast and ballistic attacks.

Building separation. The distance between closest points on the exterior walls of adjacent buildings or structures.

C

Cable barrier. Cable or wire rope anchored to and suspended off the ground or attached to chain-link fence to act as a barrier to moving vehicles.

Chemical agent. A chemical substance that is intended to kill, seriously injure, or incapacitate people through physiological effects. Generally separated by severity of effect (e.g., lethal, blister, and incapacitating).

Clear zone. An area that is clear of visual obstructions and landscape materials that could conceal a threat or perpetrator.

Closed circuit television (CCTV). An electronic system of cameras, control equipment, recorders, and related apparatus used for surveillance or alarm assessment.

Collateral damage. Injury or damage to assets that are not the primary target of an attack.

Combating terrorism. The full range of federal programs and activities applied against terrorism, domestically and abroad, regardless of the source or motive.

Community. A political entity that has the authority to adopt and enforce laws and ordinances for the area under its jurisdiction. In most cases, the community is an incorporated town, city, township, village, or unincorporated area of a county; however, each state defines its own political subdivisions and forms of government.

Components and cladding. Elements of the building envelope that do not qualify as part of the main wind-force resisting system.

Confidentiality. The protection of sensitive information against unauthorized disclosure and sensitive facilities from physical, technical, or electronic penetration or exploitation.

Consequence Management. Measures to protect public health and safety, restore essential government services, and provide emergency relief to governments, businesses, and individuals affected by the consequences of terrorism. State and local governments exercise the primary authority to respond to the consequences of terrorism.

Contamination. The undesirable deposition of a chemical, biological, or radiological material on the surface of structures, areas, objects, or people.

Control center. A centrally located room or facility staffed by personnel charged with the oversight of specific situations and/or equipment.

Controlled area. An area into which access is controlled or limited. It is that portion of a restricted area usually near or surrounding a limited or exclusion area. Correlates with exclusion zone.

Controlled lighting. Illumination of specific areas or sections.

Controlled perimeter. A physical boundary at which vehicle and personnel access is controlled at the perimeter of a site. Access control at a controlled perimeter should demonstrate the capability to search individuals and vehicles.

Conventional construction Building construction that is not specifically designed to resist weapons, explosives, or chemical, biological, and radiological effects. Conventional construction is designed only to resist common loadings and environmental effects, such as wind, seismic, and snow loads.

Coordinate. To advance systematically an exchange of information among principals who have or may have a need to know certain information in order to carry out their roles in a response.

Counterintelligence. Information gathered, and activities conducted to protect against: espionage, other intelligence activities, sabotage, or assassinations conducted for or on behalf of foreign powers, organizations, or persons; or international terrorist activities, excluding personnel, physical, document, and communications security programs.

Counterterrorism (CT). Offensive measures taken to prevent, deter, and respond to terrorism.

Covert entry. Attempts to enter a facility by using false credentials or stealth.

Crash bar. A mechanical egress device located on the interior side of a door that unlocks the door when pressure is applied in the direction of egress.

Crime Prevention through Environmental Design. A crime prevention strategy based on evidence that the design and form of the built environment can influence human behavior. CPTED usually involves the use of three principles: natural surveillance (by placing physical features, activities, and people to maximize visibility); natural access control (through the judicial placement of entrances, exits, fencing, landscaping, and lighting); and territorial reinforcement (using buildings, fences, pavement, signs, and landscaping to express ownership).

Crisis Management. The measures taken to identify, acquire, and plan the use of resources needed to anticipate, prevent, and/or resolve a threat or act of terrorism.

Critical assets. Those assets essential to the minimum operations of the organization, and to ensure the health and safety of the general public.

Critical infrastructure. Primary infrastructure systems (e.g., utilities, telecommunications, and transportation) whose incapacity would have a debilitating impact on the school's ability to function.

D

Damage assessment. The process used to appraise or determine the number of injuries and deaths, damage to public and private property, and the status of key facilities and services (e.g., schools, hospitals, and other health care facilities, fire and police stations, communications networks, water and sanitation systems, utilities, and transportation networks) resulting from a manmade or natural disaster.

Data gathering panel. A local processing unit that retrieves, processes, stores, and/or acts on information in the field.

Debris-catching system Blast wallpaper, fragmentation blankets, or any similar system applied to the inside of a building's exterior walls. Debris-catching systems are often made of Kevlar or geotextile material and are designed to collect wall material debris in the event of an external explosion and to shield occupants from injuries.

Decontamination. The reduction or removal of a chemical, biological, or radiological material from the surface of a structure, area, object, or person.

Defense layer. Building design or exterior perimeter barriers intended to delay attempted forced entry.

Defensive measures. Protective measures that delay or prevent attack on an asset or that shield the asset from weapons, explosives, and CBR effects. Defensive measures include site work and building design.

Design Basis Threat. The threat (e.g., tactics and associated weapons, tools, or explosives) against which assets within a building must be protected and upon which the security engineering design of the school is based.

Design constraint. Anything that restricts the design options for a protective system or that creates additional problems for which the design must compensate.

Design opportunity. Anything that enhances protection, reduces requirements for protective measures, or solves a design problem.

Design team. A group of individuals from various engineering and architectural disciplines responsible for the protective system design.

Disaster. An occurrence of a natural catastrophe, technological accident, or human-caused event that has resulted in severe property damage, deaths, and/or multiple injuries.

Domestic terrorism. The unlawful use, or threatened use, of force or violence by a group or individual based and operating entirely within the United States or Puerto Rico without foreign direction committed against persons or property to intimidate or coerce a government, the civilian population, or any segment thereof in furtherance of political or social objectives.

Dose rate (radiation). A general term indicating the quantity (total or accumulated) of ionizing radiation or energy absorbed by a person or animal, per unit of time.

Dosimeter. An instrument for measuring and registering total accumulated exposure to ionizing radiation.

Duress alarm devices. Also known as panic buttons, these devices are designated specifically to initiate a panic alarm.

E

Effective stand-off distance. A stand-off distance at which the required LOP can be shown to be achieved through analysis or can be achieved through building hardening or other mitigating construction or retrofit.

Electronic Entry Control Systems. Electronic devices that automatically verify authorization for a person to enter or exit a controlled area.

Electronic Security System. An integrated system that encompasses interior and exterior sensors, closed circuit television systems for assessment of alarm conditions, EECSs, data-transmission media, and alarm reporting systems for monitoring, control, and display of various alarm and system information.

Emergency. Any natural or human-caused situation that results in or may result in substantial injury or harm to the population or substantial damage to or loss of property.

Emergency Environmental Health Services. Services required to correct or improve damaging environmental health effects on humans, including inspection for food contamination, inspection for water contamination, and vector control; providing for sewage and solid waste inspection and disposal; cleanup and disposal of hazardous materials; and sanitation inspection for emergency shelter facilities.

Emergency Medical Services Services including personnel, facilities, and equipment required to ensure proper medical care for the sick and injured from the time of injury to the time of final disposition, including medical disposition within a hospital, temporary medical facility, or special care facility; release from the site; or declared dead. Further, EMS specifically include those services immediately required to ensure proper medical care and specialized treatment for patients in a hospital and coordination of related hospital services.

Emergency Operations Center. The protected site from which state and local civil government officials coordinate, monitor, and direct emergency response activities during an emergency.

Emergency Operations Plan. A document that describes how people and property will be protected in disaster and disaster threat situations; details who is responsible for carrying out specific actions; identifies the personnel, equipment, facilities, supplies, and other resources available for use in the disaster; and outlines how all actions will be coordinated.

Emergency Public Information. Information that is disseminated primarily in anticipation of an emergency or at the actual time of an emergency and, in addition to providing information, frequently directs actions, instructs, and transmits direct orders.

Entry control point. A continuously or intermittently manned station at which entry to sensitive or restricted areas is controlled.

Equipment closet. A room where field control equipment, such as data gathering panels and power supplies are typically located.

Evacuation. Organized, phased, and supervised dispersal of people from dangerous or potentially dangerous areas.

Evacuation, mandatory or directed. This is a warning to persons within the designated area that an imminent threat to life and property exists and individuals MUST evacuate in accordance with the instructions of local officials.

Evacuation, spontaneous. Residents or citizens in the threatened areas observe an emergency event or receive unofficial word of an actual or perceived threat and, without receiving instructions to do so, elect to evacuate the area. Their movement, means, and direction of travel are unorganized and unsupervised.

Evacuation, voluntary. This is a warning to persons within a designated area that a threat to life and property exists or is likely to exist in the immediate future. Individuals issued this type of warning or order are NOT required to evacuate; however, it would be to their advantage to do so.

Evacuees. All persons removed or moving from areas threatened or struck by a disaster.

Exclusion area. A restricted area containing a security interest. Uncontrolled movement permits direct access to the item. See controlled area and limited area.

Exclusion zone. An area around an asset that has controlled entry with highly restrictive access. See controlled area.

F

Federal Coordinating Officer. The person appointed by the FEMA Director to coordinate federal assistance in a presidentially declared emergency or major disaster.

Federal Response Plan. The FRP establishes a process and structure for the systematic, coordinated, and effective delivery of federal assistance to address the consequences of any major disaster or emergency.

Fence protection. An intrusion detection technology that detects a person crossing a fence by various methods, such as climbing, crawling, cutting, and so on.

Fence sensor. An exterior intrusion detection sensor that detects aggressors as they attempt to climb over, cut through, or otherwise disturb a fence.

Field of view. The visible area in a video picture.

First responder. Local police, fire, and emergency medical personnel who first arrive on the scene of an incident and take action to save lives, protect property, and meet basic human needs.

Forced entry. Entry to a denied area achieved through force to create an opening in fence, walls, doors, and so on, or to overpower guards.

Fragment retention film. A thin, optically clear film applied to glass to minimize the spread of glass fragments when the glass is shattered.

Frangible construction. Building components that are designed to fail to vent blast pressures from an enclosure in a controlled manner and direction.

G

Glare security lighting. Illumination projected from a secure perimeter into the surrounding area, making it possible to see potential intruders at a considerable distance while making it difficult to observe activities within the secure perimeter.

Glazing. A material installed in a sash, ventilator, or panes (e.g., glass and plastic, including material, such as thin granite installed in a curtain wall).

Governor's Authorized Representative. The person empowered by the governor to execute, on behalf of the state, all necessary documents for disaster assistance.

H

Hazard. A source of potential danger or adverse condition.

Hazard mitigation. Any action taken to reduce or eliminate the long-term risk to human life and property from hazards. The term is sometimes used in a stricter sense to mean cost-effective measures to reduce the potential for damage to a facility or facilities from a disaster event.

Hazardous material (HazMat). Any substance or material that, when involved in an accident and released in sufficient quantities, poses a risk to people's health, safety, and/or property. These substances and materials include explosives, radioactive materials, flammable liquids or solids, combustible liquids or solids, poisons, oxidizers, toxins, and corrosive materials.

High-hazard areas. Geographic locations that, for planning purposes, have been determined through historical experience and vulnerability analysis to be likely to experience the effects of a specific hazard (e.g., hurricane, earthquake, and hazardous materials accident), resulting in vast property damage and loss of life.

High-risk target. Any material resource or facility that, because of mission sensitivity, ease of access, isolation, and symbolic value, may be an especially attractive or accessible terrorist target.

Human-caused hazard. Human-caused hazards are technological hazards and terrorism. They are distinct from natural hazards primarily in that they originate from human activity. Within the military services, the term threat is typically used for human-caused hazard. See definitions of technological hazards and terrorism for further information.

I

International terrorism. Violent acts or acts dangerous to human life that are a violation of the criminal laws of the United States or any state, or that would be a criminal violation if committed within the jurisdiction of the United States or any state. These acts appear to be intended to intimidate or coerce a civilian population, influence the policy of a government by intimidation or coercion, or affect the conduct of a government by assassination or kidnapping. International terrorist acts occur outside the United States or transcend national boundaries in terms of the means by which they are accomplished, the persons they appear intended to coerce or intimidate, or the locale in which their perpetrators operate or seek asylum.

Intrusion Detection System. The combination of components, including sensors, control units, transmission lines, and monitor units, integrated to operate in a specified manner.

J

Jersey barrier. A protective concrete barrier initially and still used as a highway divider that now also functions as an expedient method for traffic speed control at entrance gates and to keep vehicles away from buildings.

L

Laminated glass. A flat lite of uniform thickness consisting of two monolithic glass plies bonded together with an interlayer material as defined in Specification C1172. Many different interlayer materials are used in laminated glass.

Landscaping. The use of plantings (shrubs and trees), with or without landforms and/or large boulders, to act as a perimeter barrier against defined threats.

Layers of protection. A traditional approach in security engineering using concentric circles extending out from an area to be protected as demarcation points for different security strategies.

Level of protection. The degree to which an asset is protected against injury or damage from an attack.

Liaison. An agency official sent to another agency to facilitate interagency communications and coordination.

Limited area. A restricted area within close proximity of a security interest. Uncontrolled movement may permit access to the item. Escorts and other internal restrictions may prevent access to the item. See controlled area and exclusion area.

Line of sight. Direct observation between two points with the naked eye or hand-held optics.

Line-of-sight sensor. A pair of devices used as an intrusion detection sensor that monitor any movement through the field between the sensors.

Local government. Any county, city, village, town, district, or political subdivision of any state, and Indian tribe or authorized tribal organization, or Alaska Native village or organization, including any rural community or unincorporated town or village or any other public entity.

M

Mail-bomb delivery. Bombs or incendiary devices delivered to the target in letters or packages.

Minimum measures. Protective measures that can be applied to all buildings regardless of the identified threat. These measures offer defense or detection opportunities for minimal cost, facilitate future upgrades, and may deter acts of aggression.

Mitigation. Those actions taken to reduce the exposure to and impact of an attack or disaster.

Motion detector. An intrusion detection sensor that changes state based on movement in the sensor's field of view.

Moving vehicle bomb. An explosive-laden car or truck driven into or near a building and detonated.

Mutual aid agreement. A prearranged agreement developed between two or more entities to render assistance to the parties of the agreement.

N

Natural hazard. Naturally occurring events, such as floods, earthquakes, tornadoes, tsunami, coastal storms, landslides, and wildfires that strike populated areas. A natural event is a hazard when it has the potential to harm people or property (FEMA 386-2 Understanding Your Risks). The risks of natural hazards may be increased or decreased as a result of human activity; however, they are not inherently human-induced.

Natural protective barriers. Natural protective barriers are mountains and deserts, cliffs and ditches, water obstacles, or other terrain features that are difficult to traverse.

Non-exclusive zone. An area around an asset that has controlled entry but shared or less restrictive access than an exclusive zone.

Non-persistent agent. An agent that, upon release, loses its ability to cause casualties after 10 to 15 minutes. It has a high evaporation rate, is lighter than air, and will disperse rapidly. It is considered to be a short-term hazard; however, in small, unventilated areas, the agent will be more persistent.

Nuclear, biological, or chemical weapons. Also called Weapons of Mass Destruction (WMD). Weapons that are characterized by their capability to produce mass casualties.

P

Passive vehicle barrier. A vehicle barrier that is permanently deployed and does not require response to be effective.

Perimeter barrier. A fence, wall, vehicle barrier, landform, or line of vegetation applied along an exterior perimeter used to obscure vision, hinder personnel access, or hinder or prevent vehicle access.

Persistent agent. An agent that, upon release, retains its casualty-producing effects for an extended period of time, usually anywhere from 30 minutes to several days. A persistent agent usually has a low evaporation rate and its vapor is heavier than air; therefore, its vapor cloud tends to hug the ground. It is considered to be a long-term hazard. Although inhalation hazards are still a concern, extreme caution should be taken to avoid skin contact.

Physical security. The part of security concerned with measures/concepts designed to safeguard personnel; to prevent unauthorized access to equipment, installations, materiel, and documents; and to safeguard them against espionage, sabotage, damage, and theft.

Planter barrier. A passive vehicle barrier usually constructed of concrete and filled with dirt (and flowers for aesthetics). Planters, along with bollards, are the usual street furniture used to keep vehicles away from existing buildings. Overall size and the depth of installation below grade determine the vehicle stopping capability of the individual planter.

Plume. Airborne material spreading from a particular source; the dispersal of particles, gases, vapors, and aerosols into the atmosphere.

Polycarbonate glazing. A plastic glazing material with enhanced resistance to ballistics or blast effects.

Preliminary Damage Assessment. A mechanism used to determine the impact and magnitude of damage and the resulting unmet needs of individuals, businesses, the public sector, and the community as a whole. Information collected is used by the state as a basis for the governor's request for a presidential declaration, and by FEMA to document the recommendation made to the president in response to the governor's request. PDAs are made by at least one state and one federal representative. A local government representative familiar with the extent and location of damage in the community often participates; other state and federal agencies and voluntary relief organizations also may be asked to participate, as needed.

Preparedness. Establishing the plans, training, exercises, and resources necessary to enhance mitigation of and achieve readiness for response to, and recovery from all hazards, disasters, and emergencies, including WMD incidents.

Primary asset. An asset that is the ultimate target for compromise by an aggressor.

Primary gathering building. Inhabited buildings routinely occupied by fifty or more personnel. This designation applies to the entire portion of a building that meets the population density requirements for an inhabited building.

Probability of detection. A measure of an intrusion detection sensor's performance in detecting an intruder within its detection zone.

Probability of intercept. The probability that an act of aggression will be detected and that a response force will intercept the aggressor before the asset can be compromised.

Progressive collapse. A chain reaction failure of building members to an extent disproportionate to the original localized damage. Such damage may result in upper floors of a building collapsing onto lower floors.

Protective barriers. Define the physical limits of a site, activity, or area by restricting, channeling, or impeding access and forming a continuous obstacle around the object.

Protective measures. Elements of a protective system that protect an asset against a threat. Protective measures are divided into defensive and detection measures.

Protective system. An integration of all of the protective measures required to protect an asset against the range of threats applicable to the asset.

R

Radiation. High-energy particles or gamma rays that are emitted by an atom as the substance undergoes radioactive decay. Particles can be either charged alpha or beta particles or neutral neutron or gamma rays.

Radiation sickness. The symptoms characterizing the sickness known as radiation injury, resulting from excessive exposure of the whole body to ionizing radiation.

Radiological monitoring. The process of locating and measuring radiation by means of survey instruments that can detect and measure (as exposure rates) ionizing radiation.

Recovery. The long-term activities beyond the initial crisis period and emergency response phase of disaster operations that focus on returning all systems in the community to a normal status or to reconstitute these systems to a new condition that is less vulnerable.

Response. Executing the plan and resources identified to perform those duties and services to preserve and protect life and property, as well as provide services to the surviving population.

Restricted area. Any area with access controls that is subject to these special restrictions or controls for security reasons. See controlled area, limited area, exclusion area, and exclusion zone.

Risk. The potential for loss of, or damage to, an asset. It is measured based upon the value of the asset in relation to the threats and vulnerabilities associated with it.

Rotating drum or rotating plate vehicle barrier. An active vehicle barrier used at vehicle entrances to controlled areas based on a drum or plate rotating into the path of the vehicle when signaled.

S

Sacrificial roof or wall. Roofs or walls that can be lost in a blast without damage to the primary asset.

Safe haven. Secure areas within the interior of the facility. A safe haven should be designed such that it requires more time to penetrate by aggressors than it takes for the response force to reach the protected area to rescue the occupants. It may be a haven from a physical attack or an air-isolated haven from CBR contamination.

Secondary asset. An asset that supports a primary asset and whose compromise would indirectly affect the operation of the primary asset.

Secondary hazard. A threat whose potential would be realized as the result of a triggering event that of itself would constitute an emergency (e.g., dam failure might be a secondary hazard associated with earthquakes).

Situational crime prevention. A crime prevention strategy based on reducing the opportunities for crime by increasing the effort required to commit a crime, increasing the risks associated with committing the crime, and reducing the target appeal or vulnerability (whether property or person). This opportunity reduction is achieved by management and uses policies, such as procedures and training, as well as physical approaches, such as alteration of the built environment.

Specific threat. Known or postulated aggressor activity focused on targeting a particular asset.

Stand-off distance. A distance maintained between a building or portion thereof and the potential location for an explosive detonation or other threat.

Stand-off weapons. Weapons, such as anti-tank weapons and mortars, that are launched from a distance at a target.

State Coordinating Officer. The person appointed by the governor to coordinate state, commonwealth, or territorial response and recovery activities with FRP-related activities of the federal government, in cooperation with the FCO.

State Liaison. A FEMA official assigned to a particular state, who handles initial coordination with the state in the early stages of an emergency.

Stationary vehicle bomb. An explosive-laden car or truck stopped or parked near a building.

Structural protective barriers. Manmade devices (e.g., fences, walls, floors, roofs, grills, bars, roadblocks, signs, or other construction) used to restrict, channel, or impede access.

Superstructure. The supporting elements of a building above the foundation.

Supplies-bomb delivery. Bombs or incendiary devices concealed and delivered to supply or material handling points, such as loading docks.

T

Tactics. The specific methods of achieving the aggressor's goals to injure personnel, destroy assets, or steal materiel or information.

Technological hazards. Incidents that can arise from human activities, such as manufacture, transportation, storage, and use of hazardous materials. For the sake of simplicity, it is assumed that technological emergencies are accidental and that their consequences are unintended.

Terrorism. The unlawful use of force and violence against persons or property to intimidate or coerce a government, the civilian population, or any segment thereof, in furtherance of political or social objectives.

Thermally tempered glass. Glass that is heat-treated to have a higher tensile strength and resistance to blast pressures, although with a greater susceptibility to airborne debris.

Threat. Any indication, circumstance, or event with the potential to cause loss of, or damage to an asset.

Threat analysis. A continual process of compiling and examining all available information concerning potential threats and human-caused hazards. A common method to evaluate terrorist groups is to review the factors of existence, capability, intentions, history, and targeting.

TNT equivalent weight. The weight of TNT (trinitrotoluene) that has an equivalent energetic output to that of a different weight of another explosive compound.

Tornado. A local atmospheric storm, generally of short duration, formed by winds rotating at very high speeds, usually in a counter-clockwise direction. The vortex, up to several hundred yards wide, is visible to the observer as a whirlpool-like column of winds rotating about a hollow cavity or funnel. Winds may reach 300 miles per hour or higher.

Toxic-free area. An area within a facility in which the air supply is free of toxic chemical or biological agents.

Toxicity. A measure of the harmful effects produced by a given amount of a toxin on a living organism.

Triple-standard concertina wire. This type of fence uses three rolls of stacked concertina. One roll will be stacked on top of two other rolls that run parallel to each other while resting on the ground, forming a pyramid.

U

Unobstructed space. Space around an inhabited building without obstruction large enough to conceal explosive devices 150 millimetre (6 inches) or greater in height.

V

Video motion detection. Motion detection technology that looks for changes in the pixels of a video image.

Visual surveillance. The aggressor uses ocular and photographic devices (such as binoculars and cameras with telephoto lenses) to monitor facility or installation operations or to see assets.

Volumetric motion sensor. An interior intrusion detection sensor that is designed to sense aggressor motion within a protected space.

Vulnerability. Any weakness in an asset or mitigation measure than can be exploited by an aggressor (potential threat element), adversary, or competitor. It refers to the organization's susceptibility to injury.

W

Warning. The alerting of emergency response personnel and the public to the threat of extraordinary danger and the related effects that specific hazards may cause.

Watch. Indication in a defined area that conditions are favorable for the specified type of severe weather (e.g., flash flood watch, severe thunderstorm watch, tornado watch, tropical storm watch).

Waterborne contamination. Chemical, biological, or radiological agent introduced into and fouling a water supply.

Crime and Safety Glossary

Aggravated assault. Attack or attempted attack with a weapon, regardless of whether or not an injury occurs, and attack without a weapon when serious injury results.

At school. In the school building, on school property, on a school bus, or going to or from school.

At school (students). Inside the school building, on school property (school parking area, play area, school bus, etc.), or on the way to or from school.

Bullied. Students were asked if any student had bullied them at school in one or more ways during the school year. Specifically, students were asked if another student had made fun of them, called them names, or insulted them; spread rumors about them; threatened them with harm; pushed, shoved, tripped, or spit on them; forced them to do something they did not want to do; excluded them from activities on purpose; or destroyed their property on purpose.

Crime. Any violation of a statute or regulation or any act that the government has determined is injurious to the public, including felonies and misdemeanors.

Such violation may or may not involve violence, and it may affect individuals or property.

Cult or extremist group. A group that espouses radical beliefs and practices, which may include a religious component, that are widely seen as threatening the basic values and cultural norms of society at large.

Cyber-bullied. Students were asked if another student did one or more of the following behaviors anywhere that made them feel bad or were hurtful. Specifically, students were asked about bullying by a peer that occurred anywhere via electronic means, including the internet, email, instant messaging, text messaging, online gaming, and online communities.

Gang. Street gangs, fighting gangs, crews, or something else. Gangs may use common names, signs, symbols, or colors. All gangs, whether or not they are involved in violent or illegal activity, are included.

Hate crime. A criminal offense or threat against a person, property, or society that is motivated, in whole or in part, by the offender's bias against a race, color, *Indicators of School Crime and Safety: 2011* 183.

Hate-related graffiti. Hate-related words or symbols written in school classrooms, school bathrooms, school hallways, or on the outside of the school building.

Hate-related words. Students were asked if anyone called them an insulting or bad name at school having to do with their race, religion, ethnic background or national origin, disability, gender, or sexual orientation.

Homicide. An act involving a killing of one person by another resulting from interpersonal violence.

Illegal drugs. Examples of illegal drugs were marijuana, cocaine, inhalants, steroids, or prescription drugs without a doctor's permission, heroin, and methamphetamines.

Incident. A specific criminal act or offense involving one or more victims and one or more offenders.

Intimidation. To frighten, compel, or deter by actual or implied threats. It includes bullying and sexual harassment.

Metropolitan Statistical Areas (MSAs). Geographic entities defined by the US OMB for use by federal statistical agencies in collecting, tabulating, and publishing federal statistics.

Multistage sampling. A survey sampling technique in which there is more than one wave of sampling. That is, one sample of units is drawn, and then another sample is drawn within that sample. For example, at the first stage, a number of Census blocks may be sampled out of all the Census blocks in the United States. At the second stage, households are sampled within the previously sampled Census blocks.

On school property. On school property is included in the question wording, but was not defined for respondents.

Physical attack or fight. An actual and intentional touching or striking of another person against his or her will, or the intentional causing of bodily harm to an individual.

Prevalence. The percentage of the population directly affected by crime in a given period. This rate is based upon specific information elicited directly from the respondent regarding crimes committed against his or her person, against his or her property, or against an individual bearing a unique relationship to him or her. It is not based upon perceptions and beliefs about, or reactions to, criminal acts.

Rape. Forced sexual intercourse including both psychological coercion as well as physical force. Forced sexual intercourse means vaginal, anal, or oral penetration by the offender(s). Includes attempts and verbal threats of rape. This category also includes incidents where the penetration is from a foreign object, such as a bottle.

Robbery. Completed or attempted theft, directly from a person, of property or cash by force or threat of force, with or without a weapon, and with or without injury.

School. An education institution consisting of one or more of grades K through 12.

School-associated violent death. A homicide or suicide in which the fatal injury occurred on the campus of a functioning elementary or secondary school in the United States, while the victim was on the way to or from regular sessions at such a school, or while the victim was attending or traveling to or from an official school-sponsored event. Victims included nonstudents as well as students and staff members.

School crime. Any criminal activity that is committed on school property.

School year. The twelve-month period of time denoting the beginning and ending dates for school accounting purposes, usually from July 1 through June 30.

Serious violent incidents. Include rape, sexual battery other than rape, physical attacks or fights with a weapon, threats of physical attack with a weapon, and robbery with or without a weapon.

Serious violent victimization. Rape, sexual assault, robbery, or aggravated assault.

Sexual assault. A wide range of victimizations, separate from rape or attempted rape. These crimes include attacks or attempted attacks generally involving unwanted sexual contact between the victim and offender. Sexual assault may or may not involve force and includes such things as grabbing or fondling. Sexual assault also includes verbal threats.

Sexual battery. An incident that includes threatened rape, fondling, indecent liberties, child molestation, or sodomy. Principals were instructed that classification of these incidents should take into consideration the age and developmentally appropriate behavior of the offenders.

Sexual harassment. Unsolicited, offensive behavior that inappropriately asserts sexuality over another person. The behavior may be verbal or nonverbal.

Simple assault. Attack without a weapon resulting either in no injury, minor injury, or an undetermined injury requiring less than two days of hospitalization. Also includes attempted assault without a weapon.

Stratification. A survey sampling technique in which the target population is divided into mutually exclusive groups or strata based on some variable or variables (e.g., metropolitan area) and sampling of units occurs separately within each stratum.

Suicide. An act of taking one's own life voluntarily and intentionally.

Theft/larceny. Taking things valued at over $10 without personal confrontation. Specifically, the unlawful taking of another person's property without personal confrontation, threat, violence, or bodily harm. Included are pocket picking, stealing purse or backpack (if left unattended or no force was used to take it from owner), theft from a building, theft from a motor vehicle or motor vehicle parts or accessories, theft of bicycles, theft from vending machines, and all other types of thefts.

Total victimization. Combination of violent victimization and theft. If a student reported an incident of either type, he or she is counted as having experienced any victimization. If the student reported having experienced both, he or she is counted once under "total victimization."

Unequal probabilities. A survey sampling technique in which sampled units do not have the same probability of selection into the sample. For example, the investigator may oversample rural students in order to increase the sample sizes of rural students. Rural students would then be more likely than other students to be sampled.

Vandalism. The willful damage or destruction of school property, including bombing, arson, graffiti, and other acts that cause property damage. Includes damage caused by computer hacking.

Victimization. A crime as it affects one individual person or household. For personal crimes, the number of victimizations is equal to the number of victims involved.

Victimization rate. A measure of the occurrence of victimizations among a specific population group. For personal crimes, the number of victimizations is equal to the number of victims involved. Each victimization that is reported by the respondents is counted, so there may be one incident with two victims, which would be counted as two victimizations. The number of victimizations may be greater than the number of incidents because more than one person may be victimized during an incident.

Violent incidents. Include rape, sexual battery other than rape, physical attacks or fights with or without a weapon, threats of physical attack with or without a weapon, and robbery with or without a weapon.

Violent victimization. Includes serious violent victimization, rape, sexual assault, robbery, aggravated assault, or simple assault.

Weapon. Examples of weapons appearing in the questionnaire include guns, knives, and clubs. Includes look-alikes if they are used to threaten others.

BIBLIOGRAPHY

Building Security Through Design: A Primer for Architects, Design Professionals, and their Clients, November 2001, The American Institute of Architects (book). http://www.aia.org/security.

Blast Effects on Buildings: Design of Buildings to Optimize Resistance to Blast Loading, 1995, G. C. Mays and P. D. Smith, London: Thomas Telford, Ltd., American Society of Civil Engineers, ISBN: 0-7277-2030-9. http://www.pubs.asce.org/BOOKdisplay.cgi?9990338.

Minimum Design Loads for Buildings and Other Structures, ASCE 7-02,2002, American Society of Civil Engineers, ISBN: 0-7844-0624-3 [Note revision of 7-98, does not include building security or antiterrorism, but covers all natural hazards]. http://www.pubs.asce.org/ASCE7.html?9991330.

Vulnerability and Protection of Infrastructure Systems: The State of the Art, An ASCE Journals Special Publication compiling articles from 2002 and earlier, available online, https://ascestore.aip.org/OA_HTML/aipCCtpItmDspRte.jsp?a=b.

American Society of Heating, Refrigerating, and Air-Conditioning Engineers, Defensive Filtration, ASHRAE Journal, December 2002, James D. Miller. http://resourcecenter.ashrae.org/store/ashrae/newstore.cgi?itemid=9346&view=item&categoryid=409&page=1&loginid=29483.

Centers for Disease Control and Prevention/National Institute for Occupational Safety and Health Publication No. 2002-139, Guidance for Protecting Building Environments from

Airborne Chemical, Biological, or Radiological Attacks, May 2002, Cincinnati, OH. http://www.cdc.gov/niosh/bldvent/2002-139.html.

Publication No. 2003-136, Guidance for Filtration and Air Cleaning Systems to Protect Building Environments from Airborne Chemical, Biological, or Radiological Attacks, April 2003, Cincinnati, OH. http://www.cdc.gov/niosh/docs/2003-136/2003-136.html.

Central Intelligence Agency. Chemical, Biological, Radiological Incident Handbook, October 1998. http://www.cia.gov/cia/publications/cbr_handbook/cbrbook.htm.

Cost Impact of ISC Security Criteria, GSA & Applied Research Associates, Inc., L. Bryant and J. Smith, Vicksburg, MS [Restricted Access]. http://www.oca.gsa.gov/specialphp/References.php.

Council on Tall Buildings and Urban Habitat Building Safety Enhancement Guidebook, 2002. http://www.ctbuh.org.

Crime Prevention Through Environmental Design and Community Policing, August 1996, Dan Fleissner and Fred Heinzelmann, Washington, DC. http://www.ncjrs.org/pdfiles/crimepre.pdf.

Crime Prevention Through Environmental Design in Parking Facilities, April 1996, Mary S. Smith, Washington, DC. http://www.ncjrs.org/pdfiles/cptedpkg.pdf.

Facility Standards for the Public Building Service (PBS-P100); Chapter 8, Security Design, Revised November 2000. http://hydra.gsa.gov/pbs/pc/facilitiesstandards/.

Federal Emergency Management Agency FEMA 152, Seismic Considerations: Apartment Buildings, Earthquake Hazards Reduction Series 37, November 1988, Washington, DC (not available online). Contact FEMA Distribution Center, P.O. Box 2012, 8231 Stayton Drive, Jessup, MD 20794-2012, Telephone: 1-800-480-2520, Fax: 301-362-5335.

FEMA 153, Seismic Considerations: Office Buildings, Earthquake Hazards Reduction Series 38, November 1988, Washington, DC (not available online). Contact FEMA Distribution Center, P.O. Box 2012, 8231 Stayton Drive, Jessup, MD 20794-2012, Telephone: 1-800-480-2520, Fax: 301-362-5335.

FEMA 154, Rapid Visual Screening of Buildings for Seismic Hazards: A Handbook (2nd Edition), 2002, 1988, Washington, DC (not available online). Contact FEMA Distribution Center, P.O. Box 2012, 8231 Stayton Drive, Jessup, MD 20794-2012, Telephone: 1-800-480-2520, Fax: 301-362-5335.

FEMA 277, The Oklahoma City Bombing: Improving Building Performance through Multi-Hazard Mitigation, August 1, 1996, Washington, DC. http://www.fema.gov/mit/bpat/bpat009.htm.

FEMA 372, Mitigation Resources for Success (CD-ROM), October 2001, Washington, DC. http://www.fema.gov/pdf/library/poster_fnl2.pdf.

FEMA 386-2, Understanding Your Risks, Identifying Hazards and Estimating Losses, August 2001. http://www.fema.gov/ima/planning_toc3.shtm.

FEMA 386-7, Integrating Human-Caused Hazards into Mitigation Planning, September 2002. http://www.fema.gov/ima/antiterrorism/resources.shtm.

FEMA 403, World Trade Center Building Performance Study: Data Collection, Preliminary Observations, and Recommendations, May 2002, Washington, DC. http://www.fema.gov/library/wtcstudy.shtm.

FEMA 424, Design Guide for Improving Schools in Earthquakes, Floods and High Winds, January 2004. http://www.fema.gov/plan/prevent/rms/rmsp424.shtm.

FEMA 428, Primer to Design Safe School Projects in case of Terrorist Attacks, December 2003. http://www.fema.gov/plan/prevent/rms/rmsp428.

Florida Department of Education: Florida Safe School Design Guidelines, 2. http://www.firn.edu/doe/edfacil/safe_schools.htm 003.

General Services Administration Balancing Security and Openness: A Thematic Summary of a Symposium on Security and the Design of Public Buildings, November 30, 1999. http://hydra.gsa.gov/pbs/pc/gd_iles/SecurityOpenness.pdf.

Healthy Building International, Inc. Vulnerability Assessments and Counter Terrorist Protocols. http://www.healthybuildings.com/s2/vacbt.pdf.

Interagency Security Committee (executive agent—GSA) ISC Security Design Criteria for New Federal Office Buildings and Major Modernization Projects, May 28, 2001. http://www.oca.gsa.gov/specialphp/References.php.

Institute of Transportation Engineers The Influence of Traffic Calming Devices upon Fire Vehicle Travel Times, Michael A. Coleman, 1997, ITE Annual Meeting Compendium, 1997, pp. 838–845. http://webservices.camsys.com/fhwa/cmn/cmn33.htm.

Lawrence Berkeley National Lab Protecting Buildings from a Biological or Chemical Attack: actions to take before or during a release. LBNL/PUB-51959, January 10, 2003. http://securebuildings.lbl.gov/images/bldgadvice.pdf.

Mail Center Manager's Security Guide – Second Edition, October 22, 2002. http://www.gsa.gov//attachments/GSA_PUBLICATIONS/extpub/MailCenterManagersSecurityGuideV2.pdf.

National Academy of Sciences Combating Terrorism: Prioritizing Vulnerabilities and Developing Mitigation Strategies, Project Identification Number: NAEP-R-02-01-A, National Academy of Engineering on-going project—results to be published. http://www4.nationalacademies.org/webcr.nsf/ProjectScopeDisplay/.

National Capital Planning Commission Designing for Security in the Nation's Capital, October 2001. http://www.ncpc.gov/planning_init/security/DesigningSec.pdf.

National Capital Planning Urban Design and Security Plan, October 2002. http://www.ncpc.gov/publications/udsp/Final%20UDSP.pdf.

National Institute of Building Sciences Whole Building Design Guide: Provide Security for Building Occupants and Assets. http://www.wbdg.org/design/index.php?cn=2.7.4&cx=0.

National Institute of Justice (NIJ) The Appropriate and Effective Use of Security Technologies in US Schools: A Guide for Schools and Law Enforcement Agencies, September 1999, with US Department of Education, Safe and Drug-Free Schools Program; and US Department of Energy, Sandia National Laboratories. http://www.ncjrs.org/school/home.html.

NIJ Guide 100-00, Guide for the Selection of Chemical Agent and Toxic Industrial Material Detection Equipment for Emergency First Responders, June 2000. http://www.ncjrs.org/pdfiles1/nij/184449.pdf.

NIJ Guide 101-00, An Introduction to Biological Agent Detection Equipment for Emergency First Responders, December 2001. http://www.ncjrs.org/pdfiles1/nij/190747.pdf.

NIJ Guide 102-00, Guide for the Selection of Personal Protective Equipment for Emergency First Responders, Volumes I–IV, November 2002. http://www.ncjrs.org/pdfiles1/nij/191518.pdf.

NIJ Guide 602-00, Guide to the Technologies of Concealed Weapon and Contraband Imaging and Detection, February 2001. http://www.ncjrs.org/pdfiles1/nij/184432.pdf.

NIJ Standard 0108.01, Blast Resistant Protective Materials, September 1985 [Subscription Required]. http://www.ccb.org.

National Research Council Protecting Buildings and People from Terrorism: Technology Transfer for Blast-effects Mitigation, 2001, National Academy Press, Washington, DC, ISBN 0-309-08286-2. http://books.nap.edu/books/0309082862/html/index.html.

Protecting Buildings from Bomb Blast, Transfer of Blast-Effects Mitigation Technologies from Military to Civilian Applications, 1995, National Academy Press, Washington, DC, ISBN 0-309-05375-7. http://books.nap.edu/books/0309053757/html/index.html.

Progressive Collapse Analysis and Design Guidelines for New Federal Office Buildings and Major Modernization Projects, November 2000. http://www.oca.gsa.gov/specialphp/References.php.

Protection of Federal Office Buildings Against Terrorism, 1988, Committee on the Protection of Federal Facilities Against Terrorism, Building Research Board, National Academy Press, Washington, DC, ISBN 0-309-07691-9. http://books.nap.edu/books/0309076463/html/index.html.

Security Reference Manual, Part 3: Blast Design and Assessment Guidelines, July 31, 2001 [For Official Use Only] [Restricted Access], http://www.oca.gsa.gov/specialphp/References.php.

Split Speed Bump, 1998, Kathy Mulder, Washington, DC, TE International Conference, 1998. http://www.ite.org/trafic/documents/CCA98A33.pdf.

Society of American Military Engineers National Symposium of Comprehensive Force Protection, October 200l, Charleston, SC, Lindbergh & Associates. For a list of participants, access, http://www.same.org/forceprot/force.htm.

State and Local Guide 101, Guide for All-Hazard Emergency Operations Planning, Chapter 6, Attachment G, Terrorism, April 2001. http://www.fema.gov/rrr/allhzpln.shtm.

Task Force on Tall Buildings: "The Future," October 15, 2001. http://www.lehigh.edu/ctbuh/htmliles/hot_links/report.pdf or http://www.ctbuh.org/htmliles/hot_links/report.pdf.

Technical Support Working Group (TSWG) Terrorist Bomb Threat Standoff Card with Explanation of Use. http://www.tswg.gov/tswg/prods_pubs/newBTSCPress.htm.

The House National Security Committee Statement of Chairman Floyd D. Spence on the Report of the Bombing of Khobar Towers, August 1996, Washington, DC. http://www.house.gov/hasc/Publications/104thCongress/Reports/saudi.pdf.

US Department of Education Practical Information on Crisis Planning: A Guide for Schools and Communities, May 2003. http://www.ed.gov/emergencyplan.

US Department of Justice, National Institute of Justice, The Appropriate and Effective Use of Security Technologies in US Schools, September 1999. http://www.ncjrs.gov/school/home.html.

US Secret Service, Threat Assessment in Schools: A Guide to Managing Threatening Situations and to Creating Safe School Climates, May 2002. http://www.ed.gov/admins/lead/safety/threatassessmentguide.pdf.

Organizations and Associations

Air-Conditioning and Refrigeration Institute, Inc. http://www.ari.org.

Alliance for Fire & Smoke Containment & Control. http://www.afscconline.org.

American Council of Engineering Companies (ACEC). http://www.acec.org.

American Institute of Architects (AIA), Security Resource Center. http://www.aia.org/security.

American Planning Association. http://www.planning.org.

American Society of Civil Engineers (ASCE). http://www.asce.org.

American Society of Heating, Refrigerating, and Air-Conditioning Engineers (ASHRAE). http://www.ashrae.org.

American Society of Mechanical Engineers (ASME). http://www.asme.org.

Architectural Engineering Institute (AEI) of ASCE. http://www.asce.org/instfound/aei.cfm.

Associated General Contractors of America. http://www.agc.org.

Building Futures Council. http://www.thebfc.com.

Building Performance Assessment Team. http://www.fema.gov/mit/bpat.

Civil Engineering Research Foundation (CERF) of ASCE. http://www.cerf.org.

Construction Industry Institute. http://construction-institute.org.

Federal Emergency Management Agency (FEMA). http://www.fema.gov.
National Crime Prevention Institute. http://www.louisville.edu/a-s/ja/ncpi/courses.htm.
National Fire Protection Association. http://www.nfpa.org.
National Institute of Standards and Technology (NIST), Building and Fire Research Laboratory. http://www.bfrl.nist.gov.
Office of Domestic Preparedness (ODP). http://www.ojp.usdoj.gov/odp.
Protective Glazing Council. http://www.protectiveglazing.org.
Sandia National Laboratories (SNL). http://www.sandia.gov.
Structural Engineering Institute (SEI) of ASCE. http://www.seinstitute.org.
Sustainable Buildings Industry Council. http://www.sbicouncil.org.
The Infrastructure Security Partnership (TISP). http://www.tisp.org.
US Department of Energy. http://www.energy.gov.
US Department of Homeland Security. http://www.whitehouse.gov/deptofhomeland.
US Green Building Council. http://www.usgbc.org.
US Marshals Service (USMS). http://www.usdoj.gov/marshals.

Index